T0295555

Artificial Intelligence for Drug Development, Precision Medicine, and Healthcare

Chapman & Hall/CRC Biostatistics Series

Series Editors
Shein-Chung Chow, Duke University School of Medicine, USA
Byron Jones, Novartis Pharma AG, Switzerland
Jen-pei Liu, National Taiwan University, Taiwan
Karl E. Peace, Georgia Southern University, USA
Bruce W. Turnbull, Cornell University, USA
Recently Published Titles

For more information about this series, please visit: https://www.crcpress.com/Chapman–Hall CRC-Biostatistics-Series/book-series/CHBIOSTATIS

Artificial Intelligence for Drug Development, Precision Medicine, and Healthcare

Mark Chang
Boston University

CRC Press
Taylor & Francis Group
Boca Raton London New York

CRC Press is an imprint of the
Taylor & Francis Group, an **informa** business
A CHAPMAN & HALL BOOK

CRC Press
Taylor & Francis Group
6000 Broken Sound Parkway NW, Suite 300
Boca Raton, FL 33487-2742

© 2020 by Taylor & Francis Group, LLC
CRC Press is an imprint of Taylor & Francis Group, an Informa business

No claim to original U.S. Government works

Printed on acid-free paper

International Standard Book Number-13: 978-0-3673-6292-8 (Hardback)

Visit the Taylor & Francis Web site at
http://www.taylorandfrancis.com

and the CRC Press Web site at
http://www.crcpress.com

Contents

Preface

There are many publications on artificial intelligence (AI) under various names: data mining, text mining, machine-learning (ML), pattern recognition, statistical learning, deep learning, bioinformatics, computational biology, computational linguistics, natural language processing, robotics. Most book publications are for those who don't have any statistical background at all. This is because the significant achievements in recent years, especially in deep learning for image and language processing and voice recognition, have nothing to do with conventional statistics at all, even though they are heavily dependent on big data. Some AI books are statistically oriented under the names of statistical learning, pattern-recognition, and machine learning. Because traditional statistics and AI both deal with data and an increasing number of statisticians or data scientists have become involved in AI research, the meaning and scope of AI are constantly evolving. Sometimes the differences are unintelligible, but AI is clearly not equal to statistics. Here are some subtle differences: (1) AI ML emphasizes learning and prediction, classical statistics focuses on type-I error rate, (2) statistics and AI both deal with uncertainty, but the former focuses on the mathematical approach and probability distributions, while the latter is mainly algorithm-based. (3) ML takes its main aim at real world experiences, while classical statistics often performs under an ideal assumed probability distribution.

My efforts in this book will focus on the balance between statistical and algorithm-based approaches to AI, discussing and revealing the relationships between the two methods, emphasizing the creative idea of each method, and on conceptual clarity, mathematical conciseness, and algorithms for implementation. I particularly consider real-world applications in precision medicine, healthcare, as well as drug discovery and development in the pharmaceutical industry. It is my belief that to have a deep understanding of AI methods (and to be creative in using AI's power), a combination of familiarity with the theories and applications and practice or experimentation with software implementations is necessary. For this reason, we also introduce software package in R, keras, kerasR, and other lR packages with simplified but real-world examples. R is a popular software package in the statistical community. The recent inclusion of the keras package makes R more powerful than ever in deep learning. Keras is an API with TensorFlow as backend, while Google TensorFlow is the most popular and powerful software for deep learning. Hopefully, the software choice makes it easy for statisticians to learn AI. Python is a popular language in the AI community, and readers may also want to learn some of the basics.

Recently, the FDA moved toward a new, tailored review framework for artificial intelligence-based medical devices (Gottlieb, April 2019). It is encouraging news. AI and ML have found great applications in drug discovery, disease diagnosis, and healthcare, but virtually no applications in clinical trials. This book will bridge the gap. It will cover AI methods and applications in drug discovery, clinical trials, disease diagnosis and prognosis, and patient health system management.

Road Map

The book includes fourteen chapters and an appendix. Chapter 1, Overview of Modern Artificial Intelligence, will provide a brief history that includes the three waves of AI and different types of AI methods, including supervised and unsupervised learning, reinforcement learning, evolutionary learning and swarm intelligence. Chapter 2, Classical Statistics and Modern Machine Learning, will discuss key concepts required for data scientists and compare modern AI modeling versus classic statistical modeling. Chapter 3, The Similarity Principle, will elaborate why the principle is essential for all sciences, how classic statistics is incapable of dealing with many statistical paradoxes, and how to construct similarity functions. Chapter 4, Similarity-Based Artificial Intelligence (SBAI), will be devoted to a unique similarity-based approach for both big and small data, such as rare disease clinical trials. We will discuss regularization, sequential, hierarchical, and recursive learning with SBAI and its implementations in R, and application examples. Chapter 5, Artificial Neural Networks (ANNs), is a typical algorithm-based AI, including single-layer and multilayer perceptrons (MLPs). An MLP is a feed-forward artificial neural network that serves as a foundation for many different deep learning networks. We will discuss important concepts such as linear separability and the backpropagation algorithm. Chapter 6, Deep Learning Neural Networks, will introduce the three most commonly used and successful deep learning networks: convolution neural networks (CNNs) for medical image processing, recurrent neural networks (RNNs), including long short-term memory networks for sequence (gene) data processing, language and voice-recognition, and deep belief networks (DBNs) for disease diagnosis and cancer detection. Chapter 7, Kernel Methods, will introduce a unique representation of objects called kernels. We will reveal the relationship between support vector machines (SVMs) and kernel methods (KMs) and how a KM can be viewed as special case of the similarity-based methods described in Chapter 4. Chapter 8, Decision Tree and Ensemble Methods, will briefly present decision tree methods for classification and regression and discuss commonly used algorithms including bagging and boosting, AdaBoost, and random forests. Chapter 9, The Bayesian Learning Approach, will discuss Bayesian learning mechanisms, posterior distributions, hierarchical models, Bayesian decision processes, and

Bayesian networks for similarity searches. Chapter 10, Unsupervised Learning, will cover popular unsupervised learning methods including link-analysis (association), K-means clustering, hierarchical clustering, self-organized maps, adversarial networks and autoencoders. Chapter 11, Reinforcement Learning, will introduce sequential decision-making for drug development, Q-learning, and Bayesian stochastic decision processes. Chapter 12, Swarm and Evolutionary Intelligence, will discuss artificial ant algorithms, genetic algorithms, and genetic programing. Chapter 13, Applications of AI in Medical Science and Drug Development, is a comprehensive review of applications of different AI methods in drug discovery, in cancer prediction using microarray data, and in medical image analysis, healthcare, clinical trials, and drug safety monitoring. Chapter 14, Future Perspectives, will present my thoughts on artificial general intelligence and my future research directions.

For most method, each chapter will provide examples of applications using R. For some of the AI methods, I present algorithms and implementations in R so that you will see how simple code can perform amazing tasks. In this way, you will also get insights into the AI algorithms, rather than just dealing a black box.

The book can be read sequentially chapter-by-chapter or by jump-starting to read chapters that appear most interesting to you. Many different AI methods can be used for the same application problems and many different application problems can be solved using the same AI method. That is why we cover broad methods and application areas, hoping that you can borrow the ideas from different fields and find more fruit in this promising field.

The *R* programs for the book are available on www.statisticians.org

Acknowledgements

I'd like to thank my doctoral student, Susan Hwang for her research into similarity-based machine learning and comments on the first two chapters. I'd also like to express my sincere thanks to Dr. Robert Pierce for his comprehensive review and valuable suggestions.

1

Overview of Modern Artificial Intelligence

1.1 Brief History of Artificial Intelligence

The term, artificial intelligence (AI), was coined by John McCarthy, Marvin Minsky, Nathaniel Rochester, and Claude Shannon in 1955. AI is tied to what we used to think of what comprised a robot's brain, or to a function of such a brain. In a general sense, AI includes robotics. The term AI often emphasizes the software aspects, while the term robot includes a physical body as an important part. The notions of AI and robotics come from a long way back.

Early in 1854, George Boole argued that logical reasoning could be performed systematically in the same manner as solving a system of equations. Thus a logical approach played an essential role in early AI studies. Examples: the Spanish engineer Leonardo Torres Quevedo (1914) demonstrates the first chess-playing machine, capable of king and rook against king endgames without any human intervention, Claude Shannon's (1950) "Programming a Computer for Playing Chess" is the first published article on developing a chess-playing computer program, and Arthur Samuel (1952) develops the first computer checkers-playing program and the first computer program to learn on its own. In 1997 Deep Blue becomes the first computer chess-playing program to beat a reigning world chess champion. Herbert Simon and Allen Newell (1955) develop the Logic Theorist, the first artificial intelligence program, which eventually would prove 38 of the first 52 theorems in Whitehead and Russell's Principia Mathematica. In 1961 James Slagle develops SAINT (Symbolic Automatic INTegrator), a heuristic program that solved symbolic integration problems in freshman calculus. This is perhaps the predecessor of powerful AI software often used in present-day mathematics (Gil Press, 2016).

In robotics, Nikola Tesla (1898) makes a demonstration of the world's first radio-controlled ("a borrowed mind" as Tesla described) vessel, an embryonic form of robots. Czech writer Karel Čapek (1921) introduces the word *robot*, a Czech word meaning forced work, in his play *Rossum's Universal Robots*. Four years later, a radio-controlled driverless car was released, travelling the streets of New York City. In 1929, Makoto Nishimura designs the first robot built in Japan, which can change its facial expression and move its head and hands using an air pressure mechanism. The first industrial robot, Unimate, starts working on an assembly line in a General Motors plant in New Jersey in 1961. In 1986, Bundeswehr University built the first driverless car, which drives up to 55 mph on empty streets. In 2000 Honda's ASIMO robot, an artificially intelligent humanoid robot, is able to walk as fast as a human, delivering trays to customers in a restaurant setting. In 2009 Google starts developing, in secret, a driverless car. In 2014 it became the first to pass, in Nevada, a U.S. state self-driving test.

In artificial neuro-network (ANN) development, Warren S. McCulloch and Walter Pitts publish (1943) "A Logical Calculus of the Ideas Immanent in Nervous Activity" to mimic the brain. The authors discuss networks of simplified artificial "neurons" and how they might perform simple logical functions. Eight years later, Marvin Minsky and Dean Edmunds build SNARC (Stochastic Neural Analog Reinforcement Calculator), the first artificial neural network, using 3000 vacuum tubes to simulate a network of 40 neurons. In 1957, Frank Rosenblatt develops the Perceptron, an early artificial neural network enabling pattern recognition based on a two-layer computer learning network. Arthur Bryson and Yu-Chi Ho (1969) describe a backpropagation learning algorithm for multi-layer artificial neural networks, an important precursor contribution to the success of deep learning in the 2010s, once big data become available and computing power was sufficiently advanced to accommodate the training of large networks. In the following year, AT&T Bell Labs successfully applies backpropagation in ANN to recognizing handwritten ZIP codes, though it took 3 days to train the network, given the hardware limitations at the time. In 2006 Geoffrey Hinton publishes "Learning Multiple Layers of Representation," summarizing the ideas that have led to "multilayer neural networks that contain top-down connections and training them to generate sensory data rather than to classify it," i.e., a new approach to deep learning. In March 2016, Google DeepMind's AlphaGo defeats Go champion Lee Sedol.

Computational linguistics originated with efforts in the United States in the 1950s to use computers to automatically translate texts from foreign languages, particularly Russian scientific journals, into English (John Hutchins, 1999). To translate one language into another, one has to understand the grammar of both languages, including morphology (the grammar of word forms), syntax (the grammar of sentence structure), the semantics, and the lexicon (or "vocabulary"), and even something of the pragmatics of language use. Thus, what started as an effort to translate between languages evolved into an entire discipline devoted to understanding how to represent and process natural languages using computers. Long before modern computational linguistics, Joseph Weizenbaum develops ELIZA in 1965, an interactive program that carries on a dialogue in English language on any topic. ELIZA surprised many people who attributed human-like feelings to the computer program. In 1988 Rollo Carpenter develops the chat-bot Jabberwacky to "simulate natural human chat in an interesting, entertaining and humorous manner." It is an early attempt at creating artificial intelligence through human interaction. In 1988, IBM's Watson Research Center publishes "A Statistical Approach to Language Translation," heralding the shift from rule-based to probabilistic methods of machine translation. This marks a broader shift from a deterministic approach to a statistical approach in machine learning. In 1995, inspired by Joseph Weizenbaum's ELIZA program, Richard Wallace develops the chatbot A.L.I.C.E. (Artificial Linguistic Internet Computer Entity) with natural language sample data collection at an unprecedented scale, enabled by the advent of the Web. In 2011, a convolutional neural network wins the German Traffic Sign Recognition competition with 99.46% accuracy (vs. humans at 99.22%). In the same year, Watson, a natural language question-answering computer, competes on Jeopardy and defeats two former champions. In 2009, computer scientists at the Intelligent Information Laboratory at Northwestern University develop Stats Monkey, a program that writes sport news stories without human intervention.

In the areas referred to today as machine learning, data mining, pattern recognition, and expert systems, progress may be said to have started around 1960. Arthur Samuel (1959) coins the term machine learning, reporting on programming a computer "so that it will learn to play a better game of checkers than can be played by the person who wrote the program." Edward Feigenbaum, et al. (1965) start working on DENDRAL at Stanford University, the first expert system for automating the decision-making process and constructing models of empirical induction in science. In 1978 Carnegie Mellon University developed the XCON program, a rule-based expert system that automatically selects computer system components based on the customer's requirements. In 1980, researchers at Waseda University in Japan built Wabot, a musician humanoid robot able to communicate with a person, read a musical score, and play tunes on an electronic organ. In 1988 Judea Pearl publishes Probabilistic Reasoning in Intelligent Systems. Pearl was the 2011 Turing Award recipient for creating the representational and computational foundation for the processing of information under uncertainty, and is credited with the invention of Bayesian networks. In reinforcement learning, Rodney Brooks (1990) publishes "Elephants Don't Play Chess," proposing a new approach to AI from the ground up based on physical interaction with the environment: "The world is its own best model... the trick is to sense it appropriately and often enough."

Bioinformatics involves AI or machine learning (ML) studies in biology and drug discovery. As an interdisciplinary field of science, bioinformatics combines biology, computer science, and statistics to analyze biological data such as the identification of candidate genes and single nucleotide polymorphisms (SNPs) for better understanding the genetic basis of disease, unique adaptations, desirable properties, or differences between populations. In the field of genetics and genomics, bioinformatics aids in sequencing and annotating genomes and their observed mutations. Common activities in bioinformatics include mapping and analyzing DNA and protein sequences, aligning DNA and protein sequences to compare them, and creating and viewing 3-D models of protein structures. Since AI methods were introduced to biotech companies in the late 1990s, supervised and unsupervised learning have contributed significantly to drug discovery.

Another unique approach in the study of AI is called genetic programming, an evolutionary AI that can program itself for improvement. The idea of GP was inspired by genetics. From the 1980s onward, genetic programming has produced many human-competitive inventions and reinventions in disparate fields, including electronics and material engineering.

1.2   Waves of Artificial Intelligence

DARPA, a well-known AI research agency in the US, has recently characterized AI development using three waves (Figure 1.1). It's dedicated to funding "crazy" projects – ideas that are completely outside the accepted norms and paradigms, including contribution to the establishment of the early internet and the Global Positioning System (GPS), as well as a flurry of other bizarre concepts, such as legged robots, prediction markets, and self-assembling work tools (Roey Tzezana, 2017).

1.2.1   First Wave: Logic-Based Handcrafted Knowledge

In the first wave of AI, domain-experts devised algorithms and software according to available knowledge. This approach led to the creation of chess-playing computers, and of delivery optimization software. Weizenbaum's 1965 ELIZA (see Section 1.1), an AI agent that can carry on grammatically correct conversations with a human, was a logical rule-based agent. Even most of the software in use today is based on AI of this kind — think of robots in assembly lines and Google Maps. In this wave AI systems are usually based on clear and logical rules or decision trees. The systems examine the most important parameters in every situation they encounter, and reach a conclusion about the most appropriate action to take in each case without any involvement of probability theory. As a result, when the tasks involve too many parameters, many uncertainties or hidden parameters or confounders affect the outcomes in a complex system, and it is very difficult for first wave systems to deal with the complexity appropriately. Determining drug effects in human and disease diagnosis and prognosis are examples of such complex biological systems that first wave AI cannot handle well.

In summary, first-wave AI systems are capable of implementing logical rules for well-defined problems but are incapable of learning, and not able to deal with problems with a large underlying uncertainty.

1.2.2   Second Wave: Statistical Machine Learning

Over nearly two decades, emphasis has shifted from logic to probabilities, or more accurately to mixing logic and probabilities, thanks to the availability of "big data", viable computer power, and the involvement of statisticians. Much great ongoing AI effort in both industrial applications as well as in academic research falls into this category. Here comes the second wave of AI. It is so statistics- focussed that Thomas J. Sargent, winner of the 2011 Nobel Prize in Economics, recently told the World Science and Technology Innovation Forum that artificial intelligence is actually statistics, but in a very gorgeous phrase, "it is statistics." Many formulas are very old, but all AI uses statistics to solve problems. I see his point but do not completely agree with him, as you will see in the later chapters

To deal with complex systems with great uncertainties, probability and statistics are naturally effective tools. However, it cannot be the exact same statistical methods we have used in classical settings. As Leo Breiman (2001) pointed out in his *Statistical Modeling: The Two Cultures:* "There are two cultures in the use of statistical modeling to reach conclusions from data. One assumes that the data are generated by a given stochastic data model. The other uses algorithmic models and treats the data mechanism as unknown. The statistical community has been committed to the almost exclusive use of data models. This commitment has led to irrelevant theory, questionable conclusions, and has kept statisticians from working on a large range of interesting current problems. Algorithmic modeling, both in theory and practice, has developed rapidly in fields outside statistics. It can be used both on large complex data sets and as a more accurate and informative alternative to data modeling on smaller data sets. If our goal as a field is to use data to solve problems, then we need to move away from exclusive dependence on data models and adopt a more diverse set of tools."

Statistical machine learning systems are highly successful at understanding the world around them: they can distinguish between two different people or between different vowels. They can learn and adapt themselves to different situations if they're properly trained. However, unlike first-wave systems, they're limited in their logical capacity: they don't rely on precise rules, but instead they go for the solutions that "work well enough, usually" (Roey Tzezana, 2017).

The poster boys of second-wave systems are the conceptualizers of artificial neural networks (ANN) and their great successes in the fields of deep learning, including voice-recognition and image recognition (Figures 1.1, 1.2). Starting in the early 2010s, huge amounts of training data together with massive computational power prompted a reevaluation of some particular 30-year-old neural network algorithms. To the surprise of many researchers the combination of big data, incredible computer power, and ANN, aided by new innovations such as deep-learning convolutional networks, has resulted in astonishing achievement in speech and image recognition, as well as in most categorization tasks. For example, the Johnson and Johnson's Sedasys system has been approved by the FDA to deliver anesthesia automatically for standard procedures such as colonoscopies. The machine can reduce cost since a doctor supervises several machines at the same time, often making the presence of a dedicated human

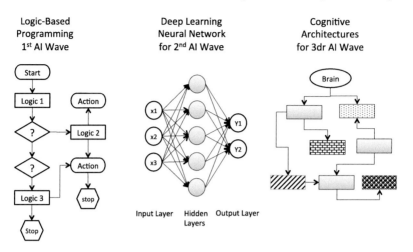

FIGURE 1.1
Architecture of Three AI Waves

anesthesiologist unnecessary. Speech-based interactive systems such as Siri of the iOS, Google Assistant, and driverless cars are well-known achievements of deep learning. Deep learning algorithms include multiple-layer perceptrons, convolution neural networks, long short-term memory networks, and deep belief networks. All these deep learning networks have been used in drug discovery, health data processing, and in disease diagnosis and prognosis.

Researchers try, usually unsuccessfully, to explain why ANN works well, even though ANN is not a black box. We are incapable of doing that consistently because we cannot only use simpler or more fundamental concepts to explain more complex or higher-level concepts. The Achilles heel of second-wave systems, that nobody is certain why they're working so well, might actually be an indication that in some aspects they are indeed much like our brains: we can throw a ball into the air and predict where it's going to fall, even without calculating Newton's equations of motion.

1.2.3   Third Wave: Contextual Adaptation

However, the third-wave AI is still nowhere near human (or even animal) intelligence in terms of general cognitive ability. In particular, today's AI is very poor at learning interactively, in the real world, capabilities such as reasoning and language understanding, how to reuse knowledge and skills, and what we call abstraction. Researchers have found that deep learning is actually less capable than some first wave approaches when it comes to certain language tasks, reasoning, planning and explaining its actions.

In the third wave, the AI systems themselves will construct models that will explain how the world works, will discover by themselves the logical

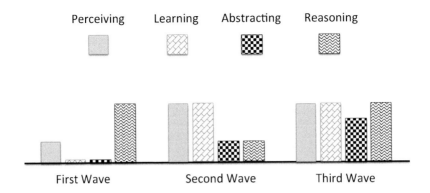

FIGURE 1.2
Characteristics of the First-, Second-, and Third-AI Waves

rules which shape their decision-making process, and will be able to train themselves as Alpha-Go did when it played a million Go games against itself, to identify commonsense rules. Third-wave systems would also be able to take information from several different sources to reach a nuanced and well-explained conclusion.

Wordnet and Conceptnet were devised initially to support AI for knowledge discovery and contextual understanding of concepts and language, though I cannot see that their approach can go anywhere further than an electronic combination of dictionary and thesaurus.

A good example of efforts towards the third wave would be genetic programming (GP). GP is essentially the creation of self-evolving programs, a patented invention from 1988 by John Koza. The series of four books by Koza, et al. (1992, 1994, 1999, 2003) fundamentally established GP and included a vast amount of GP results and example of human-competitive inventions and reinventions in different fields. Subsequently, there was an enormous expansion of the number of publications, within the Genetic Programming Bibliography, surpassing 10,000 entries (Kaza, 2010a). At the same time, industrial uptake has been significant in several areas including finance, the chemical industry, bioinformatics (Langdon and Buxton, 2004) and the steel industry (M. Kovačič).

In summary, third wave AI systems are expected to understand context and the consequences of their actions, be capable of self-improvement and be able to do most of our routine work (Figure 1.3).

1.2.4 The Last Wave: Artificial General Intelligence

It is said: in the first AI wave you had to be a programmer, in the second AI wave you have to be a data scientist, and in the third AI wave you have

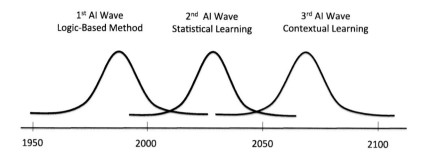

FIGURE 1.3
Architecture of Three AI Waves

to be morally better. However, third-wave AI is still nowhere near human intelligence in terms of general cognitive ability, creativity, discovery of new knowledge, emotional intelligence, or in the ability to create AI agents. Such full human-capability intelligence is called Artificial General Intelligence (AGI). AGI requires a competitively human appearance, sensors, and even organs for biological processes because knowledge acquisition depends on those sensor organs. A "strong" brain equipped in a "weak" body can only make the brain weaker and less capable.

1.3   Machine Learning Methods

1.3.1   Data Science

If we say that a statistician's job is to discover knowledge from data, it can be misleading since data alone does not tell one anything. Data and science behind data tell the whole story. Such a contemporary view of the statistician's role leads to a new name for statisticians: data scientists. Imagine a statistician without basic understanding of the science behind his data, knowledge of how the data were collected, and/or how the experiment was designed and conducted, and you can see how his blinkered analysis could lead to a nonsensical conclusion. Early in 1962 John W. Tukey writes in "The Future of Data Analysis": "For a long time I thought I was a statistician, interested in inferences from the particular to the general. But as I have watched mathematical statistics evolve, I have had cause to wonder and doubt... I have come to feel that my central interest is in data analysis... Data analysis, and the parts of statistics which adhere to it, must... take on the characteristics of science rather than those of mathematics... data analysis is intrinsically an empirical science... "

Jeff Wu, in his inaugural lecture (1997) for the H. C. Carver Chair in Statistics at the University of Michigan, calls for statistics to be renamed data science and statisticians to be renamed data scientists. Hal Varian, Google's chief economist, tells the *McKinsey Quarterly*: "I keep saying the sexy job in the next ten years will be statisticians. People think I'm joking, but who would've guessed that computer engineers would've been the sexy job of the 1990s? The ability to take data—to be able to understand it, to process it, to extract value from it, to visualize it, to communicate it—that's going to be a hugely important skill in the next decades, not only at the professional level but even at the educational level for elementary school kids, for high school kids, for college kids. Because now we really do have essentially free and ubiquitous data, the complementary scarce factor is the ability to understand that data and extract value from it....I think statisticians are part of it, but it's just a part. You also want to be able to visualize the data, communicate the data, and utilize it effectively. But I do think those skills—of being able to access, understand, and communicate the insights you get from data analysis—are going to be extremely important. Managers need to be able to access and understand the data themselves." (Interview, Napa, California, October 2008, source: Hal Varian on How the Web Challenges Managers. McKinsey & Company, 2009.)

Pete Warden from Microsoft's TensorFlow Team explained (2011) why the term data science is flawed but useful: "There is no widely accepted boundary for what's inside and outside of data science's scope. Is it just a faddish rebranding of statistics? I don't think so, but I also don't have a full definition. I believe that the recent abundance of data has sparked something new in the world, and when I look around I see people with shared characteristics who don't fit into traditional categories. These people tend to work beyond the narrow specialties that dominate the corporate and institutional world, handling everything from finding the data, processing it at scale, visualizing it and writing it up as a story. They also seem to start by looking at what the data can tell them, and then picking interesting threads to follow, rather than the traditional scientist's approach of choosing the problem first and then finding data to shed light on it."

Realizing the importance of statistics in AI, the International Association for Statistical Computing (IASC) is established in 1977, its mission being to link traditional statistical methodology, modern computer technology, and the knowledge of domain experts in order to convert data into information and knowledge.

1.3.2 Supervised Learning: Classification and Regression

Supervised learning (classification) has been used for in digital imaging recognition, signal detection, and pharmacovigilance. It has also been used in the diagnosis of disease (such as cancer) when more accurate procedures may be too invasive or expensive. In such a case, a cheaper diagnosis tool is often helpful to identify a smaller set of people who likely have the disease for further, more advanced procedures. Typically parameters in the model for predicting the disease is determined via the so-called training data set, in which the true disease status and diagnosis results are known for each individual. Supervised learning can also be nonparametric, for instance, the nearest-neighbor method. The notion of this method is that each person has compared certain characteristics with his neighbors (e.g., those in the same city or whatever is defined) who have or don't have the disease. If the person has the characteristics, to a close or similar degree, to those among the group who have the disease, we predict the person will also have the disease; otherwise, the person is predicted to not acquire this disease.

In *supervised learning*, the learner will give a response $\hat{y}$ based on input x, and will be able to compare his response $\hat{y}$ to the target (correct) response y. In other words, the "student" presents an answer $\hat{y}_i$ for each x_i in the training sample, and the supervisor provides either the correct answer and/or an error associated with the student's answer. Formally, supervised learning is to determine a decision rule that involves response $\hat{y}_i$ based on input every possible x_i so that the loss L is minimized. The loss function can be defined by simple mean squared error $L = \sum_{i=1}^{N} (\hat{y}_i - y_i)^2$ or others that fit the problem of interest. To determine the decision rule, a training set $(y_i, x_i; i = 1, ..., m)$ is obtained. In reality, the response y_i for a given x_i is often not deterministic or unique, rather the relationship is probabilistic characterized by some joint probability density $P(X, Y)$. However, in real world experiences, $P(X, Y)$ is often unknown, and thus the decision problem becomes ill-defined. When $P(X, Y)$ is known, the supervised learning can be defined statistically.

1.3.3   Unsupervised Learning: Clustering and Association

In *unsupervised learning*, the learner receives no feedback from the supervisor at all. Instead, the learner's task is to re-represent the inputs in a more efficient way, for instance, as clusters or with a reduced set of dimensions. Unsupervised learning is based on the similarities and differences among input patterns. The goal is to find hidden structure in unlabeled data without the help of a supervisor or teacher providing a correct answer or degree of error for each observation.

A typical example of unsupervised learning is a self-organizing map (SOM) in data visualization. An SOM is a type of artificial neural network (ANN) that is trained using unsupervised learning to produce a low-dimensional, discretized representation of the input space of the training samples, called a map. An SOM forms a semantic map where similar samples are mapped close together and dissimilar ones apart. The goal of learning in the self-organizing map is to cause different parts of the network to respond similarly to certain input patterns. This is partly motivated by how visual, auditory, or other sensory information is handled in separate parts of the cerebral cortex in the human brain.

Documentation classification is a typical kind of unsupervised learning. Unsupervised learning is based on the perceived similarities and differences among the input patterns. However, similarities depend on not only the input attribute x, but also the outcome (response) y of interest. Therefore, unsupervised learning has implicit outcomes of interest that might be difficult to define exactly. In this case one has a set of n observations $(x_1, ..., x_n)$ of a random vector X having joint density $P(X)$. The goal is to directly infer the properties of this probability density without the help of a supervisor or teacher providing a correct answer or degree-of error for each observation (Hastie et al., 2001).

1.3.4   Reinforcement Learning

Reinforcement learning (RL) is an active area in artificial intelligence study or machine learning that concerns how a learner should take actions in an environment so as to maximize some notion of long-term reward. Reinforcement learning emphasizes real-world experiences. The algorithm gets told how well it has performed so far or how good the answer is, but does not get told exactly how to make improvement. Reinforcement learning algorithms attempt to find a policy (or a set of action rules) that maps states of the world to the actions the learner should take in those states. In economics and game theory, reinforcement learning is considered a rational interpretation of how equilibriums may arise. A stochastic decision process (Chang, 2010) is considered a reinforcement learning model.

RL is widely studied in the field of robotics. Unlike supervised learning, in reinforcement learning, the correct input-output pairs are never presented. Furthermore, there is a focus on on-line performance, which involves finding a balance between exploration of uncharted territory and exploitation of one's current knowledge. In doing so, the agent is exploiting what it knows to receive a reward. On the other hand, trying other possibilities may produce a better reward, so exploring is sometimes the better tactic.

An example of RL would be game playing. It is difficult to determine the best move among all possible moves in the game of chess because the number of possible moves is so large that it exceeds the computational capability available today. Using RL we can cut out the need to manually specify the learning rules; agents learn simply by playing the game. Agents can be trained by playing against other human players or even other RL agents. More interesting applications are to solve control problems such as elevator scheduling or floor cleaning with a robot cleaner, in which it is not obvious what strategies would provide the most effective and/or timely service. For such problems, RL agents can be left to learn in a simulated environment where eventually they will come up with good controlling policies. An advantage of using RL for control problems is that an agent can be retrained easily to adapt to environmental changes and can be trained continuously while the system is online, improving performance all the time.

1.3.5   Swarm Intelligence

According to Mitchell (2009), a *complex system* is a system in which large networks of components (with no central control) where each individual in the system has no concept of collaboration with her peers, and simple rules of operation give rise to complex collective behavior, sophisticated information processing, and adaptation via learning or evolution. Systems in which organized behavior arises without a centralized controller or leader are often called *self-organized systems*, while the intelligence possessed by a complex system is called *swarm intelligence* (SI) or *collective intelligence*. SI is an emerging field of biologically inspired artificial intelligence, characterized by micro motives and macro behavior.

The structure of a complex system can be characterized using a network. However, the dynamics are not simply the sum of the static individual components. Instead, the system is adaptive through interactions between the individual components and the external environment. Such adaptations at the macroscopic level occur even when the simple rules governing the behaviors of each individual have not changed at all.

A good example of SI is that of ant colonies which optimally and adaptively forage for food. Ants are able to determine the shortest path leading to a food source. The process unfolds as follows. Several ants leave their nest to forage for food, randomly following different paths. Ants keep releasing *pheromones* (a chemical produced by an organism that signals its presence to other members of the same species) during the food search process. Such pheromones on the path will gradually disperse over time. Those ants reaching a food source along the shortest path are sooner to reinforce that path with pheromones, because they are sooner to come back to the nest with food; those that subsequently go out foraging find a higher concentration of pheromones on the shortest path, and therefore have a greater tendency (higher probability) to follow it. In this way, ants collectively build up and communicate information about locations, and this information adapts to changes in the environmental conditions! The SI emerges from the simple rule: follow the smell of pheromones.

Swarm intelligence is not an "accident" but rather a property of complex systems. It does not arise from a rational choice, nor from an engineered analysis. Individuals in the system have no global perspective or objective. They are not aware of what's globally happening. They are not aware how their behavior will affect the overall consequences. The behavior of swarm intelligent systems is often said to be an *emergent behavior*.

There are a lot of examples of emergent behaviors: bee colony behavior, where the collective harvesting of nectar is optimized through the waggle dance of individual worker bees, flocking of birds, which cannot be described by the behavior of individual birds, market crashes, which cannot be explained by "summing up" the behavior of individual investors, and human intelligence that cannot be explained by the behaviors of our brain cells. Likewise, a traffic jam is not just a collection of cars, but a self-organized object which emerges

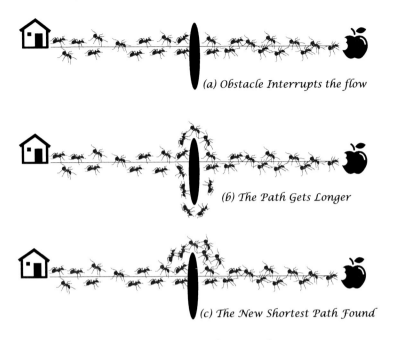

(a) Obstacle Interrupts the flow

(b) The Path Gets Longer

(c) The New Shortest Path Found

Swarm Intelligence: Ants Adapt to the Environment

FIGURE 1.4
Swarm Intelligence of Ants

from and exists at a level of analysis higher than that of the cars themselves (Xiaohui Cui, Swarm Intelligence Presentation, U.S. Department of Energy).

In a complex system, an individual agent neither has enough intelligence to solve the problem at hand nor has any goal or intention to solve it. A collection of intelligent agents can produce a better solution to a problem facing the group than the sum of the abilities of all agents when they work individually.

Interestingly, not long ago Southwest Airlines was wrestling with a difficult question: Should it abandon its long-standing policy of open seating on planes (Miller, 2010)? Using computer simulations based on artificial ant colonies, Southwest figured out that the best strategy is to assign the seats at check-in, but boarding would still be first-come, first-served. Now this strategy has become a standard for various airlines. Further applications of SI have been developed in cargo systems and in many other fields.

An SI algorithm is a kind of computer program comprising a population of individuals that interact with one another according to simple rules in order to solve problems, which may be very complex. Individuals in an SI system have mathematical intelligence (logical thought) and social intelligence (a common social mind). Social interaction thus provides a powerful problem-solving algorithm in SI.

The swarm intelligence characteristics of a human network integrate two correlated perspectives on human behavior: cognitive space and social space. In SI, we see the evolution of collective ideas, not the evolution of people who hold ideas. Evolutionary processes have costs: redundancy and futile exploration are but two. But such processes are adaptive and creative. The system parameters of SI determine the balance of exploration and exploitation. Thus uniformity in an organization is not a good sign.

An ant is simple, while a colony of ants is complex; neurons are simple, but brains are complex as a swarm. Competition and collaboration among cells lead to human intelligence; competition and collaboration among humans form a social intelligence, or what we might call the global brain. Nevertheless, such intelligence is based on a human viewpoint, and thus it lies within the limits of human intelligence. Views of such intelligence held by other creatures with a different level of intelligence could be completely different!

Swarm intelligence is different from reinforcement learning. In reinforcement learning, an individual can improve his level of intelligence over time since, in the learning process, adaptations occur. In contrast, swarm intelligence is a collective intelligence from all individuals. It is a global or macro behavior of a system. In complex systems there is a huge number of individual components, each with relatively simple rules of behavior that never change. But in reinforcement learning, there is not necessarily a large number of individuals, in fact there can just be one individual with built-in complex algorithms or adaptation rules.

A *committee machine* is a neural network that combines results from other neural networks. It is an approach to scaling artificial intelligence with multiple neural networks referred to as "experts."

The committee machine might use a variety of algorithms to assimilate expert input into a single output such as a decision. A committee machine can also be used to acquire knowledge with a technique known as boosting, whereby the committee machine learns by integrating the learning of experts.

1.3.6   Evolutionary Learning

Biological evolution can be viewed as a learning process: biological organisms adapt to improve the probabilities of survival and having offspring in their environment. This implies that we can use biological evolutionary mechanisms as artificial learning algorithms. Genetic programming (GP) is exactly such a technique. Richard Forsyth (1981) demonstrated the successful evolution of small programs, represented as trees, to perform classification of crime scene evidence. In GP, computer programs are encoded as a set of genes that are then modified (evolved) using an evolutionary algorithm. The methods used to encode a computer program in an artificial chromosome and to evaluate its fitness with respect to the predefined task are central in the GP technique. The term "genetic programming" was coined by Goldberg in 1983 in his PhD dissertation (Goldberg, 1983). In 1988 John Koza patented his invention of a genetic algorithm (GA) for program evolution. This was followed by publication in the International Joint Conference on Artificial Intelligence. The series of four books by Koza collected many GP applications in 1992 and established GP. Koza (2010b) summarized 77 results where GP was human-competitive. GP has be used for software synthesis and repair, predictive modeling, data mining, financial modeling, and image processing, cellular encoding, and for mining DNA chip data from cancer patients (Langdon and Buxton, 2004). GP has been successfully used as an automatic programming tool, a machine learning tool and an automatic problem-solving engine. GP is especially useful in the domains where the exact form of the solution is not known in advance or when an approximate solution is acceptable. Some of the applications of GP are curve fitting, data modeling, symbolic regression, feature selection, and classification.

1.4   Summary

AI and robotics have come a long way to reach their current level of development, but they are still not even close to human intelligence, and perhaps will not be within 100 years. The three-wave view is a simple way to characterize the preceding AI history, but there are large overlaps between waves in terms of the timeline. The major achievements in AI to date are centered at deep learning in the fields of image, voice, and language recognition. We have discussed five categories of learning methods: supervised, unsupervised, reinforcement, evolutionary learning, and swarm intelligence (Figure 1.5).

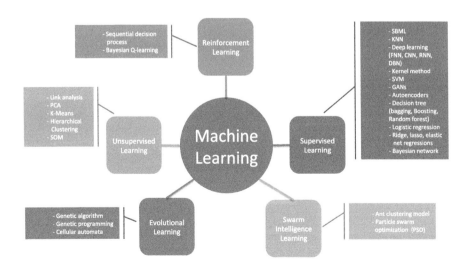

FIGURE 1.5
List of Machine Learning Methods

1.5 Problems

1.1: What is the three-wave view in AI history?

1.2: What are the major achievements in AI so far?

1.3: Describe the following terms: supervised, unsupervised, reinforcement, and evolutionary learning, and swarm intelligence

2

Classical Statistics and Modern Machine Learning

2.1 Essentials for Medical Data Scientists

2.1.1 Structured and Unstructured Data

The kinds of random variables considered here are binary, categorical (nominal and ordinal), time-to-event, vectors, matrices, sensors, sequences, trees, sets, shapes, manifolds, and functions.

Structured data refers to information with a high degree of organization. Such a typical dataset is a relational database, such as a spreadsheet, where all data have the same format, same types, same variables, and often have similar high quality. A relational database is seamless and readily searchable by simple, straightforward search engine algorithms or other search operations. In contrast, unstructured data, such as emails and social media data, are essentially the opposite. They often have mixed formats (image, text, video, sound clips), different variables, and low quality. Traditionally, classical statistics handles structured data, but they have difficulty handling massive unstructured data efficiently without manual interventions. Machine learning is expected to handle structured and unstructured data better. Since the pool of information is so large, current data mining techniques often miss a substantial amount of the information that's out there, much of which could be game-changing data if efficiently analyzed. We should develop AI to convert unstructured data into structured data or develop new AI systems that can directly handle unstructured data efficiently.

Structured data often come from surveys, epidemiological studies, or experiments such as clinical trials. Survey data are familiar to most of us. Randomized clinical trials, as the gold standard for experimentation, will be discussed in more detail later in this chapter.

A survey is a convenient and relatively inexpensive tool for empirical research in social sciences, marketing, and official statistics. Surveys usually involve a list of questions and can be conducted face-to-face, online, via phone or by mail. Most surveys are structured with a set of multiple choice questions, but unstructured surveys are also used in research. The non-randomization nature of the survey data collection can lead to biases. Understanding those biases can be helpful in survey questionnaire design, survey data analysis, result interpretation, and in the utilization of the survey results for implementation of any artificial intelligence method in survey systems.

Response bias refers to a wide range of tendencies for participants to respond inaccurately or falsely to questions. Response bias can be induced or caused by numerous factors, all relating to the idea that human subjects do not respond passively to stimuli, but rather actively integrate multiple sources of information to generate a response in a given situation. Examples of response bias include the phrasing of questions in surveys, the demeanor of the researcher, the way the experiment is conducted, or the desires of the participant to be a good experimental subject and to provide socially desirable responses. All of these can affect the response in some way.

Acquiescence bias refers to a respondent's tendency to endorse the questions in a measure. For example, participants could be asked whether they endorse the statement "I prefer to spend time with others," but then later on in the survey also endorses "I prefer to spend time alone," which is a contradictory statement. To reduce such a bias, researchers can make balanced response sets in a given measure, meaning that there are a balanced number of positively- and negatively-worded questions. Other bias, so-called question order bias, can result from a different order of questions or order of multiple choice. "Social desirability bias" is a type of response bias that influences a participant to deny undesirable traits, and ascribe to themselves traits that are socially desirable.

Online survey response rates can be very low, a few percent. In addition to refusing participation, terminating surveying during the process, or not answering certain questions, several other non-response patterns are common in online surveys. Response rates can be increased by offering some other type of incentive to the respondents, by contacting respondents several times (follow-up), and by keeping the questionnaire difficulty as low as possible. There is a drawback to using an incentive to garner a response, that is, it introduces a bias. Participation bias or non-response bias refers the potential systematic difference in response between responders and non-responders. To test for non-response bias, a common technique involves comparing the first and fourth quartiles of responses for differences in demographics and key constructs. If there is no significant difference this is an indicator that there might be no non-response bias.

Epidemiological studies as non-experimental studies can be retrospective or prospective. Suppose we want to investigate the relationship between smoking and lung cancer. We can take one of the following approaches: (1) look at hospital records, get the same number of patients with or without lung cancer, then identify them as smokers ("case") or nonsmokers ("control"); (2) look at hospital records, select about the same number of historical smokers and non-smokers, and find out if they have died of lung cancer or for other reasons, or are still alive; (3) look at hospital records, select about same number of historical smokers and non-smokers who are still alive, then follow up for a

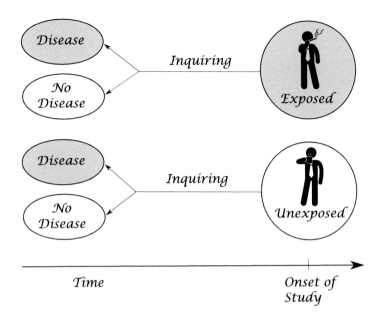

FIGURE 2.1
Retrospective Study

period, say, of 10 years to see if they die of lung cancer or not (or die for other reasons or are still alive). Approaches 1 and 2 are retrospective studies (Figure 2.1); approach 1 is also called a case-control study, while approach 2 is called a historical cohort study; approach 3 is called a prospective (cohort) study (Figure 2.2). See Chang's book (Chang, 2014), *Principles of Scientific Methods* for details.

A combination of subject matter expertise, strong statistical ability, and software engineering acumen is what people commonly expect a data scientist's skillset should comprise.

2.1.2   Random Variation and Its Causes

Variations can come from different sources. Variability within an individual is the variation in the measures of a subject's characteristics over time. Variability between individuals concerns the variation from subject to subject. *Instrumental variability* is related to the precision of the measurement tool. Other variabilities can be attributed to the difference in experimenter or other factors. Variability in measurements can be either random or systematic.

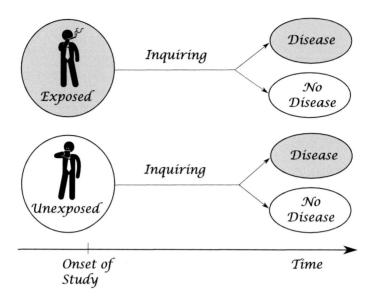

FIGURE 2.2
Prospective Study

2.1.3   Internal and External Validities

Validity is the accuracy of a lab test, which is dependent on the instrument, the person who conducts the test, and the lab environment. Reliability is how repeatable the test result is. A test can be valid, but the reliability can be very low. For example, one can measure blood pressures very accurately, but each of such measures can be unreliable if the patient's blood pressures vary greatly each time. In this case, maybe the average of the blood pressures over time is valid and more reliable. Here we are not only considering the valid lab test, but also readings of broader studies, such as clinical trials.

1. *Internal validity* is concerned with correctly concluding that an independent variable is, in fact, responsible for variation in the dependent variable. An example of violation of internal validity is an analytical method used that does not match the experiment's design.

2. *External validity* is concerned with the generalizability of research findings to and across populations of subjects and settings. An example of violation of external validity could be seen when, as usual, we recruit patients of different races in a clinical trial. If the drug has different effects on different races and the race-composition in the trial is different from the target population, the simple average drug effect on the trial patients will misrepresent the mean effect on

the target population, This is a phenomenon we have seen in current drug labeling practices.

2.1.4 Placebo and Nocebo Effects

A *placebo* is a substance or procedure that is objectively lacking in specific activity for the condition being treated (Moerman and Jonas, 2002). In medical experiments, placebos can be pills, creams, inhalants, and injections that do not involve any active ingredients. In studies of medical devices, ultrasound can act as a placebo. Sham surgery, sham electrodes implanted in the brain, and sham acupuncture can all be used as placebos.

The *placebo effect* is related to the perceptions and expectations of the patient. If the substance is viewed by the patient as helpful, it can help to heal, but, if it is viewed as harmful, it can cause negative effects, which is known as the *nocebo effect*.

Placebo effects are generally more significant in subjective measurements, such as with patient-reported outcomes, than in an objective laboratory measurement such as a hemoglobin level.

2.1.5 Bias, Bias, and Bias

Bias is a systematic error in a study that leads to a distortion of the results. Human perception occurs by a complex, unconscious process of abstraction, in which certain details of the incoming sense data are noticed and remembered, and the rest forgotten.

Selection bias refers to a subject selection process or sampling procedure that likely produces a sample not representative of the population, or produces samples that result in a systematical (probabilistic) imbalance of confounding variables between comparison groups.

The *Literary Digest* case is an unforgettable instance of selection bias causing one of the biggest political polling mistakes in history, despite the use of copious amount of data in the polling. The *Literary Digest*, founded by Isaac Kaufmann Funk in 1890, was an influential American Wiki-plagiarism weekly magazine. It is best-remembered today for the circumstances surrounding its demise. Before 1936, it had always correctly predicted the winner in the presidential election, five correct predictions in a row. As it had done previously, it conducted a straw poll regarding the likely outcome of the 1936 presidential election. The poll showed that the Republican candidate, Governor Alfred Landon of Kansas, was likely to be the overwhelming winner. However, in November of that year, when Landon carried only Vermont and Maine, President Roosevelt carrying the 46 other states, the magazine was so discredited by the discrepancy that it soon folded.

Although it had polled 10 million individuals (of whom about 2.4 million responded, an astronomical total for any opinion poll), the *Digest* had surveyed its own readers first, a group with disposable incomes well above the national average of the time, shown in part by their ability to still afford a magazine subscription during the depths of the Great Depression, and then voters from two other readily available lists: one of registered automobile owners and the other of telephone users, again representing wealthier-than-average Americans of that era.

This debacle led to a considerable refinement in the public's opinion of polling techniques and later came to be regarded as ushering in the era of modern scientific public opinion research.

Confirmation bias is a tendency of people to favor information that confirms their beliefs or hypotheses. Confirmation bias is best illustrated using examples in clinical trials. When knowledge of the treatment assignment can affect the objective evaluation of treatment effect and lead to systematic distortion of the trial conclusions, we have what is referred to as observer or ascertainment bias.

A patient's knowledge that they are receiving a new treatment may substantially affect the way they feel and their subjective assessment. If a physician is aware of which treatment the patient is receiving, this can affect the way the physician collects the information during the trial and can influence the way the assessor analyzes the study results.

Confirmation bias happens more often and is more severe in subjective evaluations than in "objective laboratory evaluations." However, it does exist in laboratory-related outcomes, where the knowledge of treatment assignment can impact how the test is run or interpreted, although the impact of this is most severe with subjectively graded results, such as pathology slides and radiological images. Blinding can effectively reduce ascertainment bias, as will be discussed in Section 3.3.

Human observations are biased toward confirming the observer's conscious and unconscious expectations and view of the world; we "see what we expect or want to see." This is called confirmation bias. This is not deliberate falsification of results, but can happen to good-faith researchers. As a result, people gather evidence and recall information from memory selectively and interpret it biasedly.

How much attention the various perceived data are given depends on an internal value system, by which we judge how important the data are to us. Thus two people can view the same event and come away with entirely different perceptions of it, even disagreeing about simple facts. This is why eyewitness testimony is so unreliable.

Confirmation bias is remarkably common. Whenever science meets some ideological barrier, scientists are accused of misinterpretation that can range from self-deception to deliberate fraud. We should know that one can tell all the truth, and nothing but the truth, but at the same time what one says can be still very biased. For example, someone might only talk about the things on the positive side of the ledger. In a church or other religious society, the same stories are told again and again; these stories often only fit into the society's needs (not a criticism).

Similarly, publication bias is the tendency of researchers and journals to publish results that appear "positive," but may or may not be truly positive. Some publication bias can be avoided or reduced, but other instances of it cannot. On the other hand, we cannot publish positive and negative results equally; otherwise, since they are many more negative results than positive ones, it could just be too much for our brains to handle!

Bias can also come from analysis, such as *survival bias*. In studying the association between tumor response and survival, people tend to divide cancer patients into two groups, tumor response and non-response groups, and then see if the survival time is longer in the response group than the non-response group. However, simply dividing into these two groups will lead to a biased estimate of survival time, since a patient who survives longer has more chance to respond, that is, a potential responder will only belong to the responder group if he/she survives until time of response. Individuals in the responder group are surviving for a specific time, and this gives them an unfair survival advantage: immortal time bias. Similarly, people seem to have discovered that smarter people (e.g., Nobel prize winners and Oscar winners) live longer, but forget that winning such honor often requires great achievement, which in turns requires long survival. A person who doesn't live long is unlikely to receive a Nobel prize.

In statistics, the bias of an estimator for a parameter is the difference between the estimator's expected value and the true value of the parameter being estimated. It is a common practice for a frequentist statistician to look for the minimum unbiased estimator. However it is not always feasible to do so, and unbiased estimators may not even exist. Therefore, we have to balance between variability and bias. It is also controversial regarding the importance of unbiasedness of an estimator because (1) repeated same experiments generally do exist or are irrelevant, and (2) for the same magnitude, an unbiased estimate wouldn't tell you it's an overestimate or underestimate, but a biased estimate tells us the direction in a probability sense.

Finally, let me quote myself: "Each of us is biased in one way or another. However, collectively, we as a social entity are not biased, which must be. In fact, the collective view defines 'reality' or the unbiased world."

2.1.6   Confounding Factors

In statistics, a *confounding factor* (also *confounder*, *hidden variable*, or *lurking variable*) is an extraneous variable that correlates, positively or negatively, with both the dependent variable and the independent variable. Such a relationship between two observed variables is termed a spurious relationship (Figure 2.3).

A classic example of confounding is to interpret the finding that people who carry matches are more likely to develop lung cancer as evidence of an association between carrying matches and lung cancer. Carrying matches or not is a confounding factor in this relationship: smokers are more likely to

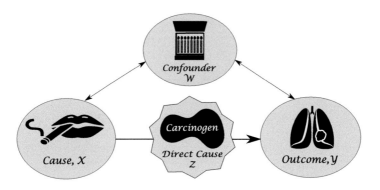

FIGURE 2.3
Direct Cause versus Confounder

carry matches and they are also more likely to develop lung cancer. However, if "carrying matches" is replaced with "drinking coffee," we may easily conclude that coffee more likely causes cancer.

For a variable to be a confounder in a clinical trial it must satisfy three conditions:

1. It must be associated with the treatment (the main factor).

2. It must be a predictor (not necessarily a cause) of the outcome being measured.

3. It must not be a consequence of the treatment (the main factor) itself.

A factor is not a confounder if it lies on the causal pathway between the variables of interest. Such a factor can be is a surrogate or a direct cause. For example, the relationship between diet and coronary heart disease may be explained by measuring serum cholesterol levels. Cholesterol is a surrogate, but not a confounder because it may be the causal link between diet and coronary heart disease. In cigarette-smoke caused lung cancer, carcinogens in the cigarette is the direct cause of the cancer.

Bias creates an association that is not true, while confounding describes an association that is true, but potentially misleading. Confounders are more usually a problem in observational studies, where the exposure of a risk factor is not randomly distributed between groups. In evaluating treatment effects from observational data, prognostic factors may influence treatment decisions, producing the so-called confounding-by-indication bias. Controlling for known prognostic factors using randomization can reduce such a bias, but it is always possible that unknown confounding factors are not balanced even when a randomization procedure is used.

In a clinical trial, some adverse reactions (ARs) may occur randomly and depend on the length of the follow-up period. The longer the follow-up is, the more ARs one is likely to see. Two drugs in evaluation having different dose schedules (e.g., one is given weekly, the other is given biweekly) can lead to different observational periods and/or frequencies for ARs, and can artificially make one drug appear to be safer than the other, even if in fact they have a similar safety profile. Likewise, consider the case when drug effectiveness is measured by the maximum response during the treatment period. If two drugs in evaluation, one given once a week, the other given five times a week, have their effects on patients measured in a clinic, the difference in observational frequency can create an artificial treatment difference simply because drug effects include a random component, so that more frequent measurements of response are more likely to capture a maximum value.

2.1.7 Regression to the Mean

The phenomenon of *regression to the mean* was first identified by Sir Francis Galton in the 19th century. Galton, a half-cousin of Charles Darwin, was a geographer, meteorologist, and tropical explorer, a founder of differential psychology, the inventor of scientific fingerprint identification, a pioneer of statistical correlation and regression, a convinced hereditarian and eugenicist, and a best-selling author.

Galton discovered that sons of very tall fathers tended to be shorter than their fathers and sons of very short fathers tended to be taller than their fathers. A similar phenomenon is noticed: A class of students takes two editions of the same test on two successive days; it has frequently been observed that the worst performers on the first day will tend to improve their scores on the second day, and the best performers on the first day will tend to do worse on the second day. This phenomenon is called "regression to the mean," and is explained as follows. Exam scores are a combination of skill and luck. The subset of students scoring above average would be composed of those who were skilled and did not have especially bad luck, together with those who were unskilled but were extremely lucky. On a retest of this subset, the unskilled will be unlikely to repeat their lucky performance, while the skilled will be unlikely to have bad luck again. In other words, their scores will likely go back (regress) to values close to their mean scores.

The phenomenon of regression to the mean holds for almost all scientific observations. Thus, many phenomena tend to be attributed to the wrong causes when regression to the mean is not taken into account. What follows are some real-life examples.

The calculation and interpretation of "improvement scores" on standardized educational tests in Massachusetts provides a good example of the regression fallacy. In 1999, schools were given improvement goals. For each school the Department of Education tabulated the difference in the average score achieved by students in 1999 and in 2000. It was quickly noted that most of the worst-performing schools had met their goals, which the Department of Education took as confirmation of the soundness of their policies. However, it was also noted that many of the supposedly best schools in the Commonwealth, such as Brookline High School (with 18 National Merit Scholarship finalists), were declared to have failed. As in many cases involving statistics and public policy, the issue was debated, but "improvement scores" were not announced in subsequent years, and the findings appear to be a case of regression to the mean.

Regression to the mean is a common phenomenon in clinical trials. Because we only include patients who meet certain criteria, e.g., hemoglobin levels lower than 10 among patients who enter the study, some subjects' levels will accidentally be below 10 and will regress to the mean later on in the trial. Therefore, the overall treatment effect usually includes the treatment effect due to the drug, the placebo effect, and a part due to regression to the mean.

2.2 Revolutionary Ideas of Modern Clinical Trials

2.2.1 Innovative and Adaptive Development Program

In recent years, the cost for drug development has increased dramatically, but the success rate of new drug applications (NDAs) remains low. The pharmaceutical industry devotes great efforts in innovative approaches, especially on adaptive design. An adaptive clinical trial design is a clinical trial design that allows adaptations or modifications to aspects of the trial after its initiation without undermining the validity and integrity of the trial (Chang, 2014). Adaptive design can also allow a combination of trials from different phases. The adaptation can be based on internal or external information to the trial. The purposes of adaptive trials are to increase the probability of success of reducing both the cost and the time to market, and to deliver the right drug to the right patient.

In modern clinical trials, there are three essential elements that generally fit most experiments: (1) control, (2) randomization, and (3) blinding (Table 2.1). They are effective ways to control confounders and reduce bias in the design and conduct of a trial. The combination of these three comprises the gold standard for a scientific experiment.

The primary purpose of a control as comparison group is to remove the placebo effect and any effect caused by regression to the mean as might be observed in a single-group study; the blinding procedure is critical for removing or reducing the confirmation and operational bias. The purpose of randomization is to reduce the selection bias and bias caused by the imbalance

TABLE 2.1
Key Concepts of Modern Clinical Trials

	Purpose (to reduce)
Control	placebo effect and effect due to regression to the mean
Blinding	confirmation and operational bias
Randomization	selection and confounder bias
Sample Size	random variation and chance of false negative funding

of potential and hidden confounders between the two treatment groups. Adequate sample size is an effective way to reduce the random variabilities, ensure the precision of statistical results, and reduce the probability of false negative findings or increase the power to detect treatment difference.

2.2.2 Control, Blinding, and Randomization

We now know there is a placebo effect. An important task is to obtain the "true" treatment effect from the total effect by subtracting the placebo effect. This can be done in an experiment with two groups in a clinical trial, one group treated with the test drug and the other treated with a placebo. More generally, the placebo group can be replaced with any treatment group (reference group) for comparison. Such a general group is called the control or control group. By subtracting the effect in the placebo group from the test group, we can tell the "pure" effect of the drug candidate. However, such a simple consideration of getting a pure treatment effect appears to be naïve, because knowledge of which treatment a patient is assigned can lead to subjective and judgmental bias by the patients and investigators. For example, patients who know they are taking a placebo may have the nocebo effect, and those who are aware that they are taking the test drug may have over-reported the response. To avoid such bias we can implement another experimental technique called blinding.

Blinding can be imposed on the investigator, the experimental subjects, the sponsor who finances the experiment, or any combination of these actors. In a *single-blind* experiment, the individual subjects do not know whether they have been assigned to the experimental group or the control group. Single-blind experimental design is used where the experimenters either must know the full facts (for example, when comparing sham to real surgery). However, there is a risk that subjects are influenced by interaction with the experimenter—known as the experimenter's bias. In *double-blind* experiments, both the investigator and experimental subjects have no knowledge of the group to which they are assigned. A double-blind study is usually better than a single-blind study in terms of bias reduction. In a *triple-blind* experiment, the patient, investigator, and sponsor are all blinded from the treatment group.

Blind experiments are an important tool of the scientific method, in many fields of research. These include medicine, psychology and the social sciences, the natural sciences such as physics and biology, the applied sciences such as market research, and many others.

It is controversial as to how to determine a drug's pure treatment benefit. If a safe drug is approved for marketing, the benefit a patient gets is the effect of the drug without the subtraction of the placebo effect, because companies cannot market a placebo anyway under current health authority regulation in the US. Also, the actual medication administration is unblinded. The benefit of a treatment is expected to be more than that from the blind trials.

To control confounding in an experiment design, we can use randomization or stratified randomization. The latter will provide a better balance of confounders between intervention groups, and confounding can be further reduced by using appropriate statistical analyses, such as the so-called analysis of covariance.

Randomization is a procedure to assign subjects or experimental units to a certain intervention group based on an allocation probability, rather than by choice. For example, in a clinical trial, patients can be assigned one of two treatments available with an equal chance (probability 0.5) when they are enrolled in the trial.

The utilization of randomization in an experiment minimizes selection bias, balancing both known and unknown confounding factors, in the assignment of treatments (e.g., placebo and drug candidate). In a clinical trial, appropriate use of randomization procedures not only ensures an unbiased and fair assessment regarding the efficacy and safety of the test drug, but also improves the quality of the experiments and increases the efficiency of the trial.

Randomization procedures that are commonly employed in clinical trials can be classified into two categories: conventional randomization, and adaptive randomization.

2.3 Hypothesis Test and Modeling in Classic Statistics

Classical statistics (frequentist) focuses on the hypothesis testing with type-I error rate control. The factors included in a model must be statistically significant. Therefore, their model predictions are also constructed on the basis of statistical significance. Readers may be familiar with the frequentist paradigm. Therefore, we will just do a quick review.

2.3.1 Statistical Hypothesis Testing

A typical hypothesis test in the frequentist paradigm can be written as

$$H_o : \delta \in \Omega_0 \text{ or } H_a : \delta \in \Omega_a, \tag{2.1}$$

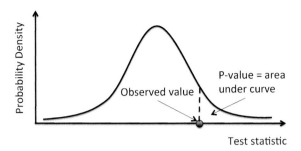

A p-value is the probability of an observed (or more extreme) result assuming the null hypothesis is true.

FIGURE 2.4
Interpretation of p-value

where δ is a parameter such as the effect of medical intervention, domain Ω_0 can be, e.g., a set of non-positive values, and the domain Ω_a can be the negation of Ω_0. In such a case, (2.1) becomes

$$H_o : \delta \leq 0 \text{ or } H_a : \delta > 0. \tag{2.2}$$

The probability of erroneously rejecting H_o, when H_o is true, is called the type-I error rate. When we make a decision to reject H_o or H_a, mistakes can happen. The allowable type-I error rate in a hypothesis test is called the size of the test or the level of significance, and is denoted by α. Similarly, the probability of erroneously rejecting H_a when H_a is true is called the type-II error rate. The probability of rejecting H_o is called the power of the hypothesis test, which is dependent on the value of the true δ in the population When H_o is true, the power is numerically equal to type-I error rate α.

The p-*value* is the probability of finding the observed or more extreme value of the test statistic under repeated experiments when the null hypothesis is true. A p-value, p, is often used to compare to the significance level α for decision-making. That is, if $p \leq \alpha = \alpha$, we reject the null hypothesis H_o; otherwise we don't reject H_o.

The extremeness here is informed by the hypothesis test. As an example, for the one-sided test in (2.2), "more extreme" means the value of the test statistic is larger than the observed one (Figure 2.4). For a two-sided test (2.3), "more extreme" means either smaller or larger than the observed test statistic. In the next section, we will give the p-value a very different interpretation, that is, in terms of similarity among subjects.

Multiplicity problem refers to the type-I error rate inflation when we use the same dataset to test different null hypotheses. For example, the error rate will be inflated from 5% to 9.8% for 2 hypothesis tests, and to 40.1% for 10 hypothesis tests (Table 2.2). .

TABLE 2.2

Error Inflation Due to Multiple Hypothesis Tests

Number of Tests	1	2	3	5	10
Familywise Error Rate	0.050	0.098	0.143	0.226	0.401

2.3.2   Generalized Linear Model

The simplest model that links K predictors $\mathbf{X} = (X_1, X_2, ..., X_K)$ to the response variable Y is perhaps a linear model:

$$Y = \mathbf{X}\boldsymbol{\beta} + \varepsilon, \tag{2.3}$$

where ε is random error due to hidden confounders that are not included in the model. Given are the observed data of K predictors from N subjects, $\mathbf{x} = [x_{ij}], i = 1, ..., N$ and $j = 1, ..., K$, and the observed responses $\mathbf{y} = (y_1, ..., y_N)$ for N subjects.

Given a normal error $\varepsilon \sim N(0, \Sigma)$, the maximum *likelihood estimate* of β is found as a solution to the normal equation

$$\mathbf{x}^T \Sigma^{-1} \mathbf{y} = \mathbf{x}^T \Sigma^{-1} \mathbf{x} \hat{\boldsymbol{\beta}}. \tag{2.4}$$

If $\mathbf{x}$ has full rank, the solution is uniquely given by

$$\hat{\boldsymbol{\beta}} = \left(\mathbf{x}^T \Sigma^{-1} \mathbf{x} \right)^{-1} \mathbf{x}^T \Sigma^{-1} \mathbf{y} \tag{2.5}$$

$$\hat{\boldsymbol{\beta}} \sim N_K \left(\boldsymbol{\beta}, \sigma^2 \left(\left(\mathbf{x}^T \Sigma^{-1} \mathbf{x} \right)^{-1} \right) \right) \tag{2.6}$$

where Σ is assumed known. The distribution of $\hat{\boldsymbol{\beta}}$ allows us to perform the hypothesis test on the parameters. Here the term "linear" means linear in parameters not the predictor. For example,

$$\theta = \beta_1 (x_1 + x_1^2) + \beta_2 x_2^2 \tag{2.7}$$

is a linear model. People also call it a polynomial model in terms of $\mathbf{X}$. $\mathbf{X}$ in the model can be any transformation of other variables. For example, we can use the data transform $\ln(x)$ before performing linear regression. In this scenario, the formulations for the linear regression are valid but apply regression instead to the transformed data.

A linear model can the generalized by applying transformations to the predictors (independent variables) and a transform to the predicted parameter μ. Such a generalized model is called a *generalized linear model* (GLM) that is characterized by (Nelder & Wedderburn, 1972):

(1) A dependent variable Y from an exponential family with density function

$$\pi(Y; \theta, \phi) = \exp[\alpha(\phi)\{Y\theta - f(\theta) + h(Y)\} + \beta(\phi, Y)], \tag{2.8}$$

where $\alpha(\phi) > 0$ so that for fixed ϕ we have an exponential family. The parameter ϕ could stand for a certain type of nuisance parameter such as the variance σ^2 of a normal distribution or the parameter p of a gamma distribution.

(2) A set of independent variables $\mathbf{x} = (x_1, x_2, .., x_K)$ and a linear predictor,

$$\eta = \mathbf{x}\boldsymbol{\beta} = \beta_0 + \beta_1 x_1 + \cdots \beta_K x_K. \tag{2.9}$$

(3) A link function $g(\cdot)$ that describes how the predicted parameter θ of Y, depends on the linear predictor,

$$g(\theta) = \eta. \tag{2.10}$$

When Y is normally distributed with mean θ and variance σ^2 and when g is the identity function, we have ordinary linear models with normally distributed errors.

From Eqs. (2.7) and (2.8), the parameter is obtained as

$$\theta = g^{-1}(\mathbf{x}\boldsymbol{\beta}) \tag{2.11}$$

Maximum-likelihood estimation of β remains popular (**to confirm**: MLE of θ is $g^{-1}\left(\mathbf{x}\hat{\beta}\right)$ where $\hat{\beta}$ is MLE of β).

Generalized linear models were formulated by John Nelder and Robert Wedderburn as a way of unifying various other statistical models, including linear regression, logistic regression, and Poisson regression through link functions. A link function provides the relationship between the linear predictor and the mean of the distribution function. The well-defined canonical link function can be derived from the exponential of the response's density function (2.6).

General linear models: A general linear model may be viewed as a special case of the generalized linear model with identity link and responses normally distributed. Most other GLMs lack closed-form estimates. However, for the normal distribution, the generalized linear model has a closed form expression for the maximum-likelihood estimates, which is convenient.

Logistic Regression: The most typical link function for *binary outcome* is the canonical logit link:

$$g(p) = \ln \frac{p}{1-p},$$

where p is the probability of success.

Probit link function is a popular choice of inverse cumulative distribution function. Alternatively, the inverse of any continuous cumulative distribution function (CDF) can be used for the link since the CDF's range is $[0,1]$, the range of the binomial mean. The normal CDF Φ is a popular choice and yields the *probit model*. Its link is

$$g(p) = \Phi^{-1}(p).$$

The *complementary log-log function* (cloglog) can be used for *Poisson distribution*:

$$g(p) = \ln\left(-\ln\left(1-p\right)\right).$$

The number of events is assumed to follow the Poisson distribution, specifically,

$$\Pr(0) = \exp{(-\mu)},$$

where μ is a positive number denoting the inverse of the expected number of events. If p represents the proportion of observations with at least one event, its complement is

$$(1 - p) = \Pr(0) = \exp(-\mu),$$

thus,

$$-\ln(1 - p) = \mu.$$

A linear model requires the response variable to take values over the entire real line. Since μ must be positive, we can enforce that by taking the logarithm, and letting $\ln{(\mu)}$ be a linear model. This produces the cloglog transformation

$$\ln(-\ln(1 - p)) = \ln(\mu).$$

Overfitting and Model Selection Overfitting refers to the phenomenon that a statistical model fits current dataset too well by, for example, including too many parameters, so that when the future data deviate from the current dataset due to randomness of the data, the model does not predict the outcome well. To overcome overfitting and improve prediction, only parameters with significant p-values after adjusting multiplicity or multiple testing will be included in the model. The model selection methods, such as the forward, backward, and stepwise methods, are familiar to traditional statisticians. However, such model selection involves multiple tests, and a problem is that when there are many parameters, as in genomics studies, multiplicity adjustment can result in significant loss in power to detect certain parameter effects.

2.3.3   Air Quality Analysis with Generalized Linear Model

We can use the glm( ) function in R for generalized linear model fitting. The defalt link functions for different outcomes are:

Family	Default Link Function
binomial	logit
Gaussian	identity
Gamma	inverse
Poisson	log

Example 2.1: New York Air Quality Measurements

The airquality dataset in the datasets package includes daily air quality measurements in New York, May to September 1973.

The dataset has 154 observations on 6 numerical variables: **Ozone** (Mean ozone in parts per billion from 1300 to 1500 hours at Roosevelt Island), **Solar.R** (Solar radiation in Langleys in the frequency band 4000–7700

Angstroms from 0800 to 1200 hours at Central Park), **Wind** (Average wind speed in miles per hour at 0700 and 1000 hours at LaGuardia Airport), **Temp** (Maximum daily temperature in degrees Fahrenheit at La Guardia Airport), **Month** and **Day**.

We are interested in how temperature changes in relation to the Ozone, Solar.R, and Wind. Here is the R code for the regression analysis with normal outcomes. Hence, the distribution family is Gaussian with identity as the default link function.

```
data("airquality", package = "datasets") # Load data
fit <- glm(Temp~Ozone+Solar.R+Wind, data=airquality, family=gaussian())
summary(fit) # display results
confint(fit) # 95% CI for the coefficients
predict(fit, type="response") # predicted values
residuals(fit, type="deviance") # residuals
```

Women Infertility Analysis with Logistic Regression

Logistic regression is useful when one is predicting a binary outcome from a set of continuous predictor variables. It is often preferred over discriminant function analysis because of its less restrictive assumptions. This time, we use the *infert* dataset from the datasets library in R. The R *datasets* package includes the infert dataset from Trichopoulos et al. (1976) with a total of 8 variables: Infertility after Spontaneous and Induced Abortion. This is a matched case-control study dates from before the availability of conditional logistic regression. There are 8 variables. (1) education: 0 = 0-5 years, 1 = 6-11 years, 2 = 12+ years; age: age in years of case; (3) parity: count; (4) number of prior: 0 = 0 induced abortions, 1 = 1, 2 = 2 or more; (5) case status: 1 = case, 0 = control; (6) number of prior: 0 = 0 spontaneous abortions, 1 = 1, 2 = 2 or more; (7) matched set number: 1-83; (8) stratum number: 1-63. Note: One case with two prior spontaneous abortions and two prior induced abortions is omitted.

```
# Logistic regression
library(datasets)
library(stats)
fit <- glm(case~age+parity+induced+spontaneous, data=infert,
family=binomial())
summary(fit) # display results
confint(fit) # 95% CI for the coefficients
exp(coef(fit)) # exponentiated coefficients
exp(confint(fit)) # 95% CI for exponentiated coefficients
predict(fit, type="response") # predicted values
```

Readers can try code. The outputs include the coefficients of parameters for the associated model. Everything else is self-explanatory

2.3.4 Lung Cancer Survival Analysis with Cox's Model

We use NCCTG Lung Cancer Data (cancer dataset) as example for the survival analysis in R. The dataset from the package survival in R includes the survival data in patients with advanced lung cancer from the North Central Cancer Treatment Group. Performance scores rate how well the patient can perform usual daily activities (Loprinzi, et al., 1994). The variables are explained as follows:

inst: Institution code, **time**: Survival time in days, **status**: censoring status 1=censored, 2=dead, **age**: Age in years, **sex**: Male=1 Female=2, **ph.ecog**: ECOG performance score (0=good 5=dead), **ph.karno**: Karnofsky performance score (bad=0-good=100) rated by physician, **pat.karno**: Karnofsky performance score as rated by patient, **meal.cal**: Calories consumed at meals, **wt.loss**: Weight loss in last six months.

```
# Survival Analysis for Mayo Clinic Lung Cancer Data
library(survival)
# learn about the dataset
help(lung)
# create a Surv object
survobj <- with(lung, Surv(time,status))
# Plot survival distribution of the total sample
# Kaplan-Meier estimator
fit0 <- survfit(survobj~1, data=lung)
summary(fit0)
plot(fit0, xlab="Survival Time in Days",
 ylab="% Surviving", yscale=100,
 main="Survival Distribution (Overall)")
# Compare the survival distributions of men and women
fit1 <- survfit(survobj~sex,data=lung)
# plot the survival distributions by sex
plot(fit1, xlab="Survival Time in Days",
 ylab="% Surviving", yscale=100, col=c("red","blue"),
 main="Survival Distributions by Gender")
 legend("topright", title="Gender", c("Male", "Female"),
 fill=c("red", "blue"))
# test for difference between male and female
# survival curves (logrank test)
survdiff(survobj~sex, data=lung)
# predict male survival from age and medical scores
```

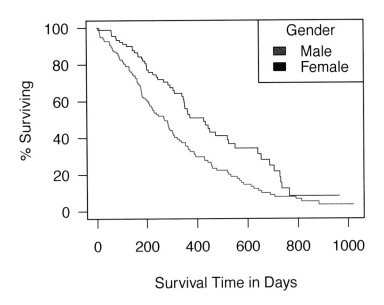

FIGURE 2.5
Kaplan-Meier Survival Curves

```
MaleMod <- coxph(survobj~age+ph.ecog+ph.karno+pat.karno,
 data=lung, subset=sex==1)
# display results
MaleMod
# evaluate the proportional hazards assumption
cox.zph(MaleMod)
```

The survival curves are shown in Figure 2.5.

2.3.5   Propensity Score Matching

In observational studies without randomization, the baseline characteristics
can be very different between two groups. The analysis of such data can lead
to biased results. Propensity score matching (PSM) is a statistical matching
technique that attempts to estimate the effect of an intervention, such as
a medical treatment for the covariates that predict receiving the treatment
in classical statistical analysis. Rosenbaum and Rubin (1983) introduced the
technique.

In a nonrandomized experiment, propensity scores (the probability of assigning a treatment given the covariates) may be used for adjusting the treatment imbalance. In this case, we need to create a treatment indicator instead of missingness indicator and model it using, e.g., a logistic model. After the propensity scores are estimated, the weight is defined as, if treatment = 1, wt=1/ (PS for treatment 1); if treatment = 0, wt=1/(PS for treatment 0).

Missing data can cause unbalance of confounders between treatment groups and bias the estimate of the treatment effect. Approaches to confounding control include propensity score control, matching, and stratification, multivariate regression, and inverse probability weighting. The propensity score (PS) method is a weight technique that is based on causes of the treatment (aka mimic randomization). A weight is assigned to each person according to the conditional probability of exposure given the confounder(s). After estimating the PS for each person, use it to control confounding by: (1) adjusting for the PS as a covariate, (2) stratifying the effect estimate by the PS, (3) matching group 1 and group 2 subjects by PS, and (4) Using PS as weight. Inverse probability weighting (IPW) is a weight technique that is based on the cause of outcome missing.

2.4 Model Selection in Machine Learning

2.4.1 Decision Approach

Modern statistical learning and model selection shift the focus from error rate control to the impact of error (decision problem) or the prediction.

A *statistical decision problem* is to determine an action rule $\wp$ and according to this rue, the action taken, $a \in A$, for given data x will minimize the defined loss function

$$V(a; x),$$

where A is the admissible action domain. The random variables, or input data are ideally from a certain (known or unknown) distribution, but in the real world of machine learning, the distribution of x or source of x is not well-defined.

Common loss functions in model selection include *ridge*, *lasso* (least absolute shrinkage and selection operator), and *elastic net* loss functions. These loss functions actually impose function smooth requirements and result in shrinkage effect, and make a trade-off between bias versus variance. When the decision approach is applied to model fitting, we offten use the term regularization. Before we discuss different learning models in detail, let's introduce some simple and commonly used model selection methods that are not p-value-based.

2.4.2 Regularization

Regularization imposes a penalty on the complexity of a model f on the basis of the principle of Occam's razor (Chang, 2014). A regularization term $R(f)$ is added to a loss function V:

$$\min_f \sum_{i=1}^{n} V\left(f\left(x_i\right), y_i\right) + \lambda R\left(f\right),$$

where V is an underlying loss function that describes the cost of predicting $f(x)$ when the observed value (label) is y, such as the square loss or hinge loss; and λ is a parameter which controls the importance of the regularization term.

The learning problem with the least squares loss function and *Tikhonov regularization* (also call *ridge regression*) can be solved analytically. Written in matrix form, the optimal parameters β will be the one for which the gradient of the loss function with respect to β is 0. Taking the linear mode $\hat{Y} = \mathbf{x}\beta$ as an example, the decision problem for the model selection is

$$\min_\beta \frac{1}{N}\left(\mathbf{x}\beta - Y\right)^T \left(\mathbf{x}\beta - Y\right) + \lambda \left\|\beta\right\|_2^2,$$

where $\mathbf{x}$ is the design matrix.

Letting derivative

$$\nabla_\beta = \frac{2}{N}\mathbf{x}^T\left(\mathbf{x}\beta - Y\right) + 2\lambda\beta = 0,$$

We obtain

$$\beta = \left(\mathbf{x}^T\mathbf{x} + \lambda N I\right)^{-1}\left(\mathbf{x}^T Y\right).$$

During training, this algorithm takes $O(K^3 + NK^2)$ computation time, where $K = $ number of parameters and $N = $ number of observations. The terms correspond to the matrix inversion and calculating $X^T X$, respectively. Testing takes $O(NK)$ time.

Regularization for learning enforces a sparsity constraint on β, which can lead to simpler and more interpretable models. This is useful in many real-life applications, such as in computational biology. An example is developing a simple predictive test for a disease in order to minimize the cost of performing medical tests while maximizing predictive power.

A sensible sparsity constraint is the L_0 norm $\left\|\beta\right\|_0$, defined as the number of non-zero elements in β. Solving a L_0 regularized learning problem, however, has been demonstrated to be NP-hard. The L_1 norm can be used to approximate the optimal L_0 norm via convex relaxation. It can be shown that the L_1 norm induces sparsity. In the case of least squares, this problem is known as *lasso* in statistics and *basis pursuit* in signal processing:

$$\min_{\beta \in R^p} \frac{1}{N}\left\|\mathbf{x}\beta - Y\right\|^2 + \lambda \left\|\beta\right\|_1.$$

L_1 regularization can occasionally produce non-unique solutions. This can be problematic for certain applications, and is overcome by combining L_1 with L_2 regularization in *elastic net regularization*, which takes the following form:

$$\min_{\beta \in R^p} \frac{1}{N} \|\mathbf{x}\boldsymbol{\beta} - Y\|^2 + \lambda \left(\alpha \|\boldsymbol{\beta}\|_1 + (1 - \alpha) \|\boldsymbol{\beta}\|_2^2 \right).$$

Elastic net regularization tends to have a grouping effect, where as correlated input features are assigned equal weights. Elastic net regularization is commonly used in practice and is implemented in many machine learning libraries.

2.4.3   Subset Selection

Best subset selection: For $k = 1, 2, ...K$ and for fit all ($\binom{K}{k}$) models that contain exactly k predictors, pick the best model among $\sum_{k=0}^{K} \binom{K}{k} = 2^p$ models using predefined criterion, such as C_K (*AIC*), *BIC*, or adjusted R_2. These criteria will be discussed later.

Forward stepwise selection: Let M_0 denote the null model, which contains no predictors. Add one more predictor, and pick the best model M_1, among K models. Add one more predictor into model M_1 and select the best model M_2 among $K - 1$ models with 2 predictors, ..., add the last predictor in the best model M_{K-1} to obtain model M_K. Finally, choose the best model among $M_0, M_1, ...,$ and M_K based on C_K (*AIC*), BIC, or adjusted R_2.

Backward stepwise selection: Start with the full model M_K, remove one predictor, identify the best model M_{k-1} among K models with exact $K - 1$ predictors; remove one predictor M_{K-1} and identify the best model M_{K-2} among the $K - 1$ models,..., remove the last predictor from M_1. Select the best model among $M_0, M_1, ...,$ and M_K.

The best subset model has 2^K models to compare, while the forward and backward stepwise model has $1 + K (K + 1) /2$ models to compare. However, the latter two methods might not be able to identify the global optimal model like the best subset method does.

Selection Criteria: C_k, *AIC*, *BIC*, and Adjusted R^2

$$C_k = MSE + \frac{2k}{N} \hat{\sigma}^2,$$

where $\hat{\sigma}^2$ is an estimate of the variance of the error associated with the outcome, k is the number of predictors in the model, and N is the number of observations. Essentially, the C_k statistic adds a penalty of $\frac{2k}{N}\hat{\sigma}^2$ to the training MSE in order to adjust for the fact that the training error tends to underestimate the test error. Clearly, the penalty increases as the number of predictors in the model increases.

The AIC criterion is defined for a large class of models fit by maximum likelihood. In the case of the model (6.1) with Gaussian errors, maximum likelihood and least squares are the same thing. In this case, AIC is given by

$$AIC = \frac{MSE}{\hat{\sigma}^2} + \frac{2k}{N},$$

where, for simplicity, we have omitted an additive constant.

For the least squares model with k predictors, the BIC is, up to irrelevant constants, given by

$$BIC = \frac{MSE}{\hat{\sigma}^2} + \frac{\ln(N)k}{N},$$

The R^2 value measures the proportion of error that can be explained by the model based on training data. To reduce overfitting, an adjusted R^2 is used in practice to penalize the model complexity:

$$\text{adjusted } R^2 = 1 - \frac{RSS/(N-k-1)}{TSS/(N-1)}.$$

Maximizing R^2 is equivalent to maximizing $RSS/(N-k-1)$.

A lower value of C_k, AIC and BIC or a larger value of adjusted R^2 indicates a better model. *Lasso* has a major advantage over *ridge* regression in that it produces simpler and more interpretable models which involve only a subset of the predictors.

2.4.4 Real-World Examples

We illustrate the best subset of linear model and *ridge* regression in R with the New York Air Quality example.

```
# Best subset of Linear Model:
NYair <- na.omit(airquality) # remove NAs
LMnull = lm(Temp ~1, data=NYair) # fit a constant
LMfull=lm(Temp~Ozone+Solar.R+Wind, data=NYair) # full Regression model
LMopt = step(LMfull, scope=list(lower=LMnull,upper=LMfull),
      data=NYair, direction="both", criterion=c("BIC"), trace=0)
LMnull; LMfull; LMopt
```

Readers can try code. The outputs include the coefficients of parameters for the associated model. Everything else is self-explanatory

```
# Ridge Regression
library(ridge)
lmRidge = linearRidge(Temp~Ozone+Solar.R+Wind, data=NYair)
predtemp = predict(lmRidge, NYair) # predicted values
```

```
MSEtrain = mean((predtemp-NYair$Temp)^2) # the mean squared error
lmRidge; MSEtrain
```

The *ridge regression* outputs indicate mean squared error is 45.06549.

Diabetes Predictive Model with Logistic Model

So far, we have developed a model based on a dataset (training dataset or training set) and test the prediction in terms of mean-squared error (MSE) based on the same dataset. Such prediction evaluation will be over-optimistic. In AI and ML, training data and test data are usually different datasets. We use a diabetes analysis as an example. Dataset *PimaIndiansDiabetes2* in *R* package mlbench includes 768 observations (392 after removing NA observations) and 9 variables.

```
library(mlbench)
data("PimaIndiansDiabetes2", package = "mlbench") # Load data
Diabetes2 <- na.omit(PimaIndiansDiabetes2) # remove NAs
# create a dummy variable for diabetes
Diabetes2$Diabetes01=ifelse(Diabetes2$diabetes=="pos", 1, 0 )
# Randomly sample 70% observations to training and the rest to testing
n = nrow(Diabetes2)
trainIndex = sample(1:n, size = round(0.7*n), replace=FALSE)
trainset=Diabetes2[trainIndex, ]
testset=Diabetes2[-trainIndex, ]
DiabetesModel <- glm(Diabetes01~pregnant+glucose+pressure+triceps+
insulin+mass+pedigree+age,
 data=trainset, family=binomial()) # logistic model
PreVs=predict(DiabetesModel, testset, type="response") # predicted values
MSEtrain = mean((PreVs-testset$Diabetes01)^2) # the mean squared error
```

The output indicates the mean squared error is 0.1411858.

2.5 Process to Apply Machine Learning to Data

2.5.1 General Steps in Applying Machine Learning

In order to use machine learning methods, there are common steps involved as outlined in *the following:*

1. Purpose: Elaborate the problem to be solved clearly and your purposes of using machine learning. This will help you narrow down a small set of machine learning methods for your target.

2. Data Source: Identify data source and data format (written on paper, recorded in text files, spreadsheets, or stored in an SQL database), then process (convert, merge) them into one electronic format suitable for analysis. This data will serve as the learning material that a machine learning algorithm uses to generate actionable knowledge. The quality of any machine learning project is based largely on the quality of data it uses. Most efforts (80%) are often devoted to the data process part in an application of a machine learning project.

3. Model Training: Unless your problem has been well studied and a trained model can be used as directed, you have to train ML algorithms using your training data. You will need to identify part of your data source for training purpose. Model training is the process of training the ML algorithm using a training dataset and determining the ML parameters.

4. Performance Evaluation: Before you apply the ML algorithm, you need to evaluate its performance. Such evaluation cannot generally be done using training data because overfitting is a problem, a trend of over-optimistically assessing a model performance. Therefore, you need to identify another dataset to better the model performance. For supervised learning, you can evaluate the accuracy of the model using a test dataset. In evaluating a ML method, using a single data trained model and evaluating it with a single test dataset is not sufficient: the result can appear to be good or bad due to the random choice of the datasets. Multiple training and test datasets can be used through bootstraping from the available dataset or cross-validation.

5. Model Optimization: Depending on model complexity, we often need to recursively use training data and test data to determine the optimal ML parameters and make comparisons among different ML algorithms to identify the optimal model among several ML methods whose parameters are optimized.

6. Apply the optimal model with trained parameters to the intended task.

In short, select your ML algorithm according your clearly defined goal, use training data to train model parameters, test the trained model and retrain it if necessary, and apply the retrained model (Figure 2.6).

2.5.2 Cross-Validation

In the diabetes predictive model with logistic model, we split the data into training and test datasets to construct the predictive model. Since results from a single training-testing split will be unstable and not reliable to model evaluation, we can improve the process by multiple, different out-of-sample training-testing splits, the train function in the *caret* library. We can set

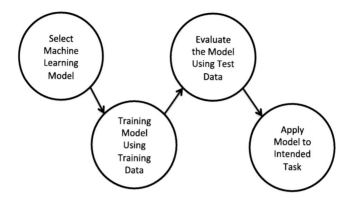

FIGURE 2.6
Steps in Application of Machine Learning

the out-of-sample training procedure to x-fold (e.g., 5-fold) cross-validation (method = "cv" and number = 5). Here is an example with a linear model:

```
library(caret)
NYair <- na.omit(airquality) # remove NAs
fit.LM <- train(Temp~Ozone+Solar.R+Wind, data=NYair, method = "lm",
          trControl = trainControl(method = "cv", number = 5))
fit.LM
```

To train the model with ridge regression using cross-validation, we can use the following *R* code:

```
library(caret)
NYair <- na.omit(airquality) # remove NAs
fit.Ridge <- train(Temp~Ozone+Solar.R+Wind, data=NYair, method = "ridge",
 trControl = trainControl(method = "cv", number = 5))
fit.Ridge
```

The output is shown below:

```
Ridge Regression
111 samples
   3 predictor
No pre-processing
Resampling: Cross-Validated (5 fold)
Summary of sample sizes: 90, 89, 89, 87, 89
```

Resampling results across tuning parameters:

lambda	RMSE	Rsquared	MAE
0e+00	7.240701	0.5318218	5.799783
1e-04	7.240694	0.5318178	5.799758
1e-01	7.247082	0.5272582	5.789384

RMSE was used to select the optimal model using the smallest value.
The final value used for the model was lambda = 1e-04.

To train the model with *lasso* regression using cross-validation, we can use the following *R* code:

```
library(caret)
library(lars)
NYair <- na.omit(airquality) # remove NAs
fit.Lasso <- train(Temp~Ozone+Solar.R+Wind, data=NYair, method = "lasso",
+ trControl = trainControl(method = "cv", number = 5))
fit.Lasso
```

The output is shown below:

```
The lasso
111 samples
  3 predictor
No pre-processing
Resampling: Cross-Validated (5 fold)
Summary of sample sizes: 89, 89, 89, 89, 88
```

Resampling results across tuning parameters:

lambda	RMSE	Rsquared	MAE
0.1	8.975250	0.5543852	7.332435
0.5	7.407286	0.5506864	6.091633
0.9	6.966473	0.5263960	5.748167

RMSE was used to select the optimal model using the smallest value.
The final value used for the model was lambda = 0.9.
The example of elastic net regression with R is represented in the following:

```
library(elasticnet)
prediction=predict.enet(fit.Lasso$finalModel, type='coefficients',
```

```
s=fit.Lasso$bestTune$fraction, mode='fraction')
##Fit models to test data with 100 observations
testset=NYair[1:100, ]
LassoTest<- predict(fit.Lasso, newdata=testset, na.action="na.pass")
postResample(pred = LassoTest, obs = testset[,c(1)])
```

The output is shown below:

RMSE	Rsquared	MAE
43.4972454	0.9881998	38.8036803

The *method* option in the **training function** is a string specifying which classification or regression model to use. Possible values are: "lm," "regLogistic," "rf," "ridge," "svmLinear," "knn," "nnet," "multinom," "bag," "cforest," "ctree," etc.. Possible values are found using names(getModelInfo()). See http://topepo.github.io/caret/train-models-by-tag.html. A list of functions can also be passed for a custom model function. See http://topepo.github.io/caret/using-your-own-model-in-train.html for details.

The function *trainControl* generates parameters that further control how models are created, with possible values for (resampling) *method* = "boot" for bootstrap and "cv" for cross-validation among others. see the details at:

https://www.rdocumentation.org/packages/caret/versions/6.0-82/topics/train.

2.6   Summary

We have discussed models for regression problems with continuous outcomes and classification problems with binary outcomes. Classical statistics focuses on a hypothesis with type-I error control. Statistical learning focuses on learning, prediction, or knowledge updating. We have discussed important concepts in data science, including internal and external validity, different bias types, confounding, regression to the mean, multiplicity, and different data sources and structures. Various predictive models, including generalized linear model, ridge, lasso, and elastic regressions are examples of models with regularizations. Readers should download an R package and play around with the example code provided to get a good understanding of the methods.

2.7 Problems

Validity

2.1: If you generate a sample $x_1, x_2, ..., x_n$ from the standard normal distribution, but it happens that it is so skewed the normality is rejected by a normality test (e.g., D'Agostino's K-squared test). However, you decide to test treatment difference based on a normality assumption will the internal validity and external validity still hold? If the distribution test should suggest something more like an exponential distribution than a normal distribution, would you use a test based on an exponential distribution?

2.2: To give an adequate examination of drug effect on a minority population, since the drug may have a different effect for different population, a clinical trial often tends to include a minimal required sample size for that minority. As a result, the minority population composition in the trial is different from the entire population. However, in the drug label, the mean effect from the trial is presented. Does this violate the internal and external validities?

Placebo Effect

2.3: However, it is interesting that a study carried out by researchers in the Program in Placebo Studies at the Harvard Medical School discovered there is an *honest placebo* effect: patients with irritable bowel syndrome in one study experienced a significant beneficial effect even though they were told the pills they were taking were placebos, as compared to a control group who received no pills (Kaptchuk, et al., 2010). This is perfectly logical if we consider that many of us are afraid of walking in a graveyard in the dark even though we don't believe there are ghosts.

Placebo effects are generally more significant in subjective measurements, such as with patient-reported outcomes, than in an objective laboratory measurement such as a hemoglobin level.

There are controversies surrounding placebo use as medical treatment. A survey conducted in the United States of more than 10,000 physicians indicated that while 24% of the physicians would prescribe a treatment that is a placebo simply because the patient wanted treatment, 58% would not, and for the remaining 18% it would depend on the circumstances. Would you recommend a patient to take placebo if no medicine is avaialble?

Selection Bias and Randomization

2.4: As an example of selection bias, suppose a student is conducting an experiment to test a chemical compound's effect on mice. He knows that inclusion of a (placebo) control and randomization are important, so he randomly catches half of the mice from the cage and administers the test compound. After days of observations, he finds that the test compound has a negative effect on mice.

However, he might overlook an important fact: the mice he caught were mostly physically weaker and thus they were more easily caught. Such selection bias can also be seen in the capture-recapture method due to physical differences in animals. How do you avoid such biases in your experiments?

2.5: A clinical trial investigator might use her knowledge to assign treatments to different patients with, say, sicker patients going to the new treatment group and less sick patients to the control group. Will such treatment assignments cause any problems in assessing the new treatment effect?

2.6: Selection bias can occur in surveys too. For instance, in an internet survey, selection bias could be caused by internet accessibility or by the differences in familiarity with computers. Randomization can reduce bias but cannot completely prevent it, especially when there are a large number of potential confounding factors.

Confirmation Bias

2.7: Researchers conduct a two-group controlled trial but don't want to specify sample size; instead they want to continue the trial until the response rate in the test group is 10% better that the control. Would that be any problem?

2.8: Have you intentionally or subconsciously put more effort in collecting the data that support your hypothesis? Have you biasedly presented the evidences that support your opinion when you argue with your coworker, classmates, or even your significant other?

Publication Bias

2.9: The FDA only sees the positive results and approves significant positive drugs, will that be bias to the positive side?

Regression to the Mean

2.10: When multiple screenings are used to qualify subjects for an experiment, bias can be introduced because the qualifying screening value may just happen to be lower than it actually should be, and after subjects are entered in the study, the value can come back to the natural status without any intervention. Such a phenomenon is called "regression to the mean." How do you design the experiment such that the bias can be removed?

Randomization

2.11: A confounding factor, if measured, often can be identified and its effect can be isolated from other effects using appropriate statistical procedures. A recent example is an ABC poll (http://abcnews.go.com, 2004) that found that Democrats were less satisfied with their sex lives than Republicans. But women are also less satisfied with their sex lives than men, and more Democrats are women than Republicans. How do we know whether the correlation between happy sex lives and political affiliation is an honest relationship, or just a side effect of the gender differences between Democrats and Republicans?

Multiplicity

2.12: What is multiplicity issue? Given examples in clinical trials.

2.13: A statistician suggests to control type-I error in clinical trial data analysis, by using different significance level, alphas for different trials as long as the average is equal to the significance level. For example, for event trials use 8% alpha, for old trials with 2% alpha, the average alpha is 5%.

Missing Data

2.14: Paired versus unpaired data

In a clinical trial, if there are a baseline measure and a postbaseline measure for each patient, we can use paired t-tests for testing the treatment effect. Such a paired approach, on one hand, reduces the intera-subject variability since each subject serves his own control, on the other hand reduces sample size to half. In the so-called natural history study, the subjects in the natural history study are not the same subjects in the current single-arm trial. In such cases, matching subjects between the natural history and the current trial can be used in order to use paired t-tests. However, the power of a paired t-test is not necessarily higher than a two-sample t-test, dependent on how well the matching is. In general, each patient's postbaseline measure can match each patient's baseline measure. Thus, if there are m baseline measures and n postbaseline measures, we can create $m \times n$ data points. This significant increase in sample size (data points) can increase the power for the hypothesis test. Is this m-to-n matching statistically valid?

2.15: If we make very frequent measurements near a time point (e.g., every second) on Hgb, such repeated measured data are actually "artificial" because the Hgb measures at such short time intervals will be the same. However, from a statistical point of view, the mixed effect model for such repeatedly measured data will increase the power and reduce the p-value. How do you resolve the dilemma?

3

Similarity Principle—The Fundamental Principle of All Sciences

3.1 Scientific Paradoxes Call for a New Approach

3.1.1 Dilemma of Totality Evidence with p-value

The p-value is the probability of finding the test statistic equal to or more extreme than the observed value when the null hypothesis is true. However, a p-value does not directly answer the underlying scientific question. We ask: what is the probability the drug is effective and how effective? The p-value, p, says: "If the drug is ineffective, you will have probability p of observing data equal to or more extreme than what you have observed." In hypothesis testing, we use a predetermined significance level α and $p \leq \alpha$ to control the type-I error rate, but knowing that type-I errors (e.g., errors on different endpoints) have different impacts, some being more severe than others, why don't we control the loss due to error rather than the error rate?

There are also other controversies regarding the use of p-values: suppose we have data from two trials conducted sequentially, but with no huge time gap and geographical difference between them: a superiority trial to compare drug A to placebo C was conducted first, followed by another trial with drug B versus A. The hypothetical efficacy data are presented in Table 3.1. In the A versus C trial, p-value $= 0.0217 < \alpha = 0.025$; therefore we claim A is superior to C. In the second drug B versus A trial, the effect of drug B is 2.82 larger than the effect of drug A with a p-value of $0.0275 > \alpha = 0.025$. Therefore, we fail to conclude that B is superior to A and drug B cannot be approved for marketing. However, if we consider the totality of the efficacy evidence, i.e., the data from the two trials all together, we can find the following:

(1) Constancy: The standard deviation σ_i is the same for all the treatment groups in the two trials, and the treatment effects are consistent between the two trials regarding the active control group or drug A.

(2) Drug effects: observed placebo, $\hat{\theta}_C = 1$, $n = 100$; drug A, $\hat{\theta}_A = 2$, $n = 200$; drug B, $\hat{\theta}_B = 2.82$, $n = 200$. Clearly this shows that drug B has larger effect than drug A and they have the same sample size of 200 each.

TABLE 3.1

Dilemma of Totality Evidence

Trial	Drug	Observed Effect	Std Dev.	Sample Size n	p-value
1	Drug A	2	3.5	100	0.0217
	Placebo C	1	3.5	100	(A vs C)
2	Drug B	2.82	3.5	200	0.0279
	Drug A	2	3.5	100	(B vs A)
	Drug B	2.82	3.5	200	<0.0001
	Placebo C	1	3.5	100	(B vs C)

Note: One-sided $\alpha = 0.025$,

(3) If we test drug B versus A with 200 patients in each group, the p-value is 0.0222, so we reject the null hypothesis of no treatment difference.

(4) If we test drug B versus placebo C with 200 patients in each group, the p-value is < 0.0001, reject the null hypothesis and conclude the drug is better than the placebo.

How can we keep drug B in the market and not approve drug A for marketing? Does the frequentist hypothesis testing approach potentially prevent patients from accessing better drugs and keep worse drugs in the market?

3.1.2   Multiple-Testing Versus Multiple-Learning

Multiple Testing Issue

Suppose in a two-stage adaptive clinical trial, two hypothesis tests for drug effects were performed as the data accumulated. The interim analysis was conducted with 50% of the patients for testing no drug effect. The interim p-value was 0.01, larger than the nominal level 0.006, thus the null hypothesis of no drug effect was not rejected. The trial continued to the final stage and the null hypothesis was tested again with all data and a final p-value of 0.0249, larger than the final significance level, 0.024. Hence, we cannot reject the null hypothesis to claim that the drug is effective. For simplicity's sake, we have used the conservative Bonferroni split of α between the interim and final analyses. However, clinicians may argue that the drug effect will not change just because of the different statistical methodologies applied. Therefore, one should use $p = 0.0249$ compared against $\alpha = 0.025$, instead of the adjusted value 0.024.

Which argument is correct? Here, we can see two different concepts of the effectiveness of a drug. One is the physical properties of the test compound, which will *not* change as the hypothesis testing procedure changes (e.g., one test versus two tests). The other view is that the statistical property will change since it reflects an aggregated attribute of a group of similar things, and such a property will depend on the choice of similarity grouping.

In current practice, the type-I rate is controlled within an experiment. Thus different pharmaceutical companies test chemical compounds every day from the same or similar compound libraries, a majority of them being ineffective for their intended purpose. Because the tests are performed by different companies, no multiplicity adjustments are applied. Therefore, the false positive rate in such screening can be high. Similarly, if an ineffective compound is tested 10 times using 10 clinical trials with 10 patients each versus one trial with 100 patients, the error rate will increase from 5% to 40%.

Multiple Learning from the Same Data Source

It is interesting to anticipate the trend that, as data sources gradually become public, more researchers will perform different analyses on the same data. How can we control the error rate if the first research has already used up the error rate allowed (α)? On the other hand, the conclusions from meta-analyses (post hoc analyses using combined data from similar experiments) are usually difficult to disprove since most, if not all, available data have been used, and since with a large sample size the conclusion is difficult to change, even if we have some new data later. The question is: should we use the conclusions drawn from a meta-analysis and disregard a previous conclusion (whether positive or negative) from a subset of data, each with a smaller sample size? How can we make use of a data resource multiple times for many different questions but without inflating α, or with only a limited inflation?

Learn multiple things with the sample data and verify them later! This is completely contradictory to the multiplicity adjustment with family-wise error rate (FWER) control!

3.1.3 A Medical and Judicial Tragedy

Sally Clark was a lawyer who became the victim of an infamous miscarriage of justice when she was wrongly convicted of the murder of two of her sons in 1999. Her first son died suddenly within a few weeks of his birth in 1996. After her second son died in a similar manner, she was arrested in 1998 and tried for the murder of both sons. A professor of pediatrics testified that the chance of two children from an affluent family suffering sudden infant death syndrome, known as SIDS, was 1 in 73 million, which was arrived at by squaring 1 in 8500, the likelihood of a cot death in a family like Clark's, i.e., stable, affluent, nonsmoking, with a mother more than 26 years old (McGrayne, 2012).

The convictions were upheld at appeal in October 2000 but overturned in a second appeal in January 2003. On the day she won her freedom, Sally issued a statement in which she said: "Today is not a victory. We are not victorious. There are no winners here. We have all lost out. We simply feel relief that our nightmare is finally at an end. We are now back in the position we should have been in all along and plead that we may now be allowed some privacy to grieve for our little boys in peace and try to make sense of what has happened to us."

Following the case, another three wrongly convicted mothers were declared innocent of crimes that had never occurred. However, Sally never recovered from the experience and died in 2007 from alcohol poisoning. Clark's case is one of the great miscarriages of justice in modern British legal history (Chang, 2014).

The main criticism of the pediatrician was that the category was too specific when looking into the chance (1 in 73 million) of two children dying from SIDS: "a cot death in a family like Clark's, i.e., stable, affluent, nonsmoking, with a mother more than 26 years old." Later, a Bayesian analysis was conducted using a less specific category to support the innocent claim. *The key question is how specific is too specific?* If specific enough we can always conclude that any lottery winner is a criminal.

3.1.4 Simpson's Paradox

In probability and statistics, *Simpson's paradox*, introduced by Colin R. Blyth in 1972, points to apparently contradictory results between aggregate data analysis and analyses from data partitioning (Chang, 2012).

Suppose two drugs, A and B, are available for treating a disease. The treatment effect (in terms of remission rate) is $520/1500$ for B, better than $500/1500$ for treatment A. Thus we will prefer treatment B to A. However, after further looking into the data for males and females separately, we found that the treatment effect in males is $200/500$ with A, better than $380/1000$ with B, while the treatment effect in females is $300/1000$ with A, better than $140/500$ with B. Therefore, whether female or male, we will prefer treatment A to B (Table 3.2). Should we take treatment A or B?

The problem can be even more controversial. Suppose when we further look into the subcategories: Young Female and Old Female, and the direction of treatment effects switches again, i.e., treatment B has better effect than treatment A in both subcategories, consistent with the treatment effect for the overall population (Table 3.3). The question is: what prevents one from partitioning the data into arbitrary subcategories artificially constructed to yield wrong choices of treatments?

The key question again: how specific is too specific?

The same issue arises in the area of rare disease and precision medicine. Compounding the signal-to-noise issue calls for tailoring treatment to specific patient types, a practice also known as precision medicine. If the promise of

TABLE 3.2
Drug Responses in Males and Females

	Drug A	Drug B
Males	200/500	380/1000
Females	300/1000	140/500
Total	500/1500	520/1500

TABLE 3.3

Drug Responses in Young and Old Females

	Drug A	Drug B
Young Females	20/200	40/300
Old Females	280/800	100/200
Total	300/1000	140/500

precision medicine is to be realized, we need to stratify clinical trials into more and smaller subgroups of treatment-eligible patients. Again, the same question: *how specific is too specific?*

Today, the goal of many clinical trials is to support marketing authorization of new drugs globally. By necessity, many clinical trials became global trials. Clinical trials across multiple regions of the world have become common practice. While global trials may have increased the efficiency of clinical drug development, they present three particular challenges (Eichler and Sweeney, 2008): (1) some participating regions seek the inclusion of predefined numbers of patients from their territory and health-care environment; (2) regulators in each region often have divergent requirements for endpoints in various therapeutic areas, durations, comparators, medical practice, and other trial characteristics, making development of a single, globally acceptable trial protocol difficult; and (3) while much has indeed been harmonized, the detailed regulatory frameworks or ethical requirements governing the approval and conduct of clinical trials at the national level increase the operational complexity of running a trial across several regions.

From the statistical learning point of view, if the overall drug effect is statistically significant when all regions are combined but very different drug effects are observed in different countries/regions, how should a drug be used in different countries? A second question is: How should the region be defined, as a country, a state, a city, or even something that is not geographically defined? It indeed ends up in the same issue as before: *how specific is too specific?*

When we get into greater specificity, the sample size will become smaller and smaller and the power of rejecting the null hypothesis will become smaller and smaller. It is a challenging issue for classical statistics.

3.1.5 Bias in Predicting Drug Effectiveness

When the proportion of patients composite in the clinical trials is different from the future population, the average effect in *drug label* can be missing and fail to provide accurate information to the clinicians about drug effect on individuals. For example, if a drug has a larger effect on females than males and the proportion of females in the trial is much larger than the proportion of females in the target patient population, then using average effect observed from the pivotal trial in the *drug label* will inflate the actual benefit of the drug on the target population.

Another issue concerns the evaluation of benefit of a new drug when there is no other drug available. In this case, the benefit of approving the drug for marketing is the total benefit of the drug, not the net drug benefit beyond the placebo. Since we cannot market the placebo, without the approval of the drug patients will get nothing; if the drug is approved, the patients will get the total benefit, not just the net benefit beyond the placebo.

As to resolutions of the controversies and issues we have discussed, despite great efforts we have not yet provided any satisfactory answers. To resolve these issues, we have to think from the scientific foundation perspective. That is, we wish to understand what constitutes scientific evidence, and that is at the heart of what we call the similarity principle.

3.2   The Similarity Principle

3.2.1   Role of Similarity Principle

In ancient medicine (BELLAVITE, et al., 1997), similarity principle claimed that when a substance is able to induce a series of symptoms in a healthy living system, it would be also able under certain circumstances to cure these symptoms when applied at low doses. However, the similarity principle we will discuss here is a different, general principle in scientific discovery.

Science aims to discover causal relationships and to predict future outcomes. So does learning (human or machine learning). All science, and learning itself, is based on a fundamental principle – the *similarity principle* (Chang, 2012, 2014). The principle can be stated: similar things or individuals will likely behave similarly, and the more similar they are the more similarly they behave. For instance, people with the same (or a similar) disease, gender, and age will likely have similar responses to a particular drug or medical intervention. If they are similar in more aspects they will have more similar responses.

To qualify as a true scientific discovery, a finding must be verifiable. Otherwise, it cannot be called science. However, as history is unique, no two events are identical or repeat exactly, even the same individual (especially a living being) will change constantly. For this reason, we have to group similar things together and, considering them as approximately the same, study their common or overall behaviors. Psychologists study a group of people with similar personalities to explain why those people behave the way they do. Pharmaceutical scientists treat people with the "same" disease to study the overall effect of a drug even though individual responses to the drug may be different. Indeed, similarity grouping is the basis for scientific discoveries, and the similarity principle we believe in is the backbone behind causality. The idea of a causal relationship is a way for human beings to handle the complex world in a simple form with a reasonable approximation given the limited ability of our brains.

Here are some simple examples of people using the principle for learning in their daily lives: All objects with wheels run fast. Objects with sharp edges can be used to cut things. Many people think September 11th every year is more likely to see a terrorist attack than other days. Therefore, NYPD tightens security around the date. People use the similarity principle differently. For instance, some of my friends think sending their children's top-rated high school will increase the probability of them entering top colleges. They buy a tiny apartment in the town with a first-rank public school because they think people from the same school have similar chances for better college than people from different schools. Some of my other friends think differently. They think their kids are similarly talented to certain kids who were successful in a certain school that fit them. The school would be suitable to their kids too; therefore, they sent their kids to the same or similar school even though they may not be the best-ranked school. However, such subconscious uses of the similarity principle lack scientific rigor and thus cannot be an effective learning method. We can make the similarity principle operational so that it can be effectively used in practice. The principle is implicitly used in drug discovery, but unconsciously is used in daily life, in our work, in all the sciences, in statistics, and even in mathematics.

The *similarity principle* says that every characteristic/element of an object contributes a portion of information to real-world outcomes; therefore, the more similarity between two objects the more likely they produce the same output when they receive the same input. That is, two similar objects behave similarly (Figure 3.1).

The mathematical expression of the similarity principle is that the outcome $\hat{Y}_i$ for a new subject i can be modeled in terms of outcomes Y_j and the similarity scores:

$$\hat{Y}_i \propto \sum_j S_{ij} Y_j \tag{3.1}$$

We will use a symbol such as Y to indicate a true value (which often is an observed value, where measurement errors are ignored) and the same letter with hat $(\hat{Y})$ to represent the corresponding predicted value.

When the response is a categorical variable, a function such a rounding or majority vote may be applied. Alternatively, we can use the probability of Y_i, where

$$\Pr\left(\hat{Y}_i\right) \propto \sum_j S_{ij} Y_j \tag{3.2}$$

Here is a potential application in drug development. After a compound is tested in patients in a clinical trial, we want to know what the effect will be on future patients, and not just the overall effect but the effect on individual patients as well, since the effect of a drug varies from patient to patient. This learning method can help the doctor to predict the drug effect in individual

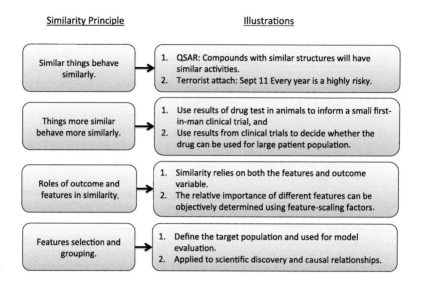

FIGURE 3.1
Illustration of the Similarity Principle

outcomes and better prescribe the medicines. It can also be used to build medical robots for personalized (precision) medicine.

Clustering (similarity grouping) and classification/regression always go hand in hand. For example, we put (cluster) subjects together based on their similarity in terms of independent variables (10–40 year old = young, >40 = elderly). In fact, because no measurement has infinite precision, or because computer memory can only maintain a limited number of significant digits, similarity grouping and the similarity principle are subconsciously applied (we think a small difference in the independent variable will not cause a larger difference in response).

Most machine learning methods require big training data. The current method does not have the requirement for big data. Therefore, this invention can be used in problems such as rare disease clinical trials where sample size is very limited.

3.2.2 The Root of Causality

According to Albert Einstein, the grand aim of all science is to explain the greatest number of empirical facts by logical deduction from the smallest number of hypotheses or axioms. The root of causality is the similarity principle, and the similarity principle is at the root of all science.

Science is a study of the reoccurrences of events. Its purpose is to reveal or interpret the "causal relationship" between paired events. One of the events is called the "cause" and the other is called the "consequence" or "effect." The reason we study history is to develop or discover a law that can be used to predict the future when the cause repeats. An event in a causal relationship can be a composite event. However, science does not simply ignore all single occurrences. Instead, one-time events are grouped into "same events" based on their similarities. In such a way we artificially construct reoccurrences of events.

According to Karl Popper, any hypothesis that does not make testable predictions is simply not science. A scientific theory is constantly tested against new observations. A new paradigm may be chosen because it does a better job of solving scientific problems than the old one. However, it is not possible for scientists to test every incidence of an action and find a reaction. You may think the brain is so amazing because it can do science, but that is also exactly due to a limitation of our brains—they are unable to deal with a huge number of separate events individually.

A cause-effect pair is the observation of a repetitive conjunction of particular events for which we postulate their linkage (causal relationship). In this philosophical sense, causality is a belief, an interpretation of what is happening, not a fact but at most a thing that cannot be verified or whose verification rests on the basis of another belief. There is a single, growing complex history. To simplify it, people try to identify the patterns or "repetitive conjunctions of particular events" and deduce "laws" or causal relationships from the process. The term "cause" relies on the recurrence of the events and the definition of "same" or "identical." The recurrence here is in regard to the recurrence of the "cause" and the "effect," both together and separately. Consider the law of factor isolation: If factors A and B exist, and if fact C exists and if C disappears when we eliminate B, then factor B is a cause of fact C. However, there is no such exact recurrence. The reoccurrence of an event is only an approximation in the real world; we ignore certain distinctive details, and therefore the characterization is always somewhat subjective.

The notion behind prediction is causality, whereas the notion of causality is the *similarity principle*: similar situations will likely result in the same outcome. When we say "the same," we ignore the differences that might be hidden and contradict the scientific law that is held. Once such a contradiction is uncovered, an exception to the law is observed, which may call for some modification of the law. Both science and statistics acknowledge the existence of exceptions, but the former doesn't act until an exception occurs, while the latter acts when the unexpected is expected.

Most of us would agree that the goal of scientific inference is to pursue the truths or laws that approximate causal relationships so that our brains can handle them. Every two things (events) are similar in one way or another, so

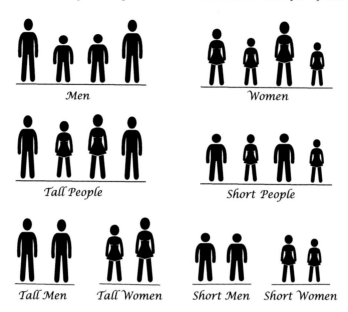

FIGURE 3.2
Similarity Grouping—Clustering

we can see the reoccurrences of the "same event" by grouping based on similarities and ignoring the differences. Importantly, such grouping is necessary for our brains to efficiently handle reality. It is this grouping that lets us see the repetition of things (both reasons and outcomes). Only under reoccurrences of "same events" can a causal relationship make sense. Such implicit grouping of similar things often creates intangible controversies in scientific inference. The differences we ignore are unspecified or hidden, but only the similarities are known and specified. The ignored differences cause a difference in defining the sets of "same things." In other words, as to what things are considered similar, it is unavoidably somewhat subjective. Different people have different opinions on different questions, as we have discussed in Simpson's paradox, rare diseases, precision medicine and clinical trial globalization. If you are a tall man, from which of the following four groups do you infer your characteristics (e.g., longevity, the likelihood of being successful in career, or the chance of having cancer): men, tall people, tall men, or the entire population (Figure 3.2)? This question needs to be answered first, before you apply any appropriate method. The answer is subjective, dependent on individual person's knowledge, experiences, or even beliefs. All scientific methods and their applications are constructed on this foundation.

We all use the similarity principle nearly every moment, throughout our lifetime. It says: A similar condition will lead to a similar consequence.

Without the similarity principle, there will be no analogy and no advancement of science.

In Simpson's paradox, with rare diseases and precision machines, and in clinical trial globalization, we face the challenging question: "how specific is too specific?" Based on the similarity principle, we will not just be taking males' information or females' information or simply pooling all patients' data for the proportion of remission rates, but each patient's response is weighted based on the similarity in predicting a new patient's response. However, we still have a key issue left, and that is how to define similarity. How the similarity principle and methods developed from this principle can resolve, exactly, the controversies about p-value and "how specific is too specific?" will be a topic of the next section.

3.3 Similarity Measures

3.3.1 Attributes Selection

Similarities and dissimilarities between objects go hand in hand. Attributes define different objects and their dissimilarity (or similarity). Depending on purpose, different types of people can be defined by their age, weight, gender, education, or other characteristics. The set of attributes selected should serve the purpose and often potentially influence the outcome. For this reason, field experts are often helpful in identifying such a concise set of attributes. For instance, in designing a clinical trial protocol, dependent on disease indication, different medical doctors are needed to identify the demographics, baseline characteristics, and inclusion/exclusion criteria for the patients to be recruited. In natural language processing (NLP), the attributes can be words at difference locations or sequences of words in a sentence. In molecular design, molecular biologists or chemists identify different molecules by molecular substructures. Such 2-D or 3-D molecular models exhibit tree-structures, which can be transferred into 1-Dimensional sequences of different substructures if one numbers the tree nodes in a particular order. For images, we can define color at different locations (bit by bit) in a two-dimensional space or a sequence of colors if we set an order rule, e.g., moving from left to right and top to bottom. The color can further be represented by 3 values (RGB) as a vector. Therefore, images can be represented by a sequence of 3-dimensional vectors, or 4-dimensional vectors if we want to include each color's brightness. A sound source or voice can be decomposed into different frequencies through a Fourier transformation and further turned into a sequence of high-dimensional vectors.

When an attribute is binary, we can use 1 or 0 to represent having or not having the property. If the attribute is categorical (more than two possible discrete values), we can use one-hot one-cold to encode the property. *One-hot* is a sequence numbers with a single 1 and all the others 0. A similar implementation in which all numbers are 1 except one 0 is sometimes called *one-cold* encoding. In statistics, *dummy variables* represent a similar technique for representing categorical data. For instance, if variable race has three categories: Asian, White, and Others, we code use dummy code $(1, 0, 0)$ for Asian, $(0, 1, 0)$ for White, and $(0, 0, 1)$ for the category Others. Here we use a 3-dimensional vector for the dummy code because there are 3 categories for the variable race.

3.3.2 Similarity Properties

Similarity (proximity, alikeness, affinity) can be measured in different ways. For example, one can use spatial distance, correlation, or comparison of local histograms, spectral properties, or even, with or without given patterns, a long data series, as in gene matching. Semantic similarity is a metric defined over a set of documents or terms, where the idea of distance between two of them is based on the likeness of their meaning or semantic content.

A similarity score is used to measure the degree of similarity. It usually ranges from 0 (no similarity at all) to 1 (identical) or ranges from -1 (completely different or opposite) to 1 (identical). For similarity score S_{AB} ranging from 0 to 1, there are a few intuitive properties of similarity:

1. If $S_{AB} > 0$, then A and B have common or similar attributes, e.g., the same color, or similarity in age.

2. If A is similar to B with score S_{AB}, then B is similar to A with the same score. In other words, $S_{AB} = S_{BA}$.

3. If A is similar to B with score S_{AB}, and B is similar to C with score S_{BC}, then it is not necessarily true that $S_{AC} > 0$. The similarity is not transitive just like friendship between people.

4. $S_{AB} > 0$, $S_{BC} > 0$, and $S_{CA} = 0$, then A, B, and C have nothing in common/are not similar.

When you want to organize your closet, kitchen, paper works, or photos or other electronic documents, you organize in terms of certain properties.

3.3.3 Cosine Similarity and Jaccard Index

Let denote the attributes by $A_1, A_2, ..., A_K$. We further use 0 and 1, respectively, to indicate that a subject or an object does not have or has a property. Examples of the objects A, B, and C with or without various properties $(K = 5)$ are:

	Attribute				
Object	A_1	A_2	A_3	A_4	A_5
A	1	0	0	1	1
B	0	1	1	0	1
C	1	1	1	0	1

Given the five properties, the objects A, B, and C can be represented by vectors.

$$
\begin{aligned}
A &= (1,0,0,1,1), \\
B &= (0,1,1,0,1), \\
C &= (1,1,1,0,1),
\end{aligned}
$$

respectively.

Cosine Similarity Score:

To measure the similarity between two objects, the most intuitive method is perhaps the proportion of common properties (0-1coding). In the above example with five-property objects, the similarity between A and B is $S_{AB} = 1/5$. Similarly $S_{AC} = 2/5$, and $S_{BC} = 4/5$. In general, the similarity is defined as

$$
S_{AB} = \frac{N_{AB}}{N}. \tag{3.3}
$$

However, *cosine similarity* has better mathematical properties in the *vector space* or *Hilbert space*. Cosine similarity is the projection of one object (vector) on the other object, defined as

$$
S_{AB} = \cos\theta = \frac{A \cdot B}{\|A\| \, \|B\|} \tag{3.4}
$$

The resulting similarity ranges from -1, meaning exactly opposite, to 1 meaning exactly the same, with 0 indicating orthogonality (decorrelation), and in-between values indicating intermediate similarity or dissimilarity. For text matching, the attribute vectors **A** and **B** are usually the term frequency vectors of the documents. Cosine similarity can be seen as a method of normalizing document length during comparison. In the above example, the similarity scores are

$$
S_{AB} = \frac{1}{3 \times 3} = \frac{1}{9}, \ S_{AC} = \frac{2}{12}, \ \text{and} \ S_{BC} = \frac{4}{12}.
$$

For documentation classification/clustering, the attribute can be keywords. For ligands (chemical compounds for drug candidates), the attributes can be substructures of chemical compounds.

From (3.4), we can see that inclusion of more common attributes to A and B will artificially reduce the similarity S_{AB}. The selection of attributes is subjective, but arguably such subjectivity is based on the individual's experience and knowledge.

In many applications, attributes often are binary or of the same type, such as physical distance in three dimensions. The question is: if attributes involve multiple things such as weight, height, disease severity, age, gender, and education, how do we define the similarity?

In our definition of cosine similarity, the value of v_j cannot be directly used since the importance of each variable in defining similarity is dependent on the problem under consideration. Therefore, we use an attribute-scaling factor R_k to rescale each variable, $u_k = R_k v_k$ and $u_{jk} = R_k v_{jk}$.

$$S_{mj} = \cos(\theta) = \frac{\sum_{k=1}^{K} u_{mk} u_{jk}}{||u_m|| \cdot ||u_j||} = \frac{\sum_{k=1}^{K} R_k v_{mk} R_k v_{jk}}{\sqrt{\sum_{k=1}^{K} R_k^2 v_{mk}^2} \sqrt{\sum_{k=1}^{K} R_k^2 v_{jk}^2}}, \quad (3.5)$$

where R_k will be determined through training. This scaling will be discussed in the next chapter.

For binary data, the *simple matching coefficient* (SMC) or Rand similarity coefficient is a statistic used for comparing the similarity and diversity of sample sets:

$$SMC = \frac{\text{number of matched attributes}}{\text{number of attributes}}. \quad (3.6)$$

The *Jaccard index (Tanimoto index)*, also known as *intersection* over *union*, is defined as the size of the intersection divided by the size of the union of the sample sets:

$$J(A, B) = \frac{|A \cap B|}{|A \cup B|} = \frac{|A \cap B|}{|A| + |B| - |A \cap B|}. \quad (3.7)$$

The Jaccard index, ranging from 0 to 1 inclusive, has been used in drug discovery. The similarity searching in molecular design is carried out to compare the representations of molecules with respect to their substructures using the *Tanimoto index*. If we use 1 to indicate that a molecule has certain attributes and 0 to indicate the molecule does not have the property, then

$$T_{AB} = \frac{n_{AB}}{n_A + n_B - n_{AB}}, \quad (3.8)$$

where n_A is the number of 1s in molecule A, n_B is the number of 1s in molecule B, and n_{AB} is the number of 1s in both A and B. The possible value of T_{AB} ranges between 0 (maximal dissimilarity) and 1 (identical bitstrings). In the previous example the Tanimoto index is

$$T_{AB} = \frac{1}{3 + 3 - 1} = \frac{1}{5} \text{ and } T_{BC} = \frac{3}{3 + 4 - 3} = \frac{3}{4}.$$

3.3.4 Distance-Based Similarity Function

A similarity measure or similarity function is a real-valued function that quantifies the similarity between two objects. Although no single definition of a similarity measure exists, usually such measures are in some sense the *inverse of distance matrices*: they take on large values for similar objects and either zero or a negative value for very dissimilar objects. For example, the *Hamming distance* between two strings of equal length is the number of positions at which the corresponding symbols are different. In other words, it measures the minimum number of substitutions required to change one string into the other, or the minimum number of errors that could have transformed one string into the other. In the context of *cluster analysis*, Frey and Dueck suggest defining a similarity measure

$$(x, y) = -||\mathbf{x} - \mathbf{y}||^2 \tag{3.9}$$

where $||\mathbf{x} - \mathbf{y}||^2$ is the squared *Euclidean distance*.

The *Mahalanobis distance* is a measure of the distance between two points, introduced by Mahalanobis in 1936. It is a multi-dimensional generalization of the idea of measuring how many standard deviations separate the two points. It is defined as

$$d_{ij} = \sqrt{(\mathbf{X}_i - \mathbf{X}_j)^T \mathbf{S}^{-1} (\mathbf{X}_i - \mathbf{X}_j)}, \tag{3.10}$$

where $\mathbf{S}$ is the covariance matrix.

A metric on a set X is a function (called the distance function or simply distance) $d : X \times X \to [0, \infty)$, where $[0, \infty)$ is the set of non-negative real numbers and for all $x, y, z \in X$, the following conditions are satisfied:

1. non-negativity or separation axiom, $d(x, y) \geq 0$
2. identity of indiscernibility, $d(x, y) = 0 \Longleftrightarrow x = y$
3. symmetry, $d(x, y) = d(y, x)$
4. subadditivity or triangle inequality, $d(x, z) \leq d(x, y) + d(y, z)$
5. Transitional invariance, $d(x, y) = d(x + z, y + z) = d(y - x, 0) = d(x - y, 0)$

Conditions 1 and 2 together define a positive-definite function. The first condition is implied by the others.

The popular L_ρ *distance* (*Chebyshev distance*) between two points $x_1 = (x_{11}, ..., x_{1n})$ and $x_2 = (x_{21}, ..., x_{2n})$ is defined as

$$L_\rho = \left(\sum_{i=1}^n |x_{2i} - x_{1i}|^\rho \right)^{1/\rho}, \text{integer } \rho > 0. \tag{3.11}$$

The special *Chebyshev distance* $L_\infty = \lim_{\rho \to \infty} L_\rho = \max_i (|x_{2i} - x_{1i}|)$, where $|x_{2i} - x_{1i}|, i = 1, ..., n$, is called the *Manhattan distance*.

In the kernel method for supervised learning, the radial-based kernel is a similarity function between two feature vectors (points) $\mathbf{x}_i$ and $\mathbf{x}_j$, defined as

$$K(\mathbf{x}_i, \mathbf{x}_j) = \exp \left(-\frac{||\mathbf{x}_i - \mathbf{x}_j||^2}{2\sigma^2} \right) \tag{3.12}$$

where $||\mathbf{x}_i - \mathbf{x}_j||^2$ may be recognized as the squared Euclidean distance between the two feature vectors. σ is a free parameter.

In similarity-based machine learning, an *exponential similarity function* S_{ij} can be used, which is defined as a function of distance d_{ij} between the two subjects, i and j,

$$S_{ij} = e^{-d_{ij}^{\eta}}; \eta > 0, \tag{3.13}$$

where

$$d_{ij} = \left[\sum_{k=1}^{K} (R_k |X_{jk} - X_{ik}|)^{\rho}\right]^{1/\rho}, R_k > 0, k = 1, 2, ..., K; \text{integer } \rho > 0. \tag{3.14}$$

We can also define similarity as a *distance-inverse function*

$$S_{ij} = \frac{1}{1 + d_{ij}^{\eta}}. \tag{3.15}$$

Logistic similarity functions are also used:

$$S_{ij} = \frac{2}{1 + \exp(d_{ij}^{\eta})}. \tag{3.16}$$

3.3.5   Similarity and Dissimilarity of String and Signal Data

Edit distance in computational linguistics is a way of quantifying how dissimilar two strings (e.g., words) are to one another by counting the minimum number of operations required to transform one string into the other. *Hamming distance* in information theory, one example of edit distance, is the number of positions at which the corresponding symbols are different for two equal-length strings.

The *Smith–Waterman algorithm* performs local sequence alignment for the purpose of determining similar regions between two strings of nucleic acid sequences or protein sequences. Instead of looking at the entire sequence, the Smith–Waterman algorithm compares segments of all possible lengths and optimizes the similarity measure. The *Needleman–Wunsch algorithm*, an algorithm used in bioinformatics to align protein or nucleotide sequences, was one of the first applications of dynamic programming to compare biological sequences.

Similarity matrices are used in sequence alignment. Higher scores are given to more similar characters, and lower or negative scores for dissimilar characters. Nucleotide similarity matrices are used to align nucleic acid sequences. Because there are only four nucleotides commonly found in DNA (adenine (A), cytosine (C), guanine (G), and thymine (T)), nucleotide similarity matrices are much simpler than protein similarity matrices. For example, a simple matrix will assign identical bases a score of $+1$ and non-identical bases a score of -1. We will discuss similarity matrixes for genomics and proteomics in the chapter on kernel method.

Dynamic time warping is often used in time series analysis. Dynamic time warping (DTW) is one of the algorithms for measuring similarity between two temporal sequences which may vary in speed. For instance, similarities in walking could be detected using DTW even if one person is walking faster than the other, or if there are accelerations and decelerations during the course of an observation. DTW has been applied to temporal sequences of video, audio, and graphics data. Indeed, any data that can be turned into a linear sequence can be analyzed with DTW. A well-known application has been automatic speech recognition, in order to cope with different speaking speeds. Other applications include speaker recognition and online signature recognition.

3.3.6   Similarity and Dissimilarity for Images and Colors

Multidimensional distributions are often used to describe and summarize different features of an image. The one-dimensional distribution of image intensities describes the overall brightness content of a gray-scale image, while a three-dimensional distribution (RGB) can play a similar role for color images. The texture content of an image can be described by a distribution of local signal energy over frequencies corresponding to different colors. The Euclidean distance in the multidimensional space can be used to measure the dissimilarity of two images.

The Earth Mover's Distance (EMD) is based on the minimal cost that must be paid to transform one distribution into the other, in a precise sense. EMD matches perceptual similarity better than other distances used for image retrieval. The EMD is based on a solution to the transportation problems from linear optimization. It is more robust than histogram matching techniques in that it can operate on variable-length representations of the distributions that avoid quantization and other binning problems typical of histograms (Rubner, Tomasi, and Guibas, 1998).

3.3.7   Similarix

A network can be graphically represented by nodes (vertices) and links (edges). If we use nodes to represent individuals (persons, objects, events, etc.) and links to represent the associated similarities, a network of similarities is formulated (Figure 3.3). We call this similarity network a similarix. A similarix is a weighted network with similarity scores as the weights of the edges (links). Similarity is a directionless or bidirectional relationship ($S_{AB} = S_{BA}$), therefore, a similarix is an undirected network. Similarices can be used to study the local and global properties of the community or the populations from the perspective of network science.

A network such as a similarix can be characterized by topological properties such as size, degree, centrality, density, geodesics, diameter, and modularity.

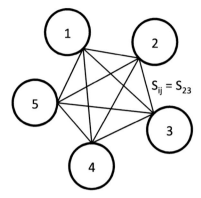

FIGURE 3.3
A Similarix–Similarity Network

Size of a network refers to the number of nodes N. *Degree* (V_i) of a node is the number of links (edges) that are directly connected to the node (vertex) i. *Average degree* of a network is $\bar{n} = \frac{1}{N} \sum_{i=1}^{N} V_i$. In a completed network, in which every node is directly connected to all other nodes, the number of edges E reaches its maximum, $E_{\max} = \frac{N(N-1)}{2}$. The factor $1/2$ in the formulation eliminates double counting the links. *Density* D of a network is defined as a ratio of the total number of edges E to the number of possible edges in a network with N nodes. Therefore, for an undirected network, $D = \frac{2E}{N(N-1)}$. Since similarix is a complete graph, it has density $D = 1$.

Centrality identifies the most important nodes in a network model. Different centrality indices encode different contexts for the word "importance." *Degree centrality*, the simplest type of centrality, is defined as the number of links or edges (E) incident upon a node. There are many other centrality measures. Betweenness centrality, for example, considers a node highly important if it forms bridges between many other nodes. Eigenvalue centrality, in contrast, considers a node highly important if many other highly important nodes link to it. In a similarix we can measure the centrality by the sum of the associated similarities, that is, the strength of the weighted network. *Node strength* is the sum of weights attached to links of a node (Barrat et al., 2004). That is, we define the *similarity centrality* of the ith node by

$$c_i = \sum_{j=1}^{n} S_{ij} \qquad (3.17)$$

Since the similarity centrality c_i is the representativeness of the node i in a network, the average of a quantity Y (such as response to a drug) of the network can be estimated by

$$\bar{Y} = \frac{1}{C} \sum_{i=1}^{n} c_i Y_i = \frac{1}{C} \sum_{i=1}^{n} \sum_{j=1}^{n} S_{ij} Y_i. \tag{3.18}$$

Here the *global closeness* $C = \sum_{j=1}^{n} c_i = \sum_{i=1}^{n} \sum_{j=1}^{n} S_{ij}$. We will come to the same conclusion from a slightly different perspective in the next chapter.

A *geodesic* between two nodes (i and j), G_{ij}, is the shortest path from i to j. The diameter of a network is the maximum of all geodesics. For a similarix, we define the geodesic between two nodes (i and j), G_{ij} as the dissimilarity-weighted shortest path from i to j. Here dissimilarity $=1/$similarity.

Clustering Coefficient

The clustering coefficient is used to measure the strength of the connectivity among members within a cluster or a community. More precisely, the clustering coefficient of a node is the ratio of the number of existing links connecting a node's neighbors to each other to the maximum possible number of such links. The clustering coefficient of the ith node is

$$C_i = \frac{e_i}{k_i (k_i - 1)/2}, \tag{3.19}$$

where k_i is the number of neighbors (nodes) of the ith node, and e_i is the number of connections between these neighbors. $k_i (k_i - 1)/2$ is the maximum possible number of connections between neighbors. The clustering coefficient for the entire network is the average of the clustering coefficients of all the nodes. A high clustering coefficient for a network is another indication of a small world. See Chapter 10 for more information.

3.3.8 Adjacency Matrix of Network

A graph or similarix can be equivalently represented by an *adjacency matrix*. For a similarix with five nodes, the adjacency matrix is

$$S = \begin{bmatrix} S_{11} & S_{12} & S_{13} & S_{14} & S_{15} \\ S_{21} & S_{22} & S_{23} & S_{24} & S_{25} \\ S_{31} & S_{32} & S_{33} & S_{34} & S_{35} \\ S_{41} & S_{42} & S_{43} & S_{44} & S_{45} \\ S_{51} & S_{52} & S_{53} & S_{54} & S_{55} \end{bmatrix}.$$

For an unweighted network, the similarity score is 1 if two nodes are directly linked and 0 otherwise. We can study the properties of a network/graph through its adjacency matrix. The adjacency matrix of an undirected simple graph is symmetric, and therefore has a complete set of real eigenvalues and an orthogonal eigenvector basis. The set of eigenvalues of a graph is the spectrum of the graph.

If A is the adjacency matrix of the directed or undirected graph G, then the matrix A^n has an interesting interpretation: the element (i, j) gives the number of *walks* of length n from vertex i to vertex j. If n is the smallest nonnegative integer such that for some i, j, the element (i, j) of A^n is positive, then n is the distance between vertex i and vertex j. This implies, for example, that the number of triangles in an undirected graph G is exactly the trace of A^3 divided by 6. The adjacency matrix can be used to determine whether or not the graph is connected.

Calculating degree centrality for all the nodes in a graph takes $O\left(V^2\right)$ time in a dense adjacency matrix representation of the graph, and for edges takes $O\left(E\right)$ time in a sparse matrix representation.

3.3.9   Biological and Medical Similarices

For examples of similarices, we can look to biological and social networks, the Internet, friends-networks, authorship-networks, metabolic networks, food webs, neural networks, pathological networks, and transportation networks, to name a few. When assigning similarity scores to the links, each of these real-world networks becomes a similarix. For example, if we assign a similarity score between two cities based on the traffic, 0 for no traffic or no direct highway, and 1 otherwise.

For a clinical trial, nodes can represent patients, and links are similarities determined by baseline characteristics such as race, age, gender, and baseline disease severity. We can also construct patient-symptom (adverse events) networks using data within a clinical trial, for a drug, for a disease indication, or all approved drugs using data sources such as www.approveddrugs.com. Similarly, drug-symptom networks, disease-symptom networks, drug-drug networks, disease-disease networks, AE-AE networks, etc., all these similarices can be constructed to study the relationships between drugs, between adverse events.

These similarices can be used for clustering and calculating the benefit-risk ratio for each cluster. Such clustering can be used for personalized medicine.

3.4   Summary

The similarity principle is a fundamental principle that we constantly use in our daily lives, and to make causality inferences and scientific discoveries. The principle asserts that each attribute contains some (≥ 0) information on the outcome of events, that similar things should have similar outcome, and that outcome similarity is positively related to the overall similarity of attributes.

Similarity is context dependent: (1) outcome-dependent, (2) attribute-dependent, and (3) scope-dependent (a difference of 2 in an attribute can be relatively small if the maximum difference in the underlying set is 100 or relatively big if the maximum difference is 2).

There are different measures of similarities, each having its own characteristics and applicability to particular problems. We will see more in the next chapter.

Choice of attributes and hierarchical attributes. The trade-off is using individual letters (only 26 English tokens) makes a long string for input, using phrases and paragraphs (millions of tokens) makes for much shorter input strings.

3.5 Problems

3.1: What is the similarity principle? Why it is important?

3.2: Provide examples where you have used similarity principles in your daily life or work.

3.3: Why can we say the following? "Causality and prediction require repeated events, as a result, similarity grouping cannot be avoided."

3.4: How do you define causality? Is causality a discovery or invention?

3.5: Identify two friends or classmates of yours. Define the similarities between you and them and compare your achievements (or grades) to see if yours are more similar to those of the one who is more similar to you.

4

Similarity-Based Artificial Intelligence

4.1 Similarity-Based Machine Learning

4.1.1 Nearest-Neighbors Method for Supervised Learning

When we buy a product or seek advice on some matter, we often seek out close neighbors or friends for their opinions since they are similar to us (in many ways), and it is a convenient way to get helpful information. This is the basic idea behind the simplest machine learning algorithm, called the *k-nearest neighbors* (KNN or kNN) algorithm. In KNN, an object is classified by a majority vote of its neighbors, with the object being assigned to the class most common among its *k*-nearest neighbors. The kNN method is a type of instance-based learning, or lazy learning, where the function is only approximated locally, and all computation is deferred until classification. Despite its simplicity, kNN has been used in many classification problems, such as the deciphering of handwritten digits and satellite image scenes. Thomas and Mathew (2016) used kNN for ECG pattern analysis and classification (Thomas and Mathew, 2016). The authors obtained a measure of similarity between two waveforms using the cross-correlation between two time-domain signals. A heuristically determined mathematical equation gives the parameters for classifying signals as normal and abnormal cardiac patterns. The kNN algorithm can be effective when the decision boundary is very irregular (Hastie et al., 2001, 2009). Jadhav and Kulkarni (2006) and Tropsha (2006) studied a 3-D QSAR using the *k*-nearest neighbors method. Nigsch et al., (2006) employed *k*-nearest neighbors algorithms for genetic parameter optimization.

The weighted nearest neighbor classifier: The kNN classifier can be viewed as assigning the *k*-nearest neighbors a weight $1/k$ and all others 0 weight. This can be generalized to weighted nearest-neighbor classifiers. That is, where the *i*th nearest neighbor is assigned a weight w_{ni}, with $\sum_{i=1}^{n} w_{ni} = 1$.

The kNN method can be used for regression, by simply assigning the property value for the object to be the average of the values of its *k*-nearest neighbors.

$$\hat{y}_i = Ave\left(y_j | x_j \in N_k\left(x_i\right)\right), \tag{4.1}$$

where $N_k\left(x_i\right)$ is the set of k points nearest to x_i and y_j is the attribute value at x_j. We may weigh neighbors differently according to the distances, e.g., perhaps we'll choose the weight to be the reciprocal of the distance to the neighbor. Therefore, a kNN algorithm can be viewed as utilizating the similarity principle, in which, a similarity score can be defined as a function of distance, for example, 1 for the each of the closest neighbors and 0 otherwise. The question is how "neighbor" should be defined, or what the definition of distance should be when multiple attributes are involved, because the distance definition will determine the neighbors. For instance, if we want to classify a person as diseased or not diseased based on symptoms and demographic characteristics such as age, race, and gender, how do we determine the similarity or distance (dissimilarity) between subjects? If we have patient A, a 60-year old male, patient B, a female aged 58 years, and patient C, a male aged 30 years, how will we determine whether B or C is the nearest neighbor of patient A? We also have to decide what unit will be used for age, year, month, or something else The second question is: how many neighbors should we use in the prediction of $\hat{y}$?

The optimal number of neighbors, k, in kNN is usually different for different problems and can be determined by trying different values of k with training data and selecting one that gives a smaller misclassification error rate.

$$Err = \frac{1}{N} \sum_{i=1}^{k} (\hat{y}_i - y_i)^2, \tag{4.2}$$

where $y_i = 0$ or 1 for two-category classification problems.

R Programming for KNN

```
    library(mlbench)
  library(class)
  data("PimaIndiansDiabetes2", package = "mlbench") # Load data
  Diabetes2 <- na.omit(PimaIndiansDiabetes2) # remove NAs
  # create a dummy variable for diabetes
  Diabetes2$Diabetes01=ifelse(Diabetes2$diabetes=="pos", 1, 0 )
  # Randomly sample 70% observations to training and the rest to testing
  n = nrow(Diabetes2)
  trainIndex = sample(1:n, size = round(0.7*n), replace=FALSE)
  train.y=Diabetes2[trainIndex,10]
  train.x=Diabetes2[trainIndex,-(9:10)]
  test.y=Diabetes2[-trainIndex,10]
  test.x=Diabetes2[-trainIndex,-(9:10)]
  set.seed (1)   # seed to randomly break the tied observation
  knn.pred=knn(train.x, test.x, train.y, k=1)
  table(knn.pred , test.y)
```

	test.y	
knn.pred	0	1
0	61	25
1	11	21

People often apply standardization to the feature variables (James, et al., 2013) so that all features have zero means and standard deviations of 1. However, such standardization may not be any better in determining the best neighbors for a reason that will be explained shortly.

We can test the effect of the number of neighbors, k, using the following code:

```
knn.pred=knn (train.x, test.x, train.y, k=3)
table(knn.pred, test.y)
```

The output for $K = 2$ is shown below:

	test.y	
knn.pred	0	1
0	60	29
1	12	17

4.1.2 Similarity-Based Learning

Define a typical person (object, event, process, etc.) by K characteristic variables (attributes), $X_k, k = 1, 2, ..., K$. The i-th person has the observed characteristics, $X_{ik}, i = 1, ..., N; k = 1, 2, ..., K$. The outcome of interest (or dependent variable) for the i-th person is Y_i. At this moment we assume the outcome variable is either continuous or of ordinal type.

A person defined by the selected attributes will not be unique. That is, even when two persons are identical in terms of the selected characteristics, they are different in other aspects and can have different responses to a stimulant such a drug. Therefore, what we have defined is actually a type of person but not an individual person. Note that the observed outcome is the same as the true value if measurement errors are ignored and we will not differentiate between the two in machine learning unless specified. According to the similarity principle, to predict the outcome $\hat{Y}_i$ of the ith type of person to a stimulant, we use the similarity-weighted outcomes Y_j $(j = 1, ...N)$, where N is the number of patients. That is,

$$\hat{Y}_i = \sum_{j=1}^{N} W_{ij} Y_j, i = 1, 2, ..., N, \tag{4.3}$$

where N is the number of subjects in the training set, and the weight, W_{ij}, depends on the similarity score S_{ij} between the i-th and j-th subjects,

$$W_{ij} = \frac{S_{ij}}{\sum_{m=1}^{N} S_{im}}. \tag{4.4}$$

The normalization of weights in (4.4) can be justified: it makes the prediction of (4.3) an unbiased one under random sampling, i.e., $\sum_{i=1}^{N} (\hat{Y}_i - Y_i) = 0$.

Note that regression models model the relationship between dependent and independent variables directly, while similarity models model the relationship indirectly through modeling the relationship among the dependent attributes (outcomes) of different subjects by using similarity scores. From (4.3) and (4.4), as the sample size n_i for the subject of type i increases, the estimation of $\hat{Y}_i$ approaches Y_i. This is an important feature of learning.

A subject is defined by the selected attributes. Therefore, for given paired subjects, a different selection of attributes can lead to a different similarity score. The key is how to determine the similarities S_{ij} when there are different attributes involved, such as age, gender, disease severity, and weight. More precisely, how do we determine the relative importance of an attribute in calculating the similarity score? We will use attribute-scaling factors to indicate the relative importance of attributes. Without scaling variables, it's difficult to intelligently and objectively determine the relative importance. To this end, we define the similarity score S_{ij} as a function of the distance d_{ij} between subjects i and j, whereas the distance is a function of the attributes and their scaling factors $R_k (k = 1, 2, \ldots, K)$. Here, K is the number of attributes. Distance-based similarity measures are translation-invariant, that is, $S(\mathbf{X}_i, \mathbf{X}_j) = S(\mathbf{X}_i - \mathbf{X}_j, 0)$.

4.1.3 Similarity Measures

Define the exponential similarity score S_{ij} as

$$S_{ij} = \exp\left(-d_{ij}^{\eta}\right), \eta > 0, \tag{4.5}$$

where

$$d_{ij} = \left(\sum_{k=1}^{K} (R_k |X_{jk} - X_{ik}|)^{\rho}\right)^{1/\rho}, k = 1, 2, \ldots, K; \rho = 1 \text{ or } 2. \tag{4.6}$$

There are reasons to choose the exponential similarity function: the sensitivity of human sense organs (visual, hearing, sense of smell, sense of weight) are all in log scale. Thus, $S = e^{-d}$ implies $d = -\ln S$. We also pay attention to similar things, and such an attention disappears exponentially as the distance increases. The common requirements for a similarity function $S_{ij}(d_{ij})$ are that

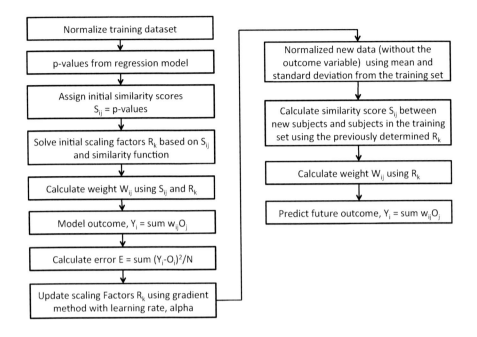

FIGURE 4.1
Flowchart of SBML

$S_{ij}(0) = 1$ and $S_{ij}(\infty) = 0$, where d_{ij} is the distance between the two subjects i and j.

We now discuss the determination of the scaling factors R_k and the SBML algorithms.

4.1.4   Algorithms for SBML

Figure 4.1 is the flowchart or algorithms of SBML as explained as follows.

(1) Data normalization

The data normalization procedure is a method used to standardize all variables in the dataset so that they all have a mean of 0 and a standard deviation of 1. The standard is achieved for each variable by subtracting its mean and then dividing by its standard deviation. The standardization makes the invented methodology robust and makes it easy to apply the SBML. After the standardization, all private information is removed and what is left is the data structure, not the actual data. Therefore, the individual's privacy is protected. For example, if original age ranges from 18 to 70, it will range from a negative value to a positive value with a mean of 0. No one can tell what the original ages are after standardization.

(2) Determination of initial scaling factors

It is difficult to determine the scaling factors R_k directly from prior knowledge or current data. It is much easier to calculate R_k through the similarity scores. The following two methods can be used

(a) When very limited training data are available, we have to use our prior knowledge in our brain, which will be vague and difficult to specify. In other words, based on prior knowledge of similarity scores between some pairs subjects, we solve for $R = (R_1, R_2, \ldots R_K)$ using the defined similarity function. That is, $R = $ function (S), where $S = (S_1, S_2, \ldots, S_K)$.

(b) p-value-based similarity scores.

A p-value can be used as a similarity measure for an attribute. If the p-value for an attribute is small, it means that the data show the subjects in the experiment are very different in terms of the attribute and the outcomes that concerns us. If the p-value for an attribute is larger, it means that the data show the subjects in the experiment are very similar in terms of the attribute and the outcome of concern.

When some training data are available, we obtain p-values from a statistical model (a linear model for continuous variables, a logistic model for binary and ordinal variables, and a log-rank test for time-to-event variables), assign these p-values to the initial similarity scores between pairs of the K selected subjects, and then determine the scaling factor R_k using the defined similarity function. Here we see that the p-value P_k can be interpreted as a similarity score. As we discussed earlier, the similarity score should relate to both attributes (independent variables) and outcome variables (dependent variables). p-values P_k from multiple regressions meet both criteria. Roughly speaking, when the two-sided $P_k = 1$, there is no difference regarding the kth attribute, but when P_k is close to 0, there is a big difference regarding the kth attribute.

For an exponential similarity function with $\eta = 1$ and $\rho = 2$, the similarity function in equation (4.5) becomes

$$S_{ij} = \exp\left(-\sqrt{\sum_{k=1}^{K}(R_k|X_{jk} - X_{ik}|)^2}\right). \tag{4.7}$$

The similarity score ranges from 0 (completely different) to 1 (identical) inclusively.

(3) Solve initial scaling factors R_k based on initial similarities

When a pair of subjects (the kth pair) are the same in terms of all variables (attributes) under consideration except one (the kth) variable, then the summation will disappear on the right-side of Eq. (4.7) and the scaling factor

R_k can be calculated explicitly using Eq. (7). For this reason, we choose the initial K (real or virtual) pairs of subjects in this way, and determine their K initial similarity scores. The corresponding K scaling factors that can be updated later through learning are

$$R_k^0 = \frac{\ln(P_k)}{|X_{jk} - X_{ik}|}, k = 1, 2, \ldots, K, \qquad (4.8)$$

where the superscript, is used to indicate the initial values, and $|X_{jk} - X_{ik}|$ is the distance between the 1st and 3rd quartiles, which can be considered to be the difference between two typical subjects in the data with respect to this variable. The p-value P_k from a regression model using the training dataset is one measure of similarity between typical subjects regarding the kth attribute when all other attributes are the same except for the kth attributes.

Note that the initial similarity score should be consistent with the similarity definition. For example, using p-value as the initial similarity score is consistent with the exponential similarity function with $\eta = 1$ because a normal cdf is an approximately exponential function. If we use the correlation coefficient matrix as the similarity function, then the initial similarity score should be the correlation matrix from the training data. The p-value as similarity score is able to take care of the problem when highly corrected variables are added to the model. When the sample size is very large, the exponential similarity function can lead to a numerical overflow and mismatch between the exponential similarity function and the multivariate normal cdf. Therefore, getting multiple small samples (50 to 100 subjects) from the training data and taking the average of the p-values as the initial similarity scores might be preferable.

For variables with large p-values (high similarity scores, not statistically significant), the corresponding Rs are large and their effects on the outcome are negligible. This means there is a low concern about overfitting with SBML.

We can also simply choose 0.5 for the initial similarity scores. That is, let $P_k = 0.5$ and use (4.8) to determine the initial R_k^0.

(4) Predicted outcomes

After the R_k are determined, we use Eq. (4.7) to calculate the similarity scores S_{ij}, Eq. (4.4) to calculate weights W_{ij}, and Eq. (4.3) to calculate the predicted outcome for each new subject m:

$$\hat{Y}_m = \sum_{j=1}^{N} W_{mj} Y_j. \qquad (4.9)$$

To calculate mean square error for the training set with N subjects:

$$E = \frac{\sum_{i=1}^{N}(\hat{Y}_i - Y_i)^2}{N} = \frac{1}{N}\sum_{i=1}^{N}(\sum_{j=1}^{N} W_{ij} Y_j - Y_i)^2 \qquad (4.10)$$

To calculate mean square error for the test set with M subjects:

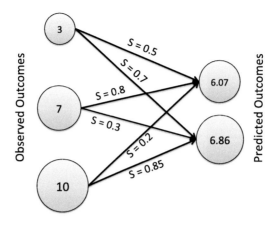

FIGURE 4.2
Similarity Principle in Action

$$E\,(Test) = \frac{\sum_{i=1}^{M}(\hat{Y}_m - Y_m)^2}{N} = \frac{1}{N}\sum_{m=1}^{M}(\sum_{j=1}^{N}W_{mj}Y_j - Y_i)^2. \qquad (4.11)$$

Note that a weight average can only be within the observed range. Figure 4.2 illustrates how the predicted outcomes are calculated from the observed values using the similarities.

4.1.5 Prediction Error Decomposition

In parametric regression and similarity-based learning methods with attribute scaling factors, we model the outcome Y by $\hat{f}(X)$, while the truth is $Y = f(X) + \varepsilon$, where X is a set of attributes or predictors and the error expectation $E(\varepsilon) = 0$ with variance σ_ε^2. From this, we can decompose the expected prediction error of $\hat{f}(X)$ at an input point $X = x_0$ using squared-error loss:

$$\begin{aligned} Err(x_0) &= E[(Y - \hat{f}(x_0))^2 | X = x_0] \\ &= \sigma_\varepsilon^2 + [E\hat{f}(x_0) - f(x_0)]^2 + E[\hat{f}(x_0) - E\hat{f}(x_0)]^2 \\ &= \sigma_\varepsilon^2 + Bias^2 + \sigma_{\hat{f}(x_0)}^2. \end{aligned} \qquad (4.12)$$

The variance σ_ε^2 is usually very difficult to reduce because it relates the hidden confounders that have not been identified or measured. In some cases, the hidden effect can be modeled to a certain extent as discussed by Chang et al., (Chang, Balser, et al., 2019). Here the bias is a statistical measure of the average accuracy of the prediction of the ideal unbiased model $f(X)$. It is dependent on the model $\hat{f}$ and the input data point x_0, which can represent a set of characteristics used to define a type of person. In practical learning problems, the true bias is usually unknown because the target population (or the similarity group to which we want to apply our predictions) is ill-defined. Only as a mathematical exercise can we assume a target population with a probability distribution and a bias that can be determined, in theory. Nevertheless, for practical problems the bias can be estimated by applying the predictive model to the test dataset. The variance of the prediction $\sigma^2_{\hat{f}(x_0)}$ reflects the average precision of the model, which also depends on the function and the set of attributes. In real-world problems, we often need to determine the trade-off between bias and precision in model selection. For example, a trial constant model has a variance of the prediction, $\sigma^2_{\hat{f}(x_0)}$, that is zero, but it is usually not a good model. (Exercise: can a model be better than $f(X)$ if we somehow can determine the hidden confounder effect?) Note that this bias is different from the bias of an estimator defined in the classic statistics.

4.1.6 Training, Validation, and Test Datasets

A ML model is generally in need of training to determine its parameters before using it in a real-world problem. Moreover, the trained model often needs to be validated or tested for its performance. The training-validation (test) processes are often done recursively to ensure that the optimally trained model is identified. Validation and testing are both used for evaluating and tuning the model; some people use the terms interchangeably, some people use "validation" in tuning a model and "testing" in evaluating a model, as we do in this book.

Cross-validation combines (averages) measures of fitness in prediction to derive a more accurate estimate of a model's predictive performance. Cross-validation can also be used in variable selection, to determine the model parameters (for example, the number of neighbors and scaling factors in the KNN method), and to compare the performances of different predictive modeling procedures.

(1) Exhaustive cross-validation methods are cross-validation methods which learn and test using all possible ways to divide the original sample into a training and a validation set.

(2) Leave-p-out cross-validation (LpO CV) involves using p observations as the validation set and the remaining observations as the training set. This is repeated for all ways to cut the original sample into a validation set of p observations and a training set.

(3) Bootstrapping is the random selection, with replacement, of m samples of size p as training sets and n samples of size q as test sets.

In general, larger scaling values R_k will lead to a smaller training error, and the gradient method is used to adjust R_k slightly in the direction of maximum reduction of the training error. We can always reduce the training error to near-zero when the R_k approach infinity. However, too much of an

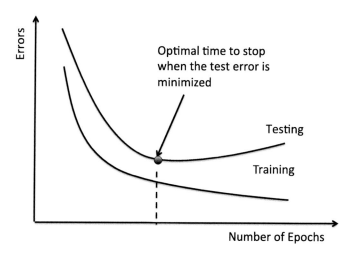

FIGURE 4.3
Training Error Versus Test Error

increase in the values R_k will lead to overfitting and increase test error (Figure 4.3). The test error is our real concern, not the training error.

4.2 Regularization and Cross-Validation

4.2.1 Learning—Updating Attribute-Scaling Factors

The gradient method is used for updating the scaling factor R_k in SBML. The scaling factor at iteration $t+1$ from iteration t is given by

$$R_k^{(t+1)} = \max(0, R_k^{(t)} - \alpha \frac{\partial E}{\partial R_k}). \tag{4.13}$$

However, larger R_s will generally lead to a smaller training error. As the R_s become very large the prediction using SBML will degenerate to a k-nearest neighbors ($k=1$) algorithm, but the relative scaling effects of different attributes are still there (the distances and neighbors of a data point are determined by the scaling factors).

4.2.2 Loss Function

If the prediction accuracy is measured by the MSE for the test set, then ridge loss function can be used and the optimization problem becomes one of finding a vector **R** that minimizes

$$\varrho_2\left(\mathbf{R};\boldsymbol{\lambda}\right) = MSE + \lambda\left\|\mathbf{R}\right\|_2^2, \tag{4.14}$$

where the tuning parameter λ can be determined by cross validation, and

$$\left\|\mathbf{R}\right\|_2 = \mathbf{R}\cdot\mathbf{R} = \sqrt{\sum_{k=1}^{K} R_k^2}. \tag{4.15}$$

If the prediction accuracy is measured by the absolute difference or L_1 for the test set, then the *lasso* loss function can be used and the optimization problem is to find a vector $\mathbf{R}$ that minimizes

$$\varrho_1\left(\mathbf{R};\boldsymbol{\lambda}\right) = MAD + \lambda\|\mathbf{R}\|_1, \tag{4.16}$$

where

$$\left\|\mathbf{R}\right\|_1 = \sum_{k=1}^{K}|R_k| = \sum_{k=1}^{K} R_k. \tag{4.17}$$

We can also use an *elastic net* loss function, and find a vector $\mathbf{R}$ that minimizes

$$\varrho_{12}\left(\mathbf{R};\boldsymbol{\lambda},\boldsymbol{\alpha}\right) = E + \lambda_1((1-\omega)\left\|\mathbf{R}\right\|_2 + \omega\lambda_2\|\mathbf{R}\|_1), \alpha \in [0,1], \tag{4.18}$$

where the parameter ω, $0 < \omega < 1$, is a weight and E can be the MSE and MAD (mean absolute difference). Other, similar optimization methods can involve different error functions, loss functions, and various similarity functions (cosine, Jacard, logistic-alike, distance-inverse functions), as we have discussed in Chapter 3.

Gradient Method Algorithm for Training:

As an example to solve the optimization problem in Eq. (4.18), we use a gradient method for a given λ:

$$\frac{\partial\varrho_i\left(\mathbf{R};\lambda\right)}{\partial R_m} = \frac{\partial E}{\partial R_m} + \frac{\lambda\partial\left\|\mathbf{R}\right\|^2}{\partial R_m}, \tag{4.19}$$

where constant λ is a constant (e.g., -0.3 for normalized data).

Since $\frac{\partial\|\mathbf{R}\|^2}{\partial R_m} = 2R_m$, (4.19) can be written as

$$\frac{\partial\varrho_i\left(\mathbf{R};\lambda\right)}{\partial R_m} = \frac{\partial E}{\partial R_m} + 2\lambda R_m. \tag{4.20}$$

The MSE (or L_2) derivative can be written as

$$\frac{\partial E}{\partial R_m} = \frac{\partial MSE}{\partial R_m} = \frac{\partial E}{\partial W_m}\frac{\partial W_{ij}}{\partial R_m} = \frac{2}{N}\sum_{i=1}^{N}\left(\left(Y_i - \hat{Y}_i\right)\sum_{j-1}^{N}\hat{Y}_j\frac{\partial W_{ij}}{\partial R_m}\right), \tag{4.21}$$

where N is the size of the training set.

For MAD (mean absolute difference or L_1), the error derivatives are

$$\frac{\partial E}{\partial R_m} = \frac{\partial MAD}{\partial R_m} = \frac{\partial E}{\partial W_m}\frac{\partial W_{ij}}{\partial R_m} = \frac{1}{N}\sum_{i=1}^{N}\frac{\hat{Y}_i - Y_i}{|\hat{Y}_i - Y_i|}\sum_{j=1}^{N}\frac{\partial W_{ij}}{\partial R_m}Y_j. \quad (4.22)$$

Here the weight derivatives are

$$\frac{\partial W_{ij}}{\partial R_m} = \frac{\frac{\partial S_{ij}}{\partial R_m}}{\sum_{n=1}^{N}S_{in}} - \frac{S_{ij}\sum_{n=1}^{N}\frac{\partial S_{in}}{\partial R_m}}{\left(\sum_{n=1}^{N}S_{in}\right)^2}. \quad (4.23)$$

The similarity derivatives are

$$\frac{\partial S_{ij}}{\partial R_m} = \frac{\partial S_{ij}}{\partial d_{ij}}\frac{\partial d_{ij}}{\partial R_m} = -\eta d_{ij}^{\eta-1}S_{ij}\frac{\partial d_{ij}}{\partial R_m}, \quad (4.24)$$

where for an exponential distance function, the distance derivatives are

$$\frac{\partial d_{ij}}{\partial R_m} = \begin{cases} \left(\frac{R_m}{d_{ij}}\right)^{\rho-1}|X_{jm}-X_{im}|^{\rho}, i \neq j \\ 0, \text{otherwise} \end{cases}. \quad (4.25)$$

If a *distance-inverse function* is used as the similarity measure

$$S_{ij} = \frac{1}{1+d_{ij}^{\eta}}, \quad (4.26)$$

then the distance derivatives are

$$\frac{\partial S_{ij}}{\partial R_m} = -\eta S_{ij}^2 d_{ij}^{\eta-1}\frac{\partial d_{ij}}{\partial R_m}. \quad (4.27)$$

If similarity is defined by the *logistic similarity function*

$$S_{ij} = \frac{2}{1+\exp(d_{ij}^{\eta})}, \quad (4.28)$$

then the distance derivatives are

$$\frac{\partial S_{ij}}{\partial R_m} = -\frac{1}{2}\eta S_{ij}^2 d_{ij}^{\eta-1}\frac{\partial d_{ij}}{\partial R_m}. \quad (4.29)$$

The gradient method will update the scaling factors using from iteration t to iteration $t+1$ using

$$R_k^{(t+1)} = R_k^{(t)} - \alpha\frac{\partial \varrho_2}{\partial R_k}. \quad (4.30)$$

To ensure the convergence, a small α (e.g., 0.6) should be used. The iteration of the R_k will continue until $\varrho_2\left(\mathbf{R}^{t+1};\lambda\right)$ is close to zero or sufficiently small.

For clarity, we summarize the formulas for a gradient learning algorithm with an exponential similarity function, as follows:

$$
\begin{cases}
R_k^{(t+1)} = R_k^{(t)} - \alpha \frac{\partial \varrho_2}{\partial R_k} \\
\frac{\partial \varrho_i(\mathbf{R};\lambda)}{\partial R_m} = \frac{\partial E}{\partial R_m} + 2\lambda R_m. \\
\frac{\partial E}{\partial R_m} = \frac{2}{N} \sum_{i=1}^{N} \left(\left(Y_i - \hat{Y}_i \right) \sum_{j=1}^{N} \hat{Y}_j \frac{\partial W_{ij}}{\partial R_m} \right) \\
\frac{\partial W_{ij}}{\partial R_m} = \frac{\frac{\partial S_{ij}}{\partial R_m}}{\sum_{n=1}^{N} S_{in}} - \frac{S_{ij} \sum_{n=1}^{N} \frac{\partial S_{in}}{\partial R_m}}{\left(\sum_{n=1}^{N} S_{in} \right)^2} \\
\frac{\partial S_{in}}{\partial R_k} = -\eta S_{in} d_{in}^{\eta-1} \frac{\partial d_{in}}{\partial R_k} \\
\frac{\partial d_{ij}}{\partial R_m} = \begin{cases} \left(\frac{R_m}{d_{ij}} \right)^{\rho-1} |X_{jm} - X_{im}|^\rho, i \neq j \\ 0, \text{otherwise} \end{cases}
\end{cases}
\tag{4.31}
$$

The suggested learning rate is $\alpha = 0.125$, $\lambda = 0.6$, but you can adjust these values using cross-validation or another method. With (4.31), we suggest the following algorithms to determine the scaling factors R_k $(k = 1, ..., K)$ based on minimization of the loss function and optimization of the similarity-based AI model (4.9):

1. Choose the initial R_k, which can be a constant 0.5 or using (4.8) with p-values P_k from multiple regression using the training dataset.

2. Calculate the derivatives $\frac{\partial \varrho_2}{\partial R_k}$ using (4.20), (4.21), (4.23) through (4.25), based on the training dataset.

3. Update the scaling factors R_k using (4.30) until the loss function ϱ_2 increases (or stops decreasing) or we reach the maximum number of iterations allowed.

4. Once the R_k are determined, calculate similarity scores S_{mj} between the training subjects and the future subjects using (4.7).

5. Calculate the weights W_{ij} using (4.4) and the predicted outcomes $\hat{Y}_i$ for the test subjects using (4.9).

6. To evaluate the performance of the ML method, calculate the error E using

$$
E\,(test) = \sum_{i=1}^{M} \frac{\left(Y_i - \hat{Y}_i \right)^2}{M} = \sum_{i=1}^{M} \frac{\left(\sum_{j=1}^{N} W_{ij} Y_j - Y_i \right)^2}{M},
\tag{4.32}
$$

where M is the number of subjects in the testset.

Alternatively, the secant algorithm can be used. This algorithm is similar to the gradient algorithm except for using the following secant to replace the gradient in (4.20):

$$
\frac{\partial \varrho_2(\mathbf{R};\lambda)}{\partial R_m} \approx \frac{\Delta \varrho_2(\mathbf{R};\lambda)}{\Delta R_m} \approx \frac{\Delta E}{\Delta R_m} + 2\lambda R_m = \frac{E(\mathbf{R} + \Delta \mathbf{R}_m)}{\Delta R_m} + 2\lambda R_m, \tag{4.33}
$$

where the incremental vector $\Delta\mathbf{R}_m$ has an mth component equal to R_m and all other components equal to zero.

Other similarity functions can be used, such as the cosine, the logistic-alike similarity function, and the distance-inverse function. The derivatives are provided in Chapter 2.

4.2.3   Computer Implementation

All the necessary R functions for SBML are provided the appendix. The following R code is an example of SBML used for simulated multivariate data.

```
# Continuous outcome
library(mvtnorm)
library(class)
getwd()
# covariance matrix for creating correlated X.
sigma = matrix(c(1, 0.0, 0.6, 0.2, 0.0, 0.0, 0.0,
                 0.0, 1, 0.0, 0.0, 0.0, 0.0, 0.0,
                 0.6, 0.0, 1, 0.5, 0.0, 0.0, 0.0,
                 0.2, 0.0, 0.5, 1, 0.0, 0.0, 0.0,
                 0.0, 0.0, 0.0, 0.0, 1, 0.0, 0.0,
                 0.0, 0.0, 0.0, 0.0, 0.0, 1, 0.0,
                 0.0, 0.0, 0.0, 0.0, 0.0, 0.0, 1), ncol=7)
mu=c(0,0,0,0,0,0,0) # 7 multivarite normal means
initS=rep(0.5, 7) # 7 intial similarities.
xTrain = rmvnorm(n=40, mean=mu, sigma=sigma)
xTest = rmvnorm(n=40, mean=mu, sigma=sigma)
# Construct y that are not linearly related to x
Ytrain=xTrain[,1]+xTrain[,2]*xTrain[,3]+xTrain[,4]^2+xTrain[,5]
Ytest=xTest[,1]+xTest[,2]*xTest[,3]+xTest[,4]^2+xTest[,5]
# Data normalization
Xtrain = as.data.frame(normalizeData(xTrain, xTrain))
Xtest = as.data.frame(normalizeData(xTest, xTrain))
# Assign p-value from LM to initial similarities
# initS = GLMPv(Y=Ytrain, X=Xtrain, outcome = "gaussian")
mySBMLtrain=SBMLtrain(Epoch=5, S0=initS, Lamda=-0.3,
LearningRate=0.125, eta=1, Xtrain, Ytrain)
PredictedY(Weight(Similarity(eta=1, mySBMLtrain$Rs, Xtrain,
Xtest)),Ytrain, Ytest)
# MSE using training mean for prediction
mean((Ytest-mean(Ytrain))^2)
```

The following is the training and testing MSEs output for a run (varies each time):

TABLE 4.1
Training MSEs with Independent Xs

N_{train}	FLM	OLM	RR	SBML
20	2.19	2.43	2.72	0.65
50	3.13	3.26	3.31	0.84
100	3.62	3.70	3.71	0.86
200	3.84	3.88	3.87	1.21
400	3.86	3.88	3.88	1.60

Note: 100 replicates, each has 50 subjects in the test set.

```
$MSE
[1] 1.191225
$MSE
[1] 2.096909
```

Simulated Multivariate Normal Data

Since the MSE varies each time, depending on the simulated data, to evaluate the SBML we use 100 datasets from the multivariate distribution. Training sample size N_{train} ranges from 20 to 400. MSE: Naive = observed, FLM = full linear model, OLM = optimal model using BIC, RR = ridge regression, and SBML = similarity-based machine learning.

$$Y = X_1 + X_2 X_3 + X_4^2 + X_5$$

Initially seven predictors $X_1, X_2, ..., X_7$ are considered in the full linear model and SBML.

(1) $X_1, X_2, ..., X_7$ are independently drawn from the standard normal distribution.

The training MESs from simulations are presented in Table 4.1. The testing MSEs are presented in Table 4.2.

(2) $x_1, x_2, ..., x_7$ are drawn from the multivariate normal with zero means and variance matrix:

$$\Sigma = \begin{bmatrix} 1 & 0.5 & 0.4 & 0.3 & 0.2 & 0.1 & 0.05 \\ 0.5 & 1 & 0.5 & 0 & 0 & 0 & 0.5 \\ 0.4 & 0.5 & 1 & 0 & 0 & 0 & 0 \\ 0.3 & 0 & 0 & 1 & 0 & 0 & 0 \\ 0.2 & 0 & 0 & 0 & 1 & 0 & 0 \\ 0.1 & 0 & 0 & 0 & 0 & 1 & 0 \\ 0.05 & 0.5 & 0 & 0 & 0 & 0 & 1 \end{bmatrix}.$$

The testing MSEs are summarized in Table 4.3. We can see that (1) the

TABLE 4.2
Test MSE with with Independent Xs

N_{train}	Naive	FLM	OLM	RR	SBML
20	6.01	5.99	5.87	5.24	3.75
50	5.67	4.52	4.48	4.43	3.54
100	5.95	4.36	4.36	4.35	3.57
200	5.65	3.93	3.96	3.94	3.38

Note: 100 replicates, each has 50 subjects in the test set

TABLE 4.3
Testing MSE with Multivariate Xs

N_{train}	Naive	FLM	OLM	RR	SBML
20	5.30	6.28	5.98	5.21	3.61
50	5.22	4.83	4.65	4.62	3.55
100	4.94	4.31	4.24	4.26	3.50
200	4.82	3.92	3.91	3.93	3.48

Note: 100 replicates, each has 50 subjects in the test set.

model errors (MSE) reduce as the sample size increases, (2) SBML is consistently better than other methods, (3) similar conclusions can be drawn when x is independent or from multivariate normal. Simulations of each set of 100 replicates require CPU time ranges of 6 minutes ($N_{train} = 20$), 2.4 hours ($N_{train} = 200$), and 8 hours ($N_{train} = 400$) on a fast MacBook Pro 2.9 GHz Intel Core i9.

In general, when the data show locally linear but distantly nonlinear behavior, SBML is more effective than a linear model. If the data show linearity in the whole range of the population then linear models are better, but given big data or the presence of heterogeneity in the field of ML/AI, SBML is better.

4.3   Case Studies

Cystic fibrosis (CF) is a rare, inherited, autosomal recessive genetic, life-threatening disorder that causes exocrine glands to work incorrectly. CF damages multiple organs and systems in the body including respiratory (sinuses, lungs), gastrointestinal, reproductive, and integumentary. In CF drug development, a clinical endpoint to evaluate the drug's efficacy is the absolute improvement in lung function (percent predicted forced expiratory volume in one second or ppFEV1) compared to baseline. The attributes of interest include treatment, age, sex, and baseline ppFEV1.

TABLE 4.4

Performance Comparisons Using Bootstraps

| Training Set | | MSE | | | |
Size	Variance	FLM	OLM	RR	SBML
25%	96.5	74.7	76.0	76.8	59.4
50%	99.3	79.1	79.6	79.8	65.0
75%	99.0	79.8	80.1	80.2	66.1

To apply SBML to the clinical trial data analysis, data from three random-ized, placebo-controlled, parallel group studies were combined. The data structures were derived from the data. Based on the data structures, 100 datasets were simulated. For each dataset, we generated MSE mean from each of the following methods: full linear model (FLM), optimal linear model (OLM), ridge regression (RR), and similarity-based machine learning (SBML). Then average the 100 MSEs from each methods. The results are presented in Table 4.4 and show that SBML outperforms all other methods in the table.

4.4 Different Outcome Variables

For a binary outcome variable coded by 0s and 1s, p-values from multivariate logistic regression can be used for the initial similarity scores. All independent variables should be standardized. Nominal outcome variables can be converted into dummy variables. Ordinal variables can be treated as continuous variables or by using dummy variables. For a time-to-event outcome, each subject has either a time with the associated probability of censoring or survival time with the associated probability. Here is exampe of R code for running SBML with a binary outcome.

```
# Bindary outcome
# covariance matrix for creating correlated X.
sigma = matrix(c(1, 0.0, 0.6, 0.2, 0.0, 0.0, 0.0,
                 0.0, 1, 0.0, 0.0, 0.0, 0.0, 0.0,
                 0.6, 0.0, 1, 0.5, 0.0, 0.0, 0.0,
                 0.2, 0.0, 0.5, 1, 0.0, 0.0, 0.0,
                 0.0, 0.0, 0.0, 0.0, 1, 0.0, 0.0,
                 0.0, 0.0, 0.0, 0.0, 0.0, 1, 0.0,
                 0.0, 0.0, 0.0, 0.0, 0.0, 0.0, 1), ncol=7)
mu=c(0,0,0,0,0,0,0) # 7 multivarite normal means
initS=rep(0.5, 7) # 7 intial similarities.
xTrain = rmvnorm(n=40, mean=mu, sigma=sigma)
xTest = rmvnorm(n=40, mean=mu, sigma=sigma)
```

```
# Construct y that are not linearly related to x
Ytrain = (sign(xTrain[,1]+xTrain[,2]*xTrain[,3]+xTrain[,4]^2+xTrain[,5])+1)/2
Ytest = (sign(xTest[,1]+xTest[,2]*xTest[,3]+xTest[,4]^2+xTest[,5])+1)/2
# Data normalization
Xtrain = as.data.frame(normalizeData(xTrain, xTrain))
Xtest = as.data.frame(normalizeData(xTest, xTrain))
# Assign p-value from logistic regression to initial similarities
# initS = GLMPv(Y=Ytrain, X=Xtrain, outcome = "binomial")
mySBMLtrain=SBMLtrain(Epoch=5, S0=initS, Lamda=-0.3,
LearningRate=0.125, eta=1, Xtrain, Ytrain)
mypred = PredictedY(Weight(Similarity(eta=1, mySBMLtrain$Rs, Xtrain,
Xtest)),Ytrain, Ytest)
# misclassification error
mean((Ytest-round (myPred$pred_y))^2)
```

The training and testing MSEs are (varies from each simulation): 0.1052001 for training and 0.1379271 for testing. The misclassification error is 0.28.

4.5   Further Development of Similarity-Based AI Approach

4.5.1   Repeated Measures

For repeated measures or longitudinal data, measures from each subject at different timepoints can be viewed as data from different individuals. In other words, the measures on the same patient at different timepoints can be treated as data from different individuals who are very similar.

4.5.2   Missing Data Handling

In a linear model, we will delete the entirety of a patient's data if the measure on one of the attributes is missing. In SBML, when we can only use partial data the missing attributes will not be considered in the similarity calculation, thus no assumption is needed about the mechanism of missing.

If some attributes are not collected for some subjects, there are two methods to deal with the issue:

(1) exclude the attributes in the similarity scores calculation for these subjects, and

(2) assign the mean, median or mode of the attributes to these subjects.

4.5.3   Multiple Outcomes

For multiple outcomes, such as survival time (t), and quality of life score (q), an outcome vector (Y) can be used, in which each component of the vector is an outcome parameter: $O = \{t, q\}$. For the ith subject, the observed outcome is denoted by Y_i, and the corresponding modeled outcome is denoted by $\hat{Y}_i$.

The predictive error E is defined as

$$E = \frac{1}{M} \sum_{i=1}^{M} \left(\hat{Y}_i - Y_i \right)^T C \left(\hat{Y}_i - Y_i \right), \tag{4.34}$$

where elements in matrix C measure the relative importance of different outcomes, whereas the scaling factors measuring the relative importance in contributing to the outcome can be the same or different for different outcomes for a given subject. The determination of the attribute-scaling factors will be similar to the case of a single outcome variable.

4.5.4   Sequential Similarity-Based Learning

Sequential similarity-based machine learning (SSBL) is an improved version of basic SBML in which the predicted outcomes are eventually observed at later times. SSBL uses the same method as for SBML, but the training will be continually performed over time as the observed outcomes accumulate. In other words, the scaling factors are continually updated as the data accumulates.

4.5.5   Ensemble Methods and Collective Intelligence

Ensemble methods can be viewed as meta-learning that combine several machine learning methods into one predictive model to decrease variance (bagging), bias (boosting), or improve predictions (stacking).

Ensemble methods can be sequential (e.g., AdaBoost in Chaper 8) or parallel (e.g., random forest in Chapter 8) approaches. A sequential method is, through exploiting the dependence between the base learners, to improve the overall performance by weighing previously mislabeled cases with higher weight. A parallel method can reduce the error dramatically by averaging the independent base learners. In terms of the difference of base learners, an ensemble method can be homogeneous ensembles (the same learner) or heterogeneous ensembles (different learners). Collective intelligence is an ensemble of intelligence from members of a society or group.

In SBML, when the sample is small, the scaling variables may become unstable or heavily dependent on the training sample. A solution would be using bootstrap to take multiple training samples and average the Rs obtained from the multiple trainings.

4.5.6   Generalized SBML

We can generalize SBML by applying a transform $g\left(\cdot\right)$ to the outcome variable before using the similarity principle:

$$g\left(\hat{Y}_m\right) = \sum_{j=1}^{N} W_{mj}Y_j \qquad (4.35)$$

and

$$\hat{Y}_m = g^{-1}\left(\sum_{j=1}^{N} W_{mj}Y_j\right). \qquad (4.36)$$

Similarly, we can also apply transforms to the features.

4.5.7   Dimension Reduction

SBML and kernel methods (see later) are "instance-based" methods, wherein all data points need to be retained in the model. When the data are big, the model becomes significantly complex in terms of size. In such cases it is desirable to reduce the dimension in the model. Here dimension refers to the number of data points, not the number of attributes or parameters.

We discuss two methods for dimension reduction. The first one is to select n representative points, and the values of the outcome variable and attributes of the representatives are the averages of the corresponding k-nearest neighbors.

The second method itself is an instance of SMBL. Suppose we have a training dataset with N observations in a vector $\mathbf{X}_i$, $(i = 1, 2, ...N)$, each having K features. The corresponding responses (labels) are $\mathbf{Y}_i$, $(i = 1, 2, ...N)$. We generate (e.g., uniformly) n data points (a grid) in feature space, $\mathbf{X}_j^*, j = 1, 2, ...n < N$. Here, the $\mathbf{X}_j^*$ are usually not the same data points as the $\mathbf{X}_i$. Treat the $\mathbf{X}_j^*$ as a test dataset and predict the associated responses (labels), $\mathbf{Y}_j^*$, by the SBML model that has been trained using the training data. Now we reverse the procedure: treat $\left(\mathbf{X}_j^*, \mathbf{Y}_j^*\right), (j = 1, 2, ...n)$ as the training set and $\left(\mathbf{X}_i^*, \mathbf{Y}_i^*\right), (i = 1, 2, ...N)$ as the test set to retrain the model to obtain new scaling factors R_k^*, $(k = 1, 2, ...K)$. At this moment, we can disregard the original data $\left(\mathbf{X}_i, \mathbf{Y}_i\right)$, $(i = 1, 2, ...N)$ since our SBML model is fully determined by the SBML with parameters $R_k^*, (k = 1, ..., K)$ and virtual data points $\left(\mathbf{X}_j^*, \mathbf{Y}_j^*\right), (j = 1, 2, ...n)$.

Because n can be much smaller than N, a dimension reduction can be achieved. Again, the dimension here is not the feature dimension, but the number of data points.

4.5.8 Recursive SBML

The hierarchical, or recursive SBML (RSBML) allows the utilization of diverse data sources to achieve a better prediction. Recursive learning resembles the natural human way of learning. It is an efficient way for learning from complicated data in which the differences are often difficult to precisely define. For instance, two trials conducted at different times or in different countries may differ in medical practice, or on account of race or other unknown characteristics.

Let's illustrate RSBML using clinical trials as an example. Recursive SBML involves hierarchical learning at multiple levels, the individual and group (trial) levels. First, SBML is applied to the individual patients within each trial to obtain the scaling factors and weights as described above. Then SBML is again applied to the group or trial levels, in which similarities between different trials are considered. To determine the similarities between groups, aggregated attributes have to be used, such as mean outcome, mean age, and the proportion of female participants in each clinical trial. New variables (often necessary) can be added to further characterize the different trials. Finally, the SBML weights at different levels are combined to predict the individual patient's results.

Specifically, in the aggregated level, the mean outcome $\hat{Y}_t$ for subject t is modeled based on the SBML:

$$\hat{Y}_t = \sum_{l=1}^{L} W_{tl}^* \hat{Y}_l(\bar{X}_l), \tag{4.37}$$

where W_{tl}^* is determined by the similarity score S_{tl} between subject t and virtual subject l, using group attributes such as mean age and mean weight, $\bar{X}_l$. Here, $(\bar{X}_l, \hat{Y}_l(\bar{X}_l))$ is the training set for determining the attribute scaling factors R^* in weights W_{tl}^*. $\hat{Y}_l(\bar{X}_l)$ is the predicted value using only the subjects in group l.

$$\hat{Y}_l(\bar{X}_l) = \sum_{j=1}^{N_l} W_{lj} Y_{lj}, \quad l = 1, 2, \dots L. \tag{4.38}$$

That is,

$$\hat{Y}_t = \sum_{l=1}^{L} W_{tl}^* \sum_{j=1}^{N_l} W_{lj} Y_{lj}, \tag{4.39}$$

where for a given group l, weight W_{lj} is determined in the same way as W_{ij} for the paired individuals in a single group/clinical trial, i.e., by using the similarity score between the ith and j^{th} subjects in the lth group. N_l is the number of subjects in group l. The computation time for SBML is $O(N^2)$, while it is $O(N^{3/2})$ for RSBML.

Sometimes only the aggregated results, instead of individual patient data, are available. These aggregated results from one trial can provide information for estimation of individual responses in another trial. For example, suppose we have individual data for one trial but can only access the means, standard deviations of the variables, and sample size (which is often the case in publications of clinical trial results) for another trial; then RSBML can also be used directly. Alternatively, the second approach to the problem is to treat the N patients in the second trial as if there are N identical patients with the same response and attributes, and then combine these data with the individual data from the first trial into a single dataset. Furthermore, add a trial ID and maybe other new variables, such as duration of the trial, etc. Finally, apply SBML to the combined data.

This recursive learning can continue on more than two levels. Recursive (hierarchical) learning is a much more efficient way of learning in the real world than learning everything from one level. It should be noted, too, that SBML and RSBML can also be used in meta-analysis.

The following is an example of RSBML in *R*:

```
# Recursive SBML
# covariance matrix for creating correlated X.
sigma = matrix(c(1, 0.0, 0.6, 0.2, 0.0, 0.0, 0.0,
                 0.0, 1, 0.0, 0.0, 0.0, 0.0, 0.0,
                 0.6, 0.0, 1, 0.5, 0.0, 0.0, 0.0,
                 0.2, 0.0, 0.5, 1, 0.0, 0.0, 0.0,
                 0.0, 0.0, 0.0, 0.0, 1, 0.0, 0.0,
                 0.0, 0.0, 0.0, 0.0, 0.0, 1, 0.0,
                 0.0, 0.0, 0.0, 0.0, 0.0, 0.0, 1), ncol=7)
mu=c(0,0,0,0,0,0,0) # 7 multivarite normal means
S0=rep(0.5, 7) # 7 intial similarities.
#
nGroups = 10; ntrain=20; ntest=20; ngvars=4
Xgtrain = matrix(nrow=1, ncol=ntrain)
Xgstrain=matrix(nrow=nGroups, ncol=ngvars)
Xgstest=matrix(nrow=nGroups, ncol=ngvars)
Ytest=matrix(nrow=nGroups, ncol=ntest)
Ygtrain=rep(0,1); Ygtest=rep(0,nGroups)
predYs(0,nGroups)
for (ig in 1:nGroups) {
xTrain = rmvnorm(n=ntrain, mean=mu, sigma=sigma)
xTest = rmvnorm(n=ntest, mean=mu, sigma=sigma)
Xstrain=xTrain[,c(1:3, 7)]; Xstest=xTest[,c(1:3, 7)]
# Xgtrain[ig,]=rowMeans(xTrain[,4:6]); Xgtest[ig,]=rowMeans(xTest[,4:6])
Xgtrain[1,]=colMeans(XsTrain)
Xgstrain[ig,]=Xgtrain
# Construct y that are not linearly related to x
Ystrain=xTrain[,1]+xTrain[,2]*xTrain[,3]
```

```
Ystest=xTest[,1]+xTest[,2]*xTest[,3]
Ygtrain[ig]=Xgtrain[1,1]+Xgtrain[1,2]*Xgtrain[1,3]

# normalize testset first before trainset and convert to dataframe
Xstest = as.data.frame(normalizeData(Xstest, Xstrain))
Xstrain = as.data.frame(normalizeData(Xstrain, Xstrain))

# Assign p-value from LM to initial similarities
initS = GLMPv(Y=Ystrain, X=Xstrain, outcome = "gaussian")
mySBMLtrain=SBMLtrain(Epoch=2, S0=initS, Lamda=-0.3,
LearningRate=0.125, eta=1, Xstrain, Ystrain)
predYgi=PredictedY(Weight(Similarity(eta=1, mySBMLtrain$Rs,
Xstrain, Xgtrain)),Ystrain, Ygtrain)
predYs[ig,]=predYgi$pred_Y
}
# perform SBML on groups
Xgstest = as.data.frame(normalizeData(Xgstest, Xgstrain))
Xgstrain = as.data.frame(normalizeData(Xgstrain, Xgstrain))

# Assign initial similarities
initS = rep(0.5,ngvars)
mygSBMLtraing=SBMLtrain(Epoch=2, S0=initS, Lamda=-0.3,
LearningRate=0.125, eta=1, Xgstrain, predYs)
Wg=Weight(Similarity(eta=1, mygSBMLtraing$Rs, Xgstrain, Xstest))
PredYrec=Wg%*%predYs
MSEtest=mean((PredYrec-Ystest)^2)
MSEtest


plot(Ystest,PredYrec)
```

The MSE produced in a simulation run is 0.87656 for a testset.

Data fusion is the process of integrating multiple data sources to produce more consistent, accurate, and useful information beyond that which any individual data source can provide. A simple sample will be integrated with the data from various clinical trials and published clinical trial data. Such data usually are a mix of individual patient data and trial summary data (such as means, medians, confidence intervals, standard errors, sample sizes and p-values). Interestingly, humans constantly use data fusion in comprehending the surrounding world. As humans, we rely heavily on our vision, smell, taste, hearing, and physical movement. We rely on a fusion of smelling, tasting, and touching food to ensure that it is edible (or not). Similarly, we rely on our ability to see, hear, and control the movement of our body to walk or drive and to perform most of our daily tasks. Our brain performs fusional processing based on individual knowledge at instants in time, and we take the appropriate action.

4.6   Similarity Principle, Filtering, and Convolution

The similarity principle can be explained by filtering processes in physics and the convolution function in math.

Humans view objects through eyes, by the filtering processes that consist of our concepts or prior knowledge and purpose in viewing the objects. These filters, as described later in our discussion of CNNs, are convolutions in mathematics. Here, we will show you that the similarity principle can be expressed as convolutions mathematically. The similarity principle when described by convolutions acts as filters, consistent with sense organs such as eyes. The similarity principle that serves as a fundamental principle for reasoning in all sciences and in our daily lives is in fact an abstract model for the behavior of our sense.

$$\hat{y}(\mathbf{x}_j) = \sum_{i,j} y(\mathbf{x}_i) S\left(\mathbf{x}_i - \mathbf{x}_j\right), \tag{4.40}$$

where $S\left(\mathbf{x}_i - \mathbf{x}_j\right)$ is a similarity function. $Y = f\left(\mathbf{x}\right) + \varepsilon, \varepsilon \sim \eta\left(0\right)$, a probability distribution with (usually) a zero mean. Let y be the observation of Y, and $\hat{y}$ is a predicted value of Y for a given x (an observed value of the predictors, X). In an extreme case, the predicted value $\hat{y}$ can be independent of the observed x.

Under the assumption of iid $\mathbf{x}_i$ and $\mathbf{x}_j$ from distribution $f\left(\mathbf{x}\right)$, the expectation of $\hat{y}$ is

$$E(\hat{y}(\mathbf{x}_j)) = \int y(\mathbf{x}_i) S\left(\mathbf{x}_i - \mathbf{x}_j\right) f\left(\mathbf{x}_i\right) f\left(\mathbf{x}_j\right) d\mathbf{x}. \tag{4.41}$$

For a given population (e.g., cancer patients in the US at a particular moment) defined by K quantifiable features, $\mathbf{X} \in \mathbb{R}^K$, given values of $\mathbf{X}$, the outcome is determined: $Y = g\left(\mathbf{X}\right)$. However, we can only measure some of the features (predictors), $\mathbf{X}_1 \in \mathbb{R}^{K_1}$, $K_1 < K$. Thus, we express Y as $Y = f\left(\mathbf{X}_1\right) + \varepsilon$, where the error term ε comes from the unmeasured part $\mathbf{X}_2$. Note that $\mathbf{X}_1$ is a subset of $\mathbf{X}$, but not a subset of the cancer patient population in the US.

4.7   Summary

1. SBML is a kind of automated learning whose development is based on the similarity principle. The principle can be stated in this way: similar things or individuals will likely behave similarly, and the

more similar they are, the more similarly they behave. The basic notion behind SBML is that in predicting outcomes we gather information from similar things and weight the source information differently based on the similarity. The similarity is determined by a set of pre-selected attributes. The relative impact of each attribute on the outcome is measured by the associated scaling factor. The scaling factors are determined through the local error-minimization with the training data set. For this reason, scaling factors will not only depend on the attributes selected, but also the outcome of interest.

2. Unlike regression models $y_i = f(\mathbf{x}_i)$ that directly model the relationship between the outcome and predictors, SBML models the relationship between the outcomes for different individuals, $y_i = \sum_j s_{ij} y_j$. This is accomplished through normalized similarity scores (or weights) as a function of the distance $s_{ij}\left(\mathbf{R}^T(\mathbf{x}_j - \mathbf{x}_i)\right)$ to indirectly establish the relationship between outcome and attributes. For this reason, SBML is a mixed parametric and memory-based (instance-based) approach. The parameters are the scaling factors R, and all training data points are still contained in the SBML algorithm. For larger training data sets, dimension reduction is desirable. We provided two different methods that can be used in SBML.

3. SBML emphasizes local relationships, meaning that the information from the subjects that are close (similar) to you will be weighted more for you. Therefore, outliers have little impact on prediction. In contrast, the outliers have a large impact on prediction when traditional regression models are used.

4. Unlike a regression model, SBML requires no assumption about data distribution. Thus, it is a robust method. SBML is most applicable when the data structure is complex, unknown, and a simple (e.g., linear) relationship between outcome and predictors is not warranted.

5. Modern learning methods such as deep learning and kernel methods require big data to be effective. SBML can be applied to small or large data and works very well. Therefore, it can be used in drug development for rare diseases and for other problems even when only small amounts of data are available.

6. In all existing machine-learning methods with similarities, the similarities between subjects are usually determined subjectively by, for example, the field experts. In SBML, the similarity score can be objectively determined by the training data through a limited number of scaling factors.

7. Regression models (e.g., linear and logistic models) often have poor predictivity for many practical problems because the assumptions about the data are often incorrect or difficult to validate. SBML has no requirements or assumptions about the distribution or the correlation structure of the data.

8. Unlike many other learning methods, such as regression approaches, SBML will not be sensitive to outliers since the similarities between an outlier and other normal data points are small. Therefore, SBML is a robust AI learning method.

9. In SBML we don't need to worry about whether a variable included is a noise variable or not, because the scaling factor for a noise variable will be very small, and consequently its effect on the outcome will be very small too.

10. Unlike other optimization problems where the global minimum error is desirable, SBML only requires an approximate local optimal solution, which makes the optimization much simpler and computationally more efficient.

11. SBML is applicable to different outcomes, multiple outcomes, and mixed types of outcomes.

12. Different types of predictors (independent variables) can be included in SBML

13. Over-fitting and multiplicity are challenging issues in statistical machine learning. SBML is virtually free of over-fitting, and there is no limitation on the number of variables to be included.

14. SSBML is an improved version of SBML in which the scaling factors are updated as the data (including the outcome) accumulate.

15. Mimicking human learning by applying SBML hierarchically at different levels, individual, group, and aggregated group-levels RSBML makes learning more efficient than one-level learning.

16. The gradient-based method is used in solving the optimization of learning problems.

17. Compared to other machine-learning methods, such as kernel methods, the results from SBML are more interpretable: the scaling factors reflect the effects of the corresponding attributes on the outcome.

18. Because SBML emphasizes the estimation of individual outcomes, it is suitable for the development of personalized medicine.

19. SBML, SSBML, and RSBML are general artificial intelligent learning methods that can be used in many different fields of scientific discovery, classification problems, prediction, and in building artificial brains for robots for artificial general intelligence.

20. SBML allows the reusability of data in training or hypothesis testing and can provide multiple-level conclusions from the same dataset.

21. Most machine-learning methods with similarity functions are used for classification problems. A kernel smoother can be used for continuous outcomes, but it is an instance-based learner. Rather than learning a fixed set of parameters corresponding to the features of their inputs, kernel methods use n weights, W_i, where n is the number of data points, which can be in the millions; this often causes over-fitting issues. In SBML, scaling factors that associate the attributes need to be determined. The number of scaling factors, K, usually ranges from a few to a few dozen.

22. In methodology comparison and validation, ideally we should define the target population. However, in reality, the target population in an AI study is unknown or hard to define. Here are two obvious options that we have: either (1) artificially define a probability distribution, or (2) verbally define a target population with a distribution never known in reality.

23. For learning parameters, initial similarity scores of 0.5, learning rate $\alpha = 0.125$, epoch=15, and $\lambda = -0.3$ work well.

4.8   Problems

4.1: Define a similarity score S_{ij} as an inverse function of distance d_{ij} between the two subjects, i and j:

$$S_{ij} = \frac{1}{1 + d_{ij}^{\rho}}$$

Prove

$$R_m = \frac{S^{-1}\left(S_m^0\right)}{|X_{qm} - X_{pm}|} = \frac{1}{|X_{qm} - X_{pm}|} \sqrt[\rho]{\frac{1}{S_m^0} - 1}$$

4.2: Use SBML to analyze the *infert* dataset (infertility after spontaneous and induced abortion). The initial similarity score can be the p-value from logistic regression:

```
infert {datasets}
require(stats)
summary(model2 <- glm(case ~age+parity+education+spontaneous+
induced, data = infert, family = binomial()))
```

4.3:   Conduct a simulation study to investigate the effects of the following factors: similarity function, training dataset size ($1/3$, $1/2$, $2/3$), initial similarity score s_0 (obtained based on prior knowledge, univariate p-values, and multivariate p-values), correlation between predictors (low, medium, high, mixed low, and high), noise variables (none, some, major), and learning rate (Note: simulations can be based on multivariate normal data).

4.4:   Prove or disprove the following statements: A drug has the same average effect as a placebo, but if we can predict who will get how much benefit, we can determine who should get the drug and who should not—truly personalized medicine in a prescription. Thus, drug approval should not be based on the average effect or frequentist's hypothesis test.

4.5:   Study how to apply SBML for a survival endpoint.

5

Artificial Neural Networks

5.1 Hebb's Rule and McCulloch-Pitts Neuronal Model

In the human brain, each neuron is typically connected to thousands of other *neurons*. A typical neuron collects signals from others through a host of fine structures called *dendrites*. The neuron sends out spikes of electrical activity through a long, thin stand known as an axon, which splits into thousands of branches. At the end of each branch, a structure called a *synapse* converts the activity from the axon into electrical effects that *inhibit* or excite activity from the axon into electrical effects that inhibit or excite activity in the connected neurons. When a neuron receives an excitatory input that is sufficiently large compared with its inhibitory input, it sends a spike of electrical activity down its axon (Figure 5.1). Learning occurs by changing the effectiveness of the synapses so that the influence of one neuron on another changes (Chang, 2010).

Hebbian theory is a neuroscientific theory claiming that an increase in synaptic efficacy arises from a presynaptic cell's repeated and persistent stimulation of a postsynaptic cell. It is an attempt to explain synaptic plasticity, the adaptation of brain neurons during the learning process. This theory was introduced by Donald Hebb in *The Organization of Behavior* (Hebb, 1949). The theory is also called Hebb's rule and is sometimes referred to as cell assembly theory.

The theory is often summarized as "Cells that fire together wire together." This actually means that cell A needs to "take part in firing" cell B, and such causality can occur only if cell A fires just before, not exactly at the same time as, cell B. Hebb's rule attempts to explain associative or *Hebbian learning*, in which simultaneous activation of cells leads to pronounced increases in synaptic strength between those cells. It also provides a biological basis for errorless learning methods for education and memory rehabilitation. If two neurons consistently fire simultaneously, then any connection between them will become stronger. Conversely, if the two neurons never fire simultaneously, the connection between them will die away. The idea is that if two neurons both respond to something then they should be connected. Pavlov used this idea, called *classical conditioning*, to train his dogs so that when food was shown to the dogs and the bell was rung at the same time, the neurons for salivating over the food and hearing the bell fired simultaneously, and so

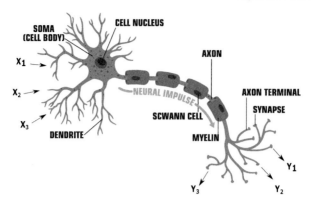

FIGURE 5.1
How a Neural Network Works (Synapse)

became strongly connected. Over time, the strength of the synapse between the neurons that responded to hearing the bell and those that caused the salivation reflex was enough that just hearing the bell caused the salivation neurons to fire in sympathy (Stephen Marsland, 2014).

An artificial neural network (ANN) is a popular artificial intelligent method due to its recent success in deep learning. An ANN mimics the mechanism of the human neural network using adaptive weights between the layers in the network to model very complicated systems. ANNs can be used in supervised learning (classification) and unsupervised learning (clustering).

Learning in biological systems involves adjustments to the synaptic connections that exist between the neurons; the same is true for ANNs. An ANN features adaptive learning (the ability to learn how to do tasks based on the data given for training or as initial experience) and self-organization (creating its own organization or representation of the information it receives at the learning phase).

We now study a simple mathematical model of a neuron that was introduced by McCulloch and Pitts in 1943 (Figure 5.2). The neural model includes:

1. A set of weighted inputs w_i $(i = 1, ..., K)$ that correspond to the synapses. A positive (negative) w_i corresponds to excitatory (inhibitory) connections that make neurons more (less) likely to fire.

2. An adder that sums the input signals (equivalent to the membrane of the cell that collects electrical charge).

3. An activation (threshold) function θ that decides whether the neuron fires (spikes) for the current inputs

$$f(x) = sign\left(\theta - \sum_{i}^{K} w_i x_i\right). \tag{5.1}$$

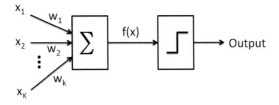

FIGURE 5.2
McCulloch-Pitts Neurona Model

The McCulloch and Pitts neuronal model (Figure 5.2) is a linear model. What it does is very limited: the neuronal model either fires (output 1) or does not fire (output 0). Real neurons are much more complicated. A noticeable difference is that real neurons do not output a single output response, but a spike train, that is, a sequence of pulses, and it is this spike train that encodes information. This means that neurons don't actually respond as threshold devices, but produce a graded output in a continuous way. This means that the threshold at which they fire changes over time. Because neurons are biochemical devices, the amount of neurotransmitter (which affects how much charge they required to spike, amongst other things) can vary according to the current state of the organism. Furthermore, the neurons are updated themselves asynchronously or somewhat randomly. But we will not discuss the asynchronous model off neurons.

5.2 The Perceptron

5.2.1 Model Construction

One neuron can only do a little, but if we put sets of neurons together into neural networks (Figure 5.3) what they can accomplish is much more interesting. Among the three things in the McCulloch-Pitts neuronal model: weights, the sum-function, and threshold, weights (modeling the synapse) are the most interesting because they can change during the learning process. That is, learning is essentially modeled by weight changes (weight changes that model the changes in the connections between neurons described in Hebb's rule). The sum-function and threshold are, for a given neural network, often chosen for simplicity. A perceptron is an extension of McCulloch-Pitts neuronal model to multiple neurons (Figure 5.3). What is it that a perceptron of n neurons can ultimately do? Since each neuron has two possible outputs, the n neurons together have 2^2 different outputs/responses, and each combination has an n-digit binary code that might be assigned different meanings. The question is how to convert a person's way of learning into a set of rules for changing the

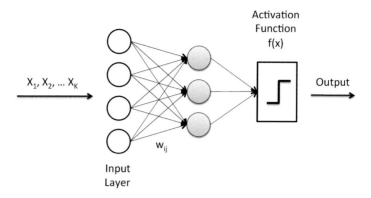

FIGURE 5.3
Perceptron with K Input Neurons

weights so that the network outputs the right answer or appropriate response more often?

$$y_i = sign\left(\sum_{j=0}^{K} w_{ij}x_j\right), i = 1, ..., M. \tag{5.2}$$

Notice that the neurons in the perceptron share the inputs, but not the weights and activation functions. The number of inputs, K, is determined by the data, and so is the number of outputs, M. You may have noticed that the subscript j starts with 0 instead of 1. This is because we often need a nonzero constant output when there is no input, which can be accomplished by a constant term. In terms of a perceptron, this can be accomplished by adding a virtual input node with unit input $x_0 \equiv -1$ and a corresponding weight denoted by w_{0j} for the jth neuron. The weights w_{0j} actually serve the role of a threshold.

For supervised learning, the rule for updating a weight w_{ij} is usually based on the gradient method with L_2 loss function:

$$w_{ij} \leftarrow w_{ij} - \eta(\hat{y}_j - y_j)x_i, \tag{5.3}$$

where the learning rate η should be stable and resistant to noise (errors) and inaccuracies in the data. A moderate learning rate, typically $0.1 < \eta < 0.4$ is recommended.

We now can outline the perceptron algorithm as follows:

(1) Set all weights w_{ij} to small random numbers.
(2) For T iterations or until all the outputs are correct:
   for each input vector x:
   compute the activation of each neuron j using activation function (5.2)
   update each of the weights individually using (5.3).

(3) After training, all weights are determined, and the activation of each neuron j will be calculated using:

$$y_i = sign\left(w_{ij}x_j\right). \tag{5.4}$$

The perceptron is implemented in R as follows:

```
w=rep(0.01,n); y=c(0,1,1,1)
x=(-1,0,0)
yhat=sign(x*T*x)
w=w-eta*(yhat-y)*xi
```

5.2.2 Perceptron Learning

Logic functions can be described in the table form:

OR Function		
x_1	x_2	y
0	0	0
0	1	1
1	0	1
1	1	1

We introduce the example of logic OR function by Marsland (2014) with one neuron and two input x_1 and x_2. We need to consider a constant input $x_0 = -1$. For the three weights, we assign them small random values, $w_0 = -0.05, w_1 = -0.02, w_2 = 0.02$. The perceptron output is determined by $\hat{y} = sign\left(\sum_{j=0}^{N} w_j x_j\right)$, and learning rate $\eta = 0.25$.

Let's feel a first input $\mathbf{x} = (x_0, x_1, x_2) = (-1, 0, 0)$, $\hat{y} = sign(0.05) = 1$, which is different from the target $y = 0$ (remember that $x_0 \equiv -1$). To make the perceptron learn, the weights need to be updated using Eq. (5.3):

$$\begin{cases} w_0 : -0.05 - 0.25 \times (1-0) \times (-1) = 0.2 \\ w_1 : -0.02 - 0.25 \times (1-0) \times 0 = -0.02 \\ w_2 : 0.02 - 0.25 \times (1-0) \times 0 = 0.02. \end{cases}$$

Now we feed in the next input $\mathbf{x} = (-1, 0, 1)$ and compute the output $\hat{y} = sign(0) = 0 \neq y$. Therefore, we apply the learning rule again:

$$\begin{cases} w_0 : 0.2 - 0.25 \times (0-1) \times (-1) = -0.05 \\ w_1 : -0.02 - 0.25 \times (0-1) \times 0 = -0.02 \\ w_2 : 0.02 - 0.25 \times (0-1) \times 1 = 0.27. \end{cases}$$

For the next two sets of inputs $x = (-1, 1, 0)$) and $x = (-1, 1, 1)$, the answer is already correct, $\hat{y} = y$, so we don't have to update the weights. However, this doesn't mean we've finished because, while updating weights means we are one step close to the correct answer, not all the answers are

correct yet. We now need to start going through the inputs again, until the
weights settle down and stop changing, which is what tells us that the algorithm has finished.

From this example we see that the training dataset can be used repeatedly
to train the perceptron. We may have the question: for real-world applications,
can the weights eventually stop changing and provide the correct answers for
all the cases? If not how long or how many iterations should the learning take?
We are going to discuss this.

The Perceptron Convergence Theorem

Rosenblatt proved in 1962 that, given a linearly separable dataset, that is,
if there is a weight vector $\mathbf{w}$ such that $f(\mathbf{w}{\cdot}p(\mathbf{q})) = t(\mathbf{q})$ for all $\mathbf{q}$, then for
any starting vector $\mathbf{w}$, the perceptron learning rule will converge to a weight
vector (not necessarily unique) that gives the correct response for all training
patterns, and it will do so in a finite number of steps.

5.2.3   Linear Separability

What the perceptron does is try to use lines to separate the different classes
(Figure 5.4). The perceptron convergence theorem requires linear separability
for the dataset. The question is: will a dataset always be linearly separable?
The answer is no. Let's investigate the simple exclusive or (XOR) function.

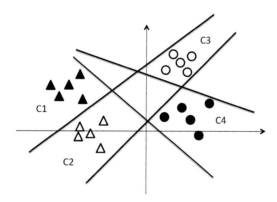

FIGURE 5.4
Different Linear Decision Boundaries Determined by a Perceptron

XOR Function			XOR in 3-D Space			
x_1	x_2	y	x_1	x_2	x_3	y
0	0	0	0	0	1	0
0	1	1	0	1	0	1
1	0	1	1	0	0	1
1	1	0	1	1	0	0

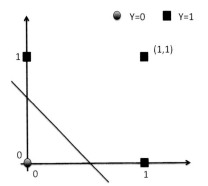
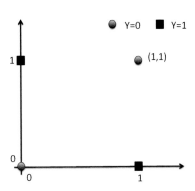

FIGURE 5.5
Linear Separability with OR and XOR Functions

A simple linearly inseparable function is the exclusive or (XOR) function (Figure 5.5). We cannot use a straight line (linear function of x_1 and x_2) to separate the four true $(y = 1)$ and false $(y = 0)$ data points.

At this point, we may think the XOR function is impossible to represent using any linear function. In fact, this is not true. If we rewrite the problem in a three-dimensional space, that is, adding the third dimension x_3, but keeping the data in the original two-dimensional space (x_1, x_2) unchanged, then it is perfectly possible to find a plane (linear function) that can separate the two classes. It is easy to see a plane separating the points with $y = 0$ from those where $y = 1$. A sufficient condition to have linear separability for XOR is that the following equation has solutions when applying the four inputs in the 3-D space:

$$y = -w_0 + w_1 x_1 + w_2 x_2 + w_3 x_3 \qquad (5.5)$$

$$\begin{cases} 0 = -w_0 \cdot 1 + w_1 \cdot 0 + w_2 \cdot 0 + w_3 \cdot 1 \\ 1 = -w_0 \cdot 1 + w_1 \cdot 0 + w_2 \cdot 1 + w_3 \cdot 0 \\ 1 = -w_0 \cdot 1 + w_1 \cdot 1 + w_2 \cdot 0 + w_3 \cdot 0 \\ 0 = -w_0 \cdot 1 + w_1 \cdot 1 + w_2 \cdot 1 + w_3 \cdot 0. \end{cases}$$

Here is a solution: $[w_0 = -2, w_1 = -1, w_2 = -1, w_3 = -2]$. In other words, the prediction is

$$y = 2 - x_1 - x_2 - 2x_3, \qquad (5.6)$$

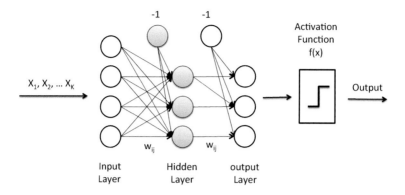

FIGURE 5.6
Multiple-Layer Perceptron

However, there is more than one solution. The perceptron output is

$$y = sign(-w_0 + w_1x_1 + w_2x_2 + w_3x_3), \qquad (5.7)$$

instead of (5.5).

5.3 Multiple-Layer Perceptron for Deep Learning

5.3.1 Model Construction

Since the learning in the neural network is due to the weights, we should be able to make an ANN learn more complicated things by increasing the number of weights. There are ways to accomplish this: (1) increase the number of hidden layers, (2) increase the number of nodes in a layer, (3) use a combination of (1) and (2). Too few layers may not be as good as we have seen in the single-layer perceptron that cannot even handle the XOR function.

Let's label the input layer Layer 0, and forward, the mth hidden layer Layer m $(m = 1, ...M - 1)$, and the output layer as Layer M. N_m is the number of nodes in Layer m. We use w_{kij} for the weight between the ith node in the kth layer and the jth node in $k + 1^{th}$ layer. We use vector $\mathbf{y}_m$ to denote the hidden output vector in Layer m. Particularly, $\mathbf{y}_0 = (y_{01}, ..., y_{0n_0}) = \mathbf{x}_0 = (x_1, ..., x_{n_0})$ is the input vector, and $\mathbf{y}_3 = (y_{31}, ..., y_{3N_3})$ is the output of the three-layer perceptron in Figure 5.6 . H_m is the activation function for Layer m, $(m = 1, ..., M)$. For convenience, as previously, we add a dummy variable y_{m0} to the mth layer. Thus, for the first layer, the hidden output is

$$y_{1t} = H_1 \left(\sum_{l=0}^{N_0} w_{0lt} y_{0l} \right), t = 1, ..., N_1, \qquad (5.8)$$

where $H_1 (\cdot)$ is the activation function for at 1st hidden layer.

For Layer 2, the hidden output is

$$y_{2j} = H_2 \left(\sum_{t=0}^{N_1} w_{1tj} y_{1t} \right), j = 1, ..., N_2. \tag{5.9}$$

In general, for the mth layer, the output is

$$y_{mi} = H_m \left(\sum_{j=0}^{N_{m-1}} w_{m-1,ji} y_{m-1,j} \right), i = 1, ..., N_m, \tag{5.10}$$

$$z_{mi} = \sum_{j=0}^{N_{m-1}} w_{m-1,ji} y_{m-1,j}, m = 1, ..., M, \tag{5.11}$$

The activation function can be the identity function $H_m(\theta) = \theta$, the sign function $H_m(\theta) = Sign(\theta)$, or the logistic function

$$H_m(\theta) = \frac{1}{1 + \exp(-b_m \theta)}, \tag{5.12}$$

where the tuning parameter b_m can depend on the layers or nodes, and can be determined using prior knowledge, through training or cross-validation. The performance of an ANN depends on the weights and the link-function. Other activation functions are also possible depending on particular needs, for example, the additive logistic function:

$$G(\theta) = \frac{1}{1 + \exp(-f(\theta))}, \tag{5.13}$$

with derivative

$$G'(\theta) = G(\theta)(1 - G(\theta)) \frac{df(\theta)}{d\theta}, \tag{5.14}$$

where $f(\cdot)$ is a real function.

This is a feed-forward architecture, in other words one having no closed directed cycles, which ensures that the outputs are deterministic functions of the inputs.

5.3.2 Gradient Method

To evaluate model performance, we can use the squared-error loss function,

$$E = \frac{1}{N_M} \sum_{i=1}^{N_M} (y_{Mi} - t_i)^2, \tag{5.15}$$

where y_i and t_i are the model and observed outputs, respectively, of the ith node at the Mth layer. The optimization problem is to determine the weights w_{mij} that minimize the squared error loss E in (5.15). We can use the gradient-descent method to adjust the weights at the rth iteration:

$$w_{mij}^{(r+1)} = w_{mij}^{(r)} - \alpha \frac{\partial E}{\partial w_{mij}}, \qquad (5.16)$$

where $\alpha > 0$ is the learning rate.

$$\frac{\partial E}{\partial w_{mij}} = \sum_{k=1}^{N_M} \frac{\partial E}{\partial y_{Mk}} \frac{\partial y_{Mk}}{\partial w_{mij}} = \frac{2}{N_M} \sum_{k=1}^{N_M} (y_{Mk} - t_k) \frac{\partial y_{Mk}}{\partial w_{mij}} \qquad (5.17)$$

We can summarize the recursive formulations (see the Appendix for derivations):

$$\begin{cases}
\text{Start propagation} \\
y_{mi} = H_m\left(\sum_{j=0}^{N_{m-1}} w_{m-1,ji} y_{m-1,j}\right), i = 1, ..., N_m \\
\text{End propagation} \\
\delta_{Mi} = (y_{Mi} - t_i), i = 1, 2, ..., N_M \\
E = \frac{1}{N_M} \sum_{i=1}^{N_M} \delta_{Mi}^2, \\
\frac{\partial E}{\partial w_{M-1,kl}} = \delta_{Ml} \cdot y_{M-1,k}, k = 1, 2, ...N_{M-1}; l = 1, 2, ..., N_M \\
\delta_{M-2,li} = H'_{M-1}(z_{M-1,l}) w_{M-1,li} \\
\text{Start backpropagation} \\
\delta_{M-m-1,li} = H'_{M-m}(z_{M-m,l}) \sum_{t=0}^{N_{M-m+1}} \delta_{M-m,ti} w_{M-m,lt}, m = 2, 3..., M-1 \\
\frac{\partial y_{Mi}}{\partial w_{M-m,kl}} = \delta_{M-m,li} y_{M-m,k}, m = 2, 3..., M \\
\frac{\partial E}{\partial w_{M-m,kl}} = \frac{2}{N_M} \sum_{i=1}^{N_M} \delta_{Mi} \frac{\partial y_{Mi}}{\partial w_{M-m,kl}}, m = 2, 3..., M \\
\text{update } w_{mij}^{(r+1)} = w_{mij}^{(r)} - \alpha \frac{\partial E}{\partial w_{mij}} \\
\text{If error } E \text{ is sufficiently small stop and take new set of inputs;} \\
\text{otherwise, start propagation with new weights, } w_{mij}^{(r+1)}.
\end{cases}$$

When the number of layers increases, the error backpropagation is a faster algorithm than the forward algorithm because the error (such as δ_{1j}) in earlier layers do not have to be recalculated for each layer in the backpropagation.

Remark: Example of deep ANN in handwriting identification: computer screen resolution 1600×1200 pixels. Each pixel is either black or white, therefore if one writes on the screen, there are $2^{1600 \times 1200}$ possible "characters." If we name each drawing as a unique thing, we have 100% accurate AI. Given $2^{1600 \times 1200}$ or fewer weights in ANN we can identify each drawing with 100% accuracy. This means that if each layer of ANN has 1000 dots, then the number of layers required to determine any character n can be obtain by solving the equation: $2^{1600 \times 1200} = 1,000,000^n$. The solution is: $n = \frac{1600(1200)\ln 2}{\ln 1,000,000} = 96,330$. If a character is only allowed to occupy a small area or a screen has 1000 characters, only $96,330/1000 = 97$ layers are needed to fully determine each character (since any possible character in any language is determined by 1000 pixels.) Given $24 \times 24 = 576$ pixels per character, the input layer will have 576 nodes, each character has then at most 576 ways of being written, so only $n = \frac{576 \ln 2}{\ln 576} = 63$ layers are needed. The same is true for voice-to-text recognition. ANN will eventually have the same results as this deterministic method, but when the data is not big enough, ANN is the better method. But ANN cannot work if the same writing can mean different characters by different people. Also, as the XOR example has shown, it is not possible because of linear inseparability without expanding into a higher dimension.

Generalization beyond linear functions in ANN:

$$y_m(x, w) = H_m \left(\sum_{j=1}^{M} w_j \phi_j(x) \right), \qquad (5.18)$$

where $H_m(\cdot)$ is a nonlinear activation function in the case of classification and is the identity in the case of regression. Our goal is to extend this model by making the basis functions $\phi_j(x)$ depend on parameters and then to allow these parameters to be adjusted, along with the coefficients w_j.

5.4   Artificial Neural Network with R

Training any *ANN* requires software. The neuralnet-package for *ANN* is developed by Stefan Fritsch and Frauke Guenther, and is supported by the German Research Foundation. Training of neural networks uses backpropagation, resilient backpropagation with (Riedmiller, 1994) or without weight backtracking (Riedmiller, 1993) or the modified globally convergent version by Anastasiadis et al. (2005). The package allows flexible settings through custom-choice of error and activation function. Furthermore, the calculation of generalized weights (Intrator O & Intrator N, 1993) is implemented. The package includes the following functions: *plot.nn* for plotting of the neural network, *gwplot* for plotting of the generalized weights, *compute* for computation of the calculated network, *confidence.interval* for calculation of a confidence interval for the weights, and *prediction* for calculation of a prediction.

Another popular package for feed-forward neural networks with a single hidden layer is *nnet*. The package also provides the capability for multinomial log-linear models. A more powerful *R* package for *ANN* is *NeuralNetTools*. The most popular *ANN* package in *R* is probably the *kearsR*. We will discuss details later.

We now provide a simple example of how to use a neuralnet.

Example 5.1: Learning Logic AND and OR Using Perceptron

```
library(neuralnet)
# create data based the definitions of logic AND and OR
AND = c(rep(0,7),1); OR = c(0,rep(1,7))
andor.data = data.frame(expand.grid(c(0,1), c(0,1), c(0,1)), AND, OR)
```

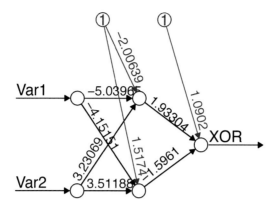

Error: 0.000177   Steps: 105

FIGURE 5.7
Two-Layer Perceptron in Learning XOR

```
print(andor.data)
# Apply a single percetron without hidden layer
print(net <- neuralnet(AND+OR~Var1+Var2+Var3, andor.data, hidden=0,
 rep=8, err.fct="ce", linear.output=FALSE))
```

The simulation results will include the entire path of updating weights.

Example 5.2: Learning XOR Using Perceptron

```
# create data based the definition of XOR
XOR <- c(0,1,1,0)
xor.data = data.frame(expand.grid(c(0,1), c(0,1)), XOR)
# Apply a single percetron with 2 hidden layers
print(net.xor <- neuralnet(XOR~Var1+Var2, xor.data, hidden=2, rep=4))
plot(net.xor, rep="best") # plot the ANN (see Fig.5.7)
```

The modeled ANN for Example 5.2 is presented in Figure 5.7. The predicted
error will vary from simulation to simulation. With two hidden layers the error
is generally small, but if only one hidden layer is used the error will be much
larger.

The neuralnet documentation can be found at https://www. rdocumen-
tation.org. The neuralnet parameter options include: *err.fct*, a differentiable

function that is used for the calculation of the error. Alternatively, the strings "sse" and "ce," which stand for the sum of squared errors and the cross-entropy can be used. *hidden*: a vector of integers specifying the number of hidden neurons (vertices) in each layer. *likelihood*: if the error function is the negative log-likelihood function, the information criteria AIC and BIC will be calculated.

5.4.1 ANN for Infertility Modeling

We now apply ANN to a real-world problem. For simplicity, we use a small dataset available in *R*. The *R datasets* package includes the dataset of *infert* dataset from Trichopoulos et al (1976). This is a matched case-control study carried out before the availability of conditional logistic regression. There are eight variables. (1) education: 0 = 0–5 years, 1 = 6–11 years, 2 = 12+ years; age: age in years of case; (3) parity: count; (4) number of prior: 0 = 0 induced abortions, 1 = 1, 2 = 2 or more; (5) case status: 1 = case, 0 = control; (6) number of prior: 0 = 0 spontaneous abortions, 1 = 1, 2 = 2 or more; (7) matched set number: 1–83; (8) stratum number: 1–63. Note: One case with two prior spontaneous abortions and two prior induced abortions is omitted.

Here is the R code for the ANN:

```
library(neuralnet)
maxs <- apply(infert[,-1], 2, max)
mins <- apply(infert[,-1], 2, min)
scaled <- as.data.frame(scale(infert[,-1], center = mins, scale = maxs - mins))
# Randomly sample 70% observations to training and the rest to testing
n = nrow(scaled)
set.seed(8)
trainIndex = sample(1:n, size = round(0.7*n), replace=FALSE)
trainset <- na.omit(scaled[trainIndex,])   # scale it and remove NAs
testset <- na.omit(scaled[-trainIndex,])   # scale it and remove NAs
infertNet1 <- neuralnet(case~parity+induced+spontaneous, trainset,
  err.fct="ce", linear.output=FALSE, likelihood=TRUE)
infertNet2 <- neuralnet(case~age+parity+induced+spontaneous, trainset,
  hidden=4, err.fct="ce", linear.output=FALSE, likelihood=TRUE)
plot(infertNet1); plot(infertNet2)
infertNet1.case <- round(predict(infertNet1, testset))
infertNet2.case <- round(predict(infertNet2, testset))
table(infertNet1.case,testset$case)
table(infertNet2.case,testset$case)
```

Figures 5.8 and 5.9 are the plot for ANN infetNet1 and inferNet2, respectvely.

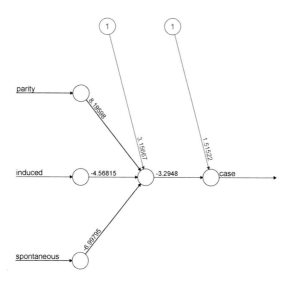

Error: 95.944132   Steps: 3297

FIGURE 5.8
Infertility ANN InfertNet1

The two corresponding confusion tables are show below:

```
infertNet1.case    0    1
             0    44   14
             1     3   11
infertNet2.case    0    1
             0    35   15
             1    12   12
```

From the predicted 2×2 confusion tables, the observed proportionate agreement is $(44 + 11)/(44 + 16 + 3 + 11) = 0.743$ for inferNet1 and $(35 + 12)/(35 + 15 + 12 + 12) = 0.635$ for inferNet2.

Example 5.4: Diabetes Modeling with ANN

We again use the diabetes dataset in the mlbench package. We first prepare the data: remove NAs, scale the variables, and split the data into training and test datasets.

```
library(mlbench)
data("PimaIndiansDiabetes2", package = "mlbench") # Load data
```

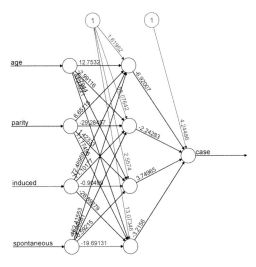

Error: 115.87938   Steps: 29272

FIGURE 5.9
Infertility ANN InfertNet2

```
Diabetes2 <- na.omit(PimaIndiansDiabetes2) # remove NAs
# create a dummy variable for diabetes
Diabetes2$Diabetes01=ifelse(Diabetes2$diabetes=="pos", 1, 0 )
# Randomly sample 70% observations to training and the rest to testing
maxs <- apply(Diabetes2[,-9], 2, max)
mins <- apply(Diabetes2[,-9], 2, min)
scaled <- as.data.frame(scale(Diabetes2[,-9], center = mins, scale = maxs - mins))
n = nrow(Diabetes)
set.seed (1)
trainIndex = sample(1:n, size = round(0.7*n), replace=FALSE)
trainset <- scaled[trainIndex,]
testset <- scaled[-trainIndex,]
```

Now we are ready to model the data using ANN, plot the ANN, and
generate the predicted outcomes and the 2×2 confusion tables.

```
diabetesNet1 <- neuralnet(Diabetes01~glucose+pressure+triceps+insulin+mass
+pedigree+age, trainset,
err.fct="ce", linear.output=FALSE, likelihood=TRUE)
diabetesNet2<- neuralnet(Diabetes01~glucose+pressure+triceps+insulin+mass
+pedigree+age, trainset,
```

```
hidden = c(3,2), linear.output=FALSE, likelihood=TRUE)
plot(diabetesNet1); plot(diabetesNet2)
diabetesNet1.case <- round(predict(diabetesNet1, testset))
diabetesNet2.case <- round(predict(diabetesNet2, testset))
table(diabetesNet1.case,testset$Diabetes01)
table(diabetesNet2.case,testset$Diabetes01)
```

The code outputs the following two confusion tables:

```
diabetesNet1.case    0    1
                0   69   18
                1    7   24

diabetesNet2.case    0    1
                0   64   18
                1   12   24
```

The ANN neuralnet1 is actually equivalent to a standard logistic regression. The neuralnet2 is a more complicated ANN, but the results are actually worse than the simple logistic model due to overfitting.

5.4.2   Feedforward Network with KarasR Package

So far we have discussed ANNs in which information flows (feed) from the previous layer to the current layer and further to the next layer. Such an ANN is called a feedforward neural network (FNN). We will discuss other types of ANNs in the next chapter. When an ANN (not necessarily an FNN) involves many layers, it is usually called a deep neural network.

Package *kerasR* provides a consistent interface to the *Keras* Deep Learning Library directly from within *R*. *Keras* provides specifications for describing dense neural networks, convolution neural networks (CNN), and recurrent neural networks (RNN) running on top of either *TensorFlow* or *Theano*. Type conversions between *Python* and *R* are automatically handled correctly, even when the default choices would otherwise lead to errors. The package includes complete *R* documentation and many working examples (https://cran.r-project.org/web/packages/kerasR/kerasR.pdf).

Let's work through the diabetes example:

Data Preparation

We need to create four datasets: *train.y* contains the label for the classification for training, *train.x* contains the attributes for the training, *test.y* contains the label for the classification for the testset, and *test.x* contains the attributes for testset.

```
library(kerasR)
library(mlbench)
data("PimaIndiansDiabetes2", package = "mlbench") # Load data
Diabetes2 <- na.omit(PimaIndiansDiabetes2) # remove NAs
# create a dummy variable for diabetes
Diabetes2$Diabetes01=ifelse(Diabetes2$diabetes=="pos", 1, 0 )
# Randomly sample 70% observations to training and the rest to testing
maxs <- apply(Diabetes2[,-9], 2, max)
mins <- apply(Diabetes2[,-9], 2, min)
scaled <- as.data.frame(scale(Diabetes2[,-9], center = mins, scale = maxs - mins))
n = nrow(Diabetes2)
set.seed (1)
trainIndex = sample(1:n, size = round(0.9*n), replace=FALSE)
trainset <- scaled[trainIndex,]
testset <- scaled[-trainIndex,]
train.y=to_categorical(trainset[,9])
train.x=as.matrix(trainset[,-9])
test.y=to_categorical(testset[,9])
test.x=as.matrix(testset[,-9])
```

Model Building

```
mod <- Sequential()
mod$add(Dense(units = 3, input_shape = dim(train.x)[2]))
# units = output units
mod$add(Dense(units = dim(train.y)[2])) # units=2 for binary outcome
mod$add(ActivityRegularization(l2=1, l1 = 0))  # using L2 for regularization
mod$add(Activation("softmax"))  # use softmax for activation function
```

Here *mod <- Sequential()* specifies we are building a model with a sequence of layers. The *mod$add(Dense(units = 3, input_shape = dim(train.x)[2]))* will add a "dense" layer with three nodes (neurons). The option *input_shape* is only needed for the first layer or the input layer. After each dense layer, an activation layer (function) can optionally be added (specified). If no activation is added, the identity activation function will be used. The *mod$add(Dense(units = dim(train.y)[2]))* adds another dense layer in the model, where units is the number of output nodes, while the *input_shape* is not specified because it will automatically take the output from the previous layer. The *mod$add(ActivityRegularization(l2=1, l1 = 0))* specifies *L2* as the loss function, and *mod$add(Activation("softmax"))* specifies the softmax activation function. At this point we have finished our model specification.

Model Compiling and Execution (Fitting Data with the ANN)

Before we can train the ANN, we need to compile the model using *keras_compile*. We specify loss = '*binary_crossentropy*' for binary classification problems and *loss = 'categorical_crossentropy'* for categorical classification problems. Several optimizers can be used for the optimization in compilation, *RMSprop()* is used here. The compiled model can be trained using *keras_fit*, where epochs = 50 indicates up to 50 interactions will be given in the gradient method during learning. The tuning parameter λ in the L_2 loss function can be obtained through the cross-validation with a 0.2/0.80 split as specified by *validation_split = 0.2*.

```
keras_compile(mod, loss = 'binary_crossentropy', optimizer = RMSprop())
keras_fit(mod, train.x, train.y, batch_size = 32, epochs = 50,
  verbose = 0, validation_split = 0.2)
```

Prediction and Confusion Tables

For the prediction and test we can generate the confusion table (2×2 contingency table) via the following code:

```
#test.predY <- keras_predict(mod, test.x)) # predict probability
test.predY <- keras_predict_classes(mod, test.x) # predict classes
table(test.predY,testset[,9])
train.predY <- keras_predict_classes(mod, train.x)
table(train.predY,trainset[,9])
```

The output confusion table is:

```
test.predY   0    1
         0   16   10
         1   7    6
```

5.4.3   MNIST Handwritten Digits Recognition

Suppose we want a neural network to classify human handwritten numbers. We will use a popular dataset called the MNIST dataset with 60,000 training examples, and 10,000 for testing. The CSV files are available at:

http://www.pjreddie.com/media/files/mnist_train.csv
http://www.pjreddie.com/media/files/mnist_test.csv

The first value (column) in the files is the label from 0 to 9, the actual digit that the handwriting is supposed to represent. It is the answer which

FIGURE 5.10
Mnist Handwritten Digits

the neural network is trying to determine. The subsequent values, all comma separated, are the pixel values of the handwritten digit (Figure 5.10). The size of the pixel array is 28×28, so there are 784 values after the label.

Before you try the following code, you should install keras (see the appendix) and download the two csv files to your R working directory. You can find out the working directory by execute the following R code: *getwd()*.

We introduce the examples given in the R documentation. You can learn more by playing around with the code to see how the accuracy changes. Note that the result will change slighty for each run even when you don't make any changes to the code.

KerasR Code

```
# Get the train and testsets
library(kerasR)
getwd()
setwd("/Users/markchang/Desktop/PharmaAI/")
mnistTrain <- read.csv("mnist_train.csv",header=T,sep=",")
mnistTest <- read.csv("mnist_test.csv",header=T,sep=",")
xTrain<-as.matrix(mnistTrain[,-1])/225
yTrain<-to_categorical(mnistTrain[,1])
xTest<-as.matrix(mnistTest[,-1])/225
yTest<-to_categorical(mnistTest[,1])
# Building ANN model
mod <- Sequential()
mod$add(Dense(units = 50, input_shape = dim(xTrain)[2])) # units
= output units
```

```
> table(test.predY,mnistTest[,1])

test.predY   0     1    2    3    4    5    6    7    8    9
         0 932     0   13   16    1   29   34    2   17   15
         1   0  1107   89   54   33   26   18   54   65   15
         2   1     1  778   22   10    7   27   12    8    4
         3   3     1   21  771    0   46    0    2   14   10
         4   2     2   11    5  801   14   17   21   14   40
         5  11     0    0   17    7  626   13    2   43    1
         6  16     5   35    5    7   26  841    1   10    1
         7   0     0   13   18    1   10    0  813    7   32
         8   8    18   64   52   19   71    8    2  749   13
         9   7     1    8   50  103   37    0  118   47  878
> mean(mnistTest[,1] == test.predY)
[1] 0.829683
```

FIGURE 5.11
FNN Prediction Versus Actual Digits

```
mod$add(Dropout(rate = 0.2))
mod$add(Dense(units = 40)) # units = output units
mod$add(Dense(units = 20)) # units = output units
mod$add(Dense(units = dim(yTrain)[2])) # units=2 for binary outcome
mod$add(ActivityRegularization(l2=1, l1 = 0))
mod$add(Activation("softmax"))
keras_compile(mod, loss = 'categorical_crossentropy', optimizer = RMSprop())
keras_fit(mod, xTrain, yTrain, batch_size = 32, epochs = 10,
 verbose = 0, validation_split = 0.2) # batch_size =32 because each image has
a size of 32x32
#test.predY <- keras_predict(mod, test.x)) # predict probability
test.predY <- keras_predict_classes(mod, xTest) # predict classes
table(test.predY,mnistTest[,1])
mean(mnistTest[,1] == test.predY)
```

The FNN provides a fair accuracy of 83.0% in identifying the handwriting
(Figure 5.11).

```
# Another model
mod <- Sequential()
```

```
> table(test.predY,mnistTest[,1])
```

test.predY	0	1	2	3	4	5	6	7	8	9
0	966	0	5	1	1	4	9	2	15	5
1	1	1121	0	0	1	0	3	0	1	1
2	1	1	968	1	2	0	0	7	2	0
3	1	3	19	985	2	15	1	2	3	3
4	2	0	5	0	941	2	7	4	7	8
5	2	0	1	4	0	852	10	0	6	1
6	2	2	2	0	4	9	921	0	2	1
7	2	1	19	10	7	2	2	995	3	4
8	2	7	11	3	2	3	4	5	927	4
9	1	0	2	6	22	5	1	12	8	982

```
> mean(mnistTest[,1] == test.predY)
[1] 0.9658966
```

FIGURE 5.12
FNN2 Prediction Versus Actual Digits

```
mod$add(Dense(units = 512, input_shape = dim(xTrain)[2]))
mod$add(LeakyReLU())
mod$add(Dropout(0.25))
mod$add(Dense(units = 512))
mod$add(LeakyReLU())
mod$add(Dropout(0.25))
mod$add(Dense(units = 512))
mod$add(LeakyReLU())
mod$add(Dropout(0.25))
mod$add(Dense(dim(yTrain)[2]))
mod$add(Activation("softmax"))
keras_compile(mod, loss = 'categorical_crossentropy', optimizer = RMSprop())
keras_fit(mod, xTrain, yTrain, batch_size = 32, epochs = 20, verbose = 1,
          validation_split = 0.1)
test.predY <- keras_predict_classes(mod, xTest) # predict classes
table(test.predY,mnistTest[,1])
mean(mnistTest[,1] == test.predY)
```

The FNN has a better result with 96.6% accuracy in identifying the hand-written digits (Figure 5.12).

5.5   Summary

Artificial neural networks (ANNs) are computing systems inspired by the biological neural networks in animal brains. ANNs take input data and output desired outcomes after training. The learning in an ANN refers to its ability of outputting more and more closely to the desired outcomes over time through training. The adjustments of weights in an ANN are what make the ANN learn.

The perceptron can model some math function such as OR exactly, but fails to mimic other functions, such as XOR, due to the issue of linear inseparability. However, the problems can be modeled in a higher dimensional space.

An ANN model includes the input layer, one or more hidden layers, and the output layer. Each layer contains input nodes and output nodes, weights, and activation functions. As the number of layers increase, the number of weights will increase exponentially. Therefore, to reduce the computational burden, we can drop some of the links (weights) between layers. The ANNs we have discussed so far are directed acyclic networks or feedforward networks (vanilla ANNs). Other networks will be discussed in the next chapter.

The numbers of layers and nodes are usually fixed. The only thing that can change is the weights in the network. Therefore, learning in ANNs occurs via updating the weights. The weight modifications are based on training data using the gradient method, more precisely, the backpropagation algorithm. There are several R packages available for building ANNs, including the popular package *keras* and *kerasR*.

A Recipe For Using The Multiple-Layer Perceptron

1. *Select inputs and outputs for your problem*
2. Loop sufficient times
    (a) *Generate training and testing datasets from entire dataset available*
    (b) *Normalize inputs*
    (c) *Select AI architecture*
    (d) *Train a network*
    (e) *Test the network*
3. End of loop

5.6   Problems

5.1: Describe the common architecture of feedforward neural network.

5.2: What is the utility of activation functions?

5.3: Oesophageal Cancer Modeling with ANN

Apply ANN to a real world case-control study about oesophageal cancer (Breslow, N. E. and Day, N. E., 1980), available in the *R* *datasets* package. The dataset includes five variables or columns and 88 observations or rows. The five variables are *agegp* for age group, *alcgp* for alcohol consumption, *tobgp* for tobacco consumption, *ncases* for the number of cases (oesophageal cancer), and *ncontrols* for the number of controls (no oesophageal cancer). Study different ANNs with different hidden layers using the data.

6

Deep Learning Neural Networks

6.1 Deep Learning and Software Packages

Deep learning (DL) is a class of machine learning algorithms that use multiple layers to progressively extract higher-level features from raw input. Deep learning architectures include (1) *Feedforward Neural Networks* (FNNs) for general classification and regression, (2) *Convolution Neural Networks* (CNNs) for image recognition, (3) *Recurrent Neural Networks* (RNNs) for speech recognition, *Natural Language Processing*, and (4) *Deep Belief Networks* (DBNs) for disease (cancer) diagnosis and prognosis. Of course, these are only examples of DL with different networks. Different problems can be solved using the same type of ANN, and different ANNs can be used to solve the same problem. We have discussed FNNs in Chapter 5. In this chapter, we will discuss CNNs, RNNs, and DBNs. To solve practical problems using these networks we need software packages. Several software packages available for deep learning are listed in Table 6.1, and there are more out there.

Keras, keras, and kerasR

Keras is a high-level neural networks API written in *Python* and capable of running on top of *TensorFlow*, *CNTK*, or *Theano*. It was developed with a focus on enabling fast experimentation. The *R* packages *keras* and *KerasR* are the two *R* version of Keras for the statistical community. How would you compare *Keras*, the original *Python* package with the R packages? The keras package uses the pipe operator (%>%) to connect functions or operations together, but you won't find this in *kerasR*: for example, to make your model with *kerasR*, you'll see that you need to make use of the $ operator. The usage of the pipe operator generally improves the readability of your code. The package *kerasR* contains functions that are named in a similar, but not identical way as the original *Keras* package. For example, the initial (*Python*) *compile()* function is called *keras_compile()*; The same holds for other functions, such as *fit()*, which becomes *keras_fit()*, or *predict()*, which is *keras_predict* when you make use of the *kerasR* package. These are all custom wrappers.

If you have not installed *keras*, you need to do so by following the instructions provided in the appendix.

123

TABLE 6.1

Machine Learning Packages in R

R Package	Description
neuralnet	Multilayer neural networks using backpropagation
keras	Inferface with tensorflow for deep learning
kerasR	Inferface with tensorflow for deep learning
nnet	Feedforward neural networks
deepnet	Deep learning toolkit in R
h2o	R scripting functionality for H2O
RSNNS	Interface to the Stuttgart Neural Network Simulator (SNNS)
tensorflow	Interface to TensorFlow
darch	Deep Architectures and Restricted Boltzmann Machines
rnn	Package to implement Recurrent Neural Networks (RRNs)
FCNN4R	Interface to the FCNN library that allows user-extensible ANNs
rcppNL	Deep Belief Nets and other deep learning methods

6.2 Convolutional Neural Network for Deep Learning

6.2.1 Ideas Behind CNN

A convolutional neural network (CNN) is a class of deep neural networks, mainly applied to image analysis. CNN architectures can also be used to detect very different lesions or pathologies without the need of manual feature design. In addition to image recognition, CNNs have been successfully used in natural language processing.

The main idea of a CNN is seen at the convolution layers, where different filters are used. Each filter is used to identify or filter out particular features or image elements such as eyes, noses, lines, etc., just as when we search for particular objects from a complex picture. The similar approaches are used when we try to identify different types of materials: we differentiate materials by running different tests (equivalent to filters). For instance, put an object on a scale to measure weight, put it into water to measure density, put it in different solutions to see if dissoluble. Filters play a similar role as those tests.

How does filtering work in computers, or mathematically? Take the identification of a picture as an example. A filter is an image of smaller size than the picture to be classified. The filter moves over the picture, searching for a particular image element. This is mathematically equivalent to convolution. A convolution (function) of two function $f(x)$ and $g(y)$ is defined as $\int f(y) \cdot g(x-y)\,dy$. Since images consist of discrete pixies, the integration becomes summation: $\sum_y f(y)g(x-y)$, that is, the sum of the products of one image $f(\cdot)$ at location y and another image, $g(\cdot)$ (called the filter) at a different location, $x - y$. That is exactly what a convolution layer does in a CNN.

6.2.2 Network Scalability Problem

The CIFAR-10 dataset (Canadian Institute For Advanced Research) is a collection of images that are commonly used to train machine learning and computer vision algorithms. It is one of the most widely used datasets for machine learning research. The CIFAR-10 dataset contains 60,000 32×32 color images in 10 different classes, including airplanes, cars, birds, cats, deer, dogs, frogs, horses, ships, and trucks. For these images of size $32 \times 32 \times 3$ (32-pixel wide, 32-pixel high, 3 color channels), a single fully-connected neuron in a first hidden layer of a regular neural network would have $32 \times 32 \times 3 = 3072$ weights. This amount might be still manageable., However, for high resolution full-screen images (computer screen resolution of 1024×768), $1024 \times 768 \times 3 = 2,359,296$ weights need to be computed. It clear that this fully-connected structure does not scale to larger images, and the huge number of hyper-parameters or weights will quickly lead to overfitting. One of the solutions to the problem is the use of local filters, which will be discussed a bit later.

Depending on the resolution and size of the image, an $n \times n$ color image will be digitalized by an $n \times n \times 3$ array of numbers (the 3 refers to RGB values). Each of these numbers, ranging from 0 to 255, represents the pixel intensity at that point. These numbers are the only inputs available to a CNN. The CNN will do the classification based on the set of numbers after training. Technically, a one-dimensional array is called a vector, a two-dimensional array is called a matrix, and a three or more dimensional array is called tensor. Therefore, given a size of $n \times n$ pixels, there are $256^{n \times n \times 3}$ possible tensors or different pictures on a computer using RGB values 0 to 255.

6.2.3 Deep Learning Architecture

A CNN consists of many layers (Figure 6.1), each of them playing a different role. The input layer takes the input from the source images or objects and converts it to data or numbers. A *convolution layer* identifies certain features (such as line segments and arches) of the images by inspecting the images piece by piece through a "filter glass" and outputting a value for each piece dependent on the filter used. A filter is a powerful tool that makes it possible to discover a feature contained in the source images. To identify different elemental features we use filters at different convolution layers. A pooling layer converts the original higher resolution images to lower resolution images, which is a way to reduce the size of the images. A *rectified linear unit* (ReLU) is an activation layer that decides whether the neuron fires ("spikes") for the current inputs. The fully connected layers take the high-level filtered images and translate them into votes in classifying the source images. Fully connected layers are the primary building blocks of traditional neural networks. However,

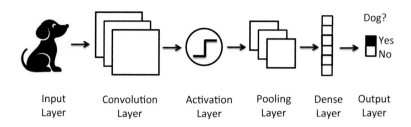

FIGURE 6.1
A Deep Learning Architecture (CNN)

the difference in a CNN is that full connection occurs at reduced images after several convolution and pooling layers. Therefore, the computational burden would have largely reduced. The classification is achieved by fully connected neural network layers that are often topped by a softmax layer that provides a probabilistic score for each class.

In summary, the CNN encodes the input image with increasing abstraction as the features are propagated through the layers of the network, arriving at a final abstract classification choice. The overlapped filters mimic connected receptive fields in human vision and result in weight-sharing among neighboring receptive fields in the CNN. Each convolution layer recognizes certain feature of the images.

6.2.4 Illustration of CNN with Example

Suppose our goal is to recognize the letters X and O from handwriting samples (Figure 6.2). We will hold small signs X and O and check again different areas of the picture/paper to see if any match. If the image to be identified is much more complicated than handwritten X's and O's, we may break down the image into smaller elements (called features in a CNN) such as lines, other geometric shapes, and blocks of simpler images. These simpler images or features will be used for classifying images in the CNN. A filter generally refers to a tool used to filter out the unwanted information or let the information of interest pass through the filter. In a CNN, when the information of interest is sized, the filter will output a larger number; otherwise it will output a small number. Let's see how the filter works using the same X's and O's example.

Convolution Layers

CNNs compare images piece by piece. Each image and feature is represented by a small 2-D array of values in the corresponding filter. In the case of X-like

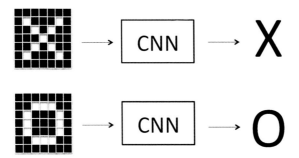

FIGURE 6.2
Recognition of X's and O's

images, we can choose as features diagonal lines and a crossing since these features will probably match up to the arms and center of any image of an X.

When presented with a new image, the CNN doesn't know exactly where these features will match, so it tries them everywhere, in every possible position. To calculate the match to a feature across the whole image (a match will output a large value, and no-match will output a small value), we are going to use a filter. A filter can be thought of as a piece of virtual glass with various transparencies at different locations according to the feature. That is, the transparency is higher at the locations of features than elsewhere, so one will see the feature when the filter moves over the target image. For the purpose of calculation, the high transparent places in the filter are assigned high values, while other places will have low values. The math we use to do this is called convolution, from which convolutional neural networks take their name.

Remark: A neuron has local connectivity. When dealing with high-dimensional inputs such as images, it is, as we saw above, computationally impractical to connect neurons to all neurons. Instead, each neuron is connected to only a local region of its input layer. The spatial extent of this connectivity is a hyperparameter called the receptive field of the neuron. The receptive field is the filter size in a CNN.

We think the image can be broken down in line segments, such as forward slashes and back-slashes. A pixel is coded 1 if an image is presented; otherwise it is coded -1 (Figure 6.3). Thus we represent an image or feature by a matrix. There are other ways of coding images. In general, we code images with higher values and whitespace with lower values. We will discuss the code effects later.

To filter the image, we place a filter over the image, starting from the upper-left corner, do the calculation (filtering) and then move to the next position by a certain distance (e.g., one or more pixels to the left, called the *stride*) and perform filtering; we continue until the filter covers all possible positions. Filtering a patch of the image, simply multiply the value at each

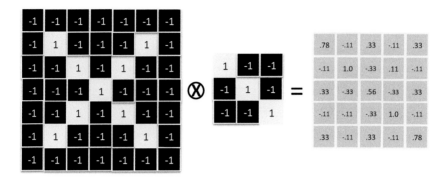

FIGURE 6.3
Convolution Produces a Shrunken Image (Fewer Pixels)

pixel in the filter by the value of the corresponding pixel in the image that is covered by the filter at the current position. Then add up the answers and divide by the total number of pixels in the filter. Let's demonstrate the calculations using an example when the filter is at the left-top corner. We use A_{ij} to represent the image and $F_{ij}(k, m)$ to represent a filter/feature at the location of the left-top corner of the filter, (k, m).

Filtering a batch of images produces a single value,

$$B_{km} = \frac{1}{KM} \sum A_{ij} F_{ij}(k, m).$$

As an example, from Figure 6.3, $B_{11} = \frac{1}{3 \times 3} \sum_{i=1}^{3} \sum_{j=1}^{3} A_{ij} F_{ij}(1, 1) = \frac{(-1 \times 1) + (-1 \times -1) + (-1 \times -1)}{3 \times 3} + \frac{(-1 \times -1) = (-1 \times -1) + (1 \times 1)}{3 \times 3} + \frac{(-1 \times -1) = (-1 \times -1) + (1 \times 1)}{3 \times 3} = 0.78$. Similarly, $B_{23} = 0.33$, and $B_{25} = -0.11$.

We can see that if the original image size is 10×10 pixes and the filter size is 3×3 pixels, then k and m range from 0 to $10 - 3 = 7$. In other words, the size of the resulting image is 8×8 pixels, which can be represented by an 8×8 matrix. In general, if the original image size is $I \times J$ pixels and filter size $K \times M$ pixels, the resulting image size after filtering is $(I - K + 1) \times (J - M + 1)$ pixels.

Pooling Layers

Pooling is a way to take large images and shrink them down while preserving the most important information in them. It consists of stepping a small window across an image and taking the maximum value from the window at each step. A window 2 or 3 pixels on a side and steps of 2 pixels works well. Other pooling methods exist, such as average pooling. Average pooling, often used in the past, has recently fallen out of favor compared to the max pooling operation, which has been shown to work better in practice. A pooling layer performs pooling on a collection of images to help manage the computational load. The visual effect of pooling is somewhat as if ones sees the image from a further distance.

With pooling, one doesn't care so much exactly where the feature fits as long as it fits somewhere within the window. The result of this is that CNNs can find whether a feature is in an image without worrying about its exact location (local transitional invariance). This helps solve the problem of computers being hyper-literal.

Rectified Linear Units

The *rectifier* is an activation function defined as the nonnegative part of its argument: $\max(0, x)$, where x is the input to a neuron. A layer or unit employing the rectifier is also called a *rectified linear unit* (*ReLU*). The output of a ReLU layer is the same size as whatever is put into it, just with all the negative values removed. ReLU is also known as a ramp function and is analogous to half-wave rectification in electrical engineering. The introduction of the activation function has strong biological motivations and mathematical justifications (Hahnloser et al., 2000; Hahnloser and Seung, 2001). It enables better training of deeper networks, compared to widely used activation functions, such as the logistic sigmoid and the hyperbolic tangent. ReLU is, as of 2018, the most popular activation function for deep neural networks (Glorot et al., 2011; Yann et al., 2015, Prajit et al., 2017).

Fully Connected Layers

A *fully connected layer* (*dense layer*) is a hidden layer like those in a regular ANN with many weights. It takes the filtered images as input and output classifications. As with a regular ANN, there can be several fully connected layers stacked together. Learning occurs in weights. By comparing the outputs from the CNN against correct outputs in the training set, we adjust weights to reduce the errors. The method for adjusting the weights is the error backpropagation we have discussed earlier. Backpropagation at the fully connected layers is another expensive computing step and a motivator for specialized computing hardware.

Hyperparameters

So far, we have have a good picture of how a CNN works, but there is still a list of questions that need to be answered:

1. How many layers of each type should there be, in what order? And how to deal with color images?

2. Some deep neural networks can have over a thousand layers, what is the trade-off between the number of layers, the size of the layer, and the complexity of filters or layers?

3. For convolution layers, what features or filters to use, and what size of each filter? How big should the stride be?

4. For each pooling layer, what window size and pooling algorithm should be used?

5. For each fully-connected layer, how many hidden neurons or weights are needed?

CNNs can be used not just for images, but also to categorize other types of data. The key is to transform them and make them look like image data, in the form of a 2-D array or matrix. For instance, audio signals can be chopped into short time chunks, and then each chunk broken up into bass, midrange, treble, or finer frequency bands. This can be represented as a two-dimensional array, where each column is a time chunk and each row is a frequency band. "Pixels" in this fake picture that are close together are closely related. CNNs work well in this way. Researchers have also used CNNs to adapt text data for natural language processing and even chemical data for drug discovery. The rule of thumb is: if your data is just as useful after swapping any of your columns with each other, then you can't use a CNN. However, if you can make your problem look like finding patterns in an image, then CNNs may be exactly what you need (Rohrer, 2019)

6.2.5 CNN for Medical Image Analysis

Medical image analysis is the science of analyzing or solving medical problems using different image analysis techniques for the effective and efficient extraction of information. Qayyuma et al. (2018) has presented a state-of-the-art review of medical image analysis using CNNs. The application area of CNN covers the whole spectrum of medical image analysis including detection, segmentation, classification, and computer-aided diagnosis.

The commonly used evaluation matrix consists of: $Sensitivity = \frac{TP}{TP+FN}$, $Specificity = \frac{TN}{FP+TN}$, $Accuracy = \frac{TP+TN}{TP+FN+FP+TN}$, $Precision = \frac{TP}{TP+FP}$, $Recall = \frac{TP}{TP+TN}$, $Dice\ Score = \frac{2|P \cap GT|}{|P|+|GT|}$, where, TP (true positive) = number of positive cases correctly recognized, FP (false positive) = number of negative cases incorrectly recognized as positives, TN (true negative) = number of negative cases correctly recognized, and FN (false negative) = number of positive cases incorrectly recognized as negatives.

The purpose of medical imaging is to aid radiologists and clinicians to make the diagnostic and treatment processes more efficient. Medical imaging is a predominant part of the diagnosis and treatment of diseases and encompasses various imaging modalities, such as computed tomography (CT), magnetic

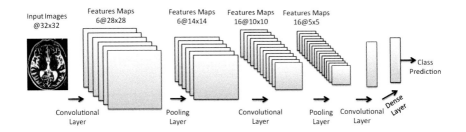

FIGURE 6.4
Typical CNN Architecture for Medical Image Analysis

resonance imaging (MRI), positron emission tomography (PET), ultrasound, X-ray, and hybrid modalities (Heidenreich, et al., 2002).

Typical CNN architecture for medical image analysis is shown in Figure 6.4. Hussain (2017) has used a brain tumor segmentation algorithm using a cascaded deep CNN. Farooq (2017) present a CNN-based method for the classification of Alzheimer's disease in MRI images having multiple classes and two networks. i.e., GoogleNet and ResNet are trained on an ADNI database. Much other research has been done on CNNs for image analysis, including a multiscale CNN-based approach for automatic segmentation of MR images for classifying voxel into brain tissue classes (Moeskops et al., 2016), a tri-planar CNN used for segmentation of tibial cartilage in knee MRI images (Prasoon et al., 2013), and segmentation of isointense brain tissue presented through a CNN using a multimodal MRI dataset by training the network on three patches that are extracted from the images (Zhang et al., 2015). Other interesting studies include "lung pattern classification for interstitial lung diseases using a deep convolutional neural network" by Anthimopoulos et al. (2016), "Predicting brain age with deep learning from raw imaging data results in a reliable and heritable biomarker," by Cole et al. (2016), and "Dermatologist-level classification of skin cancer with deep neural networks," by Esteva et al. (2017).

6.2.6   A CNN for Handwritten Digits Recognition

```
# CNN for mnist hand writing digits recognition
library(keras)
library(kerasR)
getwd()
setwd("/Users/markchang/Desktop/PharmaAI/")
# Assume mnist_train.csv, mnist_train.csv, and emdeddings have been
```

```
# dowloaded from www.kaggle to your working directory.
mnistTrain <- read.csv("mnist_train.csv",header=T,sep=",")
mnistTest <- read.csv("mnist_test.csv",header=T,sep=",")
Xtrain=mnistTrain[,-1]/225 # scaled, 225 is the maxumum value for the color
Ytrain=mnistTrain[,1]
Xtest=mnistTest[,-1]/225
Ytest=mnistTest[,1]
# Restructure data to 3-dimensional: ## [1] 60000 28 28
mnistTrain3d=array(unlist(Xtrain),dim=c(dim(Xtrain)[1],28,28))
dim(mnistTrain3d)
mnistTest3d=array(unlist(Xtest),dim=c(dim(Xtest)[1],28,28))
dim(mnistTest3d)
mnistTest4d <- array(mnistTest3d, dim = c(dim(mnistTest3d), 1))
mnistTrain4d <- array(mnistTrain3d, dim = c(dim(mnistTrain3d), 1))
dim(mnistTrain4d)
dim(mnistTest4d)
yTrain<-to_categorical(Ytrain)
yTest<-Ytest
# Not done with data restructure
# Convolutional neural networks
mod <- Sequential()
mod$add(Conv2D(filters = 32, kernel_size = c(3, 3),
 input_shape = c(28, 28, 1)))
mod$add(Activation("relu"))
mod$add(Conv2D(filters = 32, kernel_size = c(3, 3),
 input_shape = c(28, 28, 1)))
mod$add(Activation("relu"))
mod$add(MaxPooling2D(pool_size=c(2, 2)))
mod$add(Dropout(0.25))
mod$add(Flatten())
mod$add(Dense(128))
mod$add(Activation("relu"))
mod$add(Dropout(0.25))
mod$add(Dense(10))
mod$add(Activation("softmax"))
keras_compile(mod, loss = 'categorical_crossentropy', optimizer = RMSprop())
keras_fit(mod, mnistTrain4d, yTrain, batch_size = 32, epochs = 5, verbose = 1,
 validation_split = 0.1)
test.predY <- keras_predict_classes(mod, mnistTest4d) # predict classes
table(test.predY,mnistTest[,1])
mean(mnistTest[,1] == test.predY)
```

The digits recognition results are shown in Figure 6.5.

```
> table(test.predY,mnistTest[,1])

test.predY    0     1     2     3     4     5     6     7     8     9
         0  975     0     2     0     0     2     6     0     4     2
         1    0  1133     4     0     0     0     3     4     1     3
         2    0     2  1019     3     1     1     0    11     3     2
         3    0     0     0   998     0     3     0     0     1     1
         4    0     0     1     0   968     0     2     0     1     4
         5    0     0     0     5     0   881     2     1     0     3
         6    3     0     0     0     4     4   944     0     1     0
         7    1     0     6     2     0     1     0  1008     1     6
         8    1     0     0     1     1     0     1     1   958     3
         9    0     0     0     1     8     0     0     2     4   985
> mean(mnistTest[,1] == test.predY)
[1] 0.9869987
```

FIGURE 6.5
CNN for Handwritten Digits Recognition

We can see from the CNN results that the precision is 98.7%, better than the previous FNN's 96.6% accuracy.

6.2.7   Training CNN Using Keras in R

You need to install keras before you run the code in this chapter (see Appendix for keras installation). We will modify the R code example by Rai (2018) to illustrate the use of CNN for image recognition.

There are two parts in the code. The first part uses the BiocManager package to read in and standardize the images. We we only use 18 pictures (18 separate .jpg files) for the demo so that it can run fast, including 15 (5 planes, 5 cars, and 5 bikes in Figure 6.6A) for training and 3 (plane, car, bike in Figure 6.6B) for testing, all installed in the working directory. The pictures with different sizes will be difficult to handle by CNNs, thus we resize them to the same size is necessary.

The second art is to build a CNN architectures and execute the code. The general structure of CNN using keras is shown in Figure 6.7. There are 5 commonly used different layers: convolution layer (layer_conv_2d) for filtering images, pool layer (layer_max_pooling_2d) for resolution and dimension reductions, dropout layer (layer_dropout) for reducing overfitting and computational efficiency, and flatten layer (layer_flatten) to pool feature map to a single column for passing to a fully connected layer, and dense layer (layer_dense) for fully connected layer at output layer. kerner_size specifies the filter size. There are several optimization methods that can be specified using the option "optimizer" when you compile or build the model. The code is annotated for a better readability.

FIGURE 6.6A
CNN Images for Traning

FIGURE 6.6B
CNN Images for Testing

```
# Convolutional Neural Networks
library(keras)
# http://bioconductor.org/packages/release/bioc/html/EBImage.html
if (!requireNamespace("BiocManager", quietly = TRUE))
 install.packages("BiocManager")
BiocManager::install("EBImage")
library(EBImage)
BiocManager::install() # Read Images
 # set up my working directory
setwd("/Users/markchang/Desktop/pharmaAI/")
# put 15 picture file names in pic1 for training
pic1 <- c('p1.jpg', 'p2.jpg', 'p3.jpg', 'p4.jpg', 'p5.jpg',
 'c1.jpg', 'c2.jpg', 'c3.jpg', 'c4.jpg', 'c5.jpg',
 'b1.jpg', 'b2.jpg', 'b3.jpg', 'b4.jpg', 'b5.jpg')
train <- list()
```

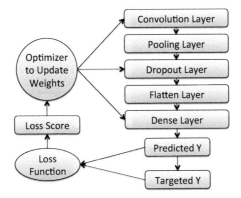

Various layers may be used repeatedly in different sequences

FIGURE 6.7
A CNN Architecture with keras

```
# Read in training pictures from the working directory and store in pic1.
for (i in 1:15) {train[[i]] <- readImage(pic1[i])}
# do the same thing for the testing
pic2 <- c('p6.jpg', 'c6.jpg', 'b6.jpg')
test <- list()
for (i in 1:3) {test[[i]] <- readImage(pic2[i])}
# Explore
print(train[[12]])
summary(train[[12]])
display(train[[12]])
plot(train[[12]])
par(mfrow = c(3, 5))
for (i in 1:15) plot(train[[i]])
par(mfrow = c(1,1))
# Resize & combine
str(train) # check size of images
for (i in 1:15) {train[[i]] <- resize(train[[i]], 100, 100)} # equalsize images
for (i in 1:3) {test[[i]] <- resize(test[[i]], 100, 100)}
str(train) # recheck size of images
train <- combine(train) # combine images into one
x <- tile(train, 5)
display(x, title='Pictures')
test <- combine(test)
y <- tile(test, 3)
display(y, title = 'Pics')
# Reorder dimension
```

```r
train <- aperm(train, c(4, 1, 2, 3))
test <- aperm(test, c(4, 1, 2, 3))
str(train)
# Response
trainy <- c(0, 0, 0, 0, 0, 1, 1, 1, 1, 1, 2, 2, 2, 2, 2)
testy <- c(0, 1, 2)
# One hot encoding
trainLabels <- keras::to_categorical(trainy)
testLabels <- keras::to_categorical(testy)
trainLabels
testLabels
# Model
model <- keras_model_sequential()
model %>%
 layer_conv_2d(filters = 32,
 kernel_size = c(3,3),
 strides = 1,
 activation = 'relu',
 input_shape = c(100, 100, 3)) %>%
 layer_conv_2d(filters = 32,
 kernel_size = c(3,3),
 activation = 'relu') %>%
 layer_max_pooling_2d(pool_size = c(2,2)) %>%
 layer_dropout(rate = 0.25) %>%
 layer_conv_2d(filters = 64,
 kernel_size = c(5,5),
 activation = 'relu') %>%
 layer_max_pooling_2d(pool_size = c(4,4)) %>%
 layer_dropout(rate = 0.125) %>%
 layer_flatten() %>%
 layer_dense(units = 256, activation = 'relu') %>%
 layer_dropout(rate=0.25) %>%
 layer_dense(units = 3, activation = 'softmax' ) %>%

 compile(loss = 'categorical_crossentropy',
 optimizer = optimizer_sgd(lr = 0.01,
 decay = 1e-6,
 momentum = 0.9,
 nesterov = T),
 metrics = c('accuracy'))
summary(model)
# Fit model
history <- model %>%
```

```
fit(train,
trainLabels,
epochs = 120,
batch_size = 32,
validation_split = 0.2,
validation_data = list(test, testLabels))
plot(history)
# Evaluation & Prediction - train data
model %>% evaluate(train, trainLabels)
pred <- model %>% predict_classes(train)
table(Predicted = pred, Actual = trainy)
prob <- model %>% predict_proba(train)
cbind(prob, Predicted_class = pred, Actual = trainy)
# Evaluation & Prediction - test data
model %>% evaluate(test, testLabels)
pred <- model %>% predict_classes(test)
table(Predicted = pred, Actual = testy)
prob <- model %>% predict_proba(test)
cbind(prob, Predicted_class = pred, Actual = testy)
```

Figure 6.8 shows histories of loss function and accuracy during training. Because the small training and testing datasets, the results can vary a lot each time you run the model. The current CNN run shows a 100% accuracy.

6.3 Recurrent Neural Networks

Location invariance and local compositionality are two key ideas behind CNNs that do not always bear fruit. They make sense for computer vision applications, but they do not make much sense in the case of natural language processing or time-series events. The location where a word lies in the whole sentence is critical to the meaning of the sentence. Words that are not close to one another in a sentence may be more connected in terms of meaning, which is quite contrary to pixels in a specific region of an image that may be a part of a certain object. Therefore, it makes sense to look for a neural network that reflects the sequence of the tokens, whether they are words, events, or something else. One such network is called a memory network.

6.3.1 Short-Term Memory Network

The idea of recurrent neural networks came from the work by William and his colleagues (William et al. 1986). A recurrent neural network (RNN) is a class

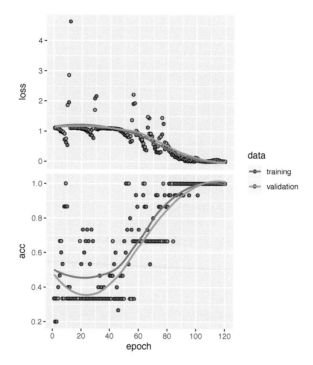

FIGURE 6.8
History of Loss Function and Accuracy

of artificial neural networks for modeling temporal dynamic behavior. Unlike feedforward neural networks, RNNs can use their internal state as memory to process sequences of inputs. In other words, they often reuse the output or hidden outputs (internal states) as input again, hence their name. RNNs are useful for tasks such as unsegmented, connected handwriting recognition (Graves, et al., 2009) or speech recognition (Sak, et al., 2014; Li and Wu, 2014). RNNs have also been implemented for stock market prediction, sequence generation, test generation, voice recognition, image captioning, poem-writing (after being trained on Shakespeare's poetry), reading handwriting from left to right, and generating music.

Suppose a home school teacher wants to schedule three routine classes, Math, Science, and English, alternating classes each week, and the teacher has decided to do this using a RNN.

Let's use a vector variable $(Math, Science, English)^T$ to represent the three subjects:

$$\mathbf{Math} = \begin{bmatrix} 1 \\ 0 \\ 0 \end{bmatrix}, \mathbf{Science} = \begin{bmatrix} 0 \\ 1 \\ 0 \end{bmatrix}, \mathbf{English} = \begin{bmatrix} 0 \\ 0 \\ 1 \end{bmatrix}.$$

The RNN model to be used is

$$
\begin{bmatrix} \text{Math} \\ \text{Science} \\ \text{English} \end{bmatrix}^T = \mathbf{W} \begin{bmatrix} \text{Science} \\ \text{English} \\ \text{Math} \end{bmatrix}^T .
$$

If we use weights to reflect the memory capability of the system:

$$
\mathbf{W} = \begin{bmatrix} 0 & 0 & 1 \\ 1 & 0 & 0 \\ 0 & 1 & 0 \end{bmatrix},
$$

the mission is accomplished. Let's check on this using the model:

$$
\mathbf{W} \cdot \mathbf{Math} = \begin{bmatrix} 0 & 0 & 1 \\ 1 & 0 & 0 \\ 0 & 1 & 0 \end{bmatrix} \begin{bmatrix} 1 \\ 0 \\ 0 \end{bmatrix} = \begin{bmatrix} 0 \\ 1 \\ 0 \end{bmatrix} = \mathbf{Science}
$$

$$
\mathbf{W} \cdot \mathbf{Science} = \begin{bmatrix} 0 & 0 & 1 \\ 1 & 0 & 0 \\ 0 & 1 & 0 \end{bmatrix} \begin{bmatrix} 0 \\ 1 \\ 0 \end{bmatrix} = \begin{bmatrix} 0 \\ 0 \\ 1 \end{bmatrix} = \mathbf{English}
$$

$$
\mathbf{W} \cdot \mathbf{English} = \begin{bmatrix} 0 & 0 & 1 \\ 1 & 0 & 0 \\ 0 & 1 & 0 \end{bmatrix} \begin{bmatrix} 0 \\ 0 \\ 1 \end{bmatrix} = \begin{bmatrix} 1 \\ 0 \\ 0 \end{bmatrix} = \mathbf{Math}.
$$

In reality, we can train the RNN using training data to get the right $\mathbf{W}$. We can use the R-Program provided earlier. But we want to use RNN, more precisely, short-term memory (STM) RNN. If the home school student doesn't do the homework well, the teacher will teach the same subject the next week; otherwise, the class subjects will rotate as usual. Therefore, to determine the next class, the teacher needs to remember the previous class (STM) and grade the homework (Figure 6.9).

6.3.2 An Example of RNN in R

In this simple example with the *rnn* package in R, we use noisy sine-wave time-series data X to predicted a time series with cosine wave Y (Figure 6.10). The key element of the package is the *trainr* function.

Arguments used in this example are listed as follows.

Y: array of output values, dim 1: samples (must be equal to dim 1 of X), dim 2: time (must be equal to dim 2 of X). X: array of input values, dim 1: samples, dim 2: time. *model*: a model trained before, used for retraining purpose. *learningrate*: learning rate to be applied for weight iteration. *hidden_dim*: dimension(s) of hidden layer(s). *network_type*: type of network, could be rnn, gru or lstm. gru and lstm are experimental. *numepochs*: number of iterations, i.e., number of times the whole dataset is presented to the network.

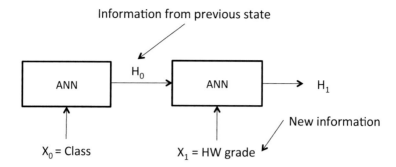

H_0 = class without consideration of HW grade

H_1 = class with consideration of HW grade

FIGURE 6.9
Recurrent Neural Network (Short-Term Memory)

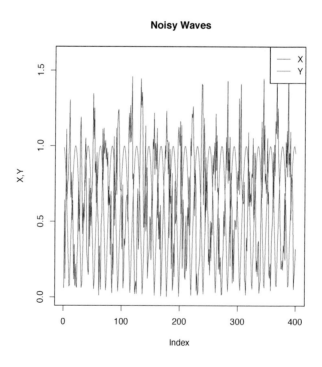

FIGURE 6.10
Noisy Sine Wave X and Cosine Time Series Y

```
# Example of RNN for simulated wave data
library(rnn)
set.seed(5)
# Create sequences of 400 observations
t <- seq(0.005,2,by=0.005)
x <- abs(sin(t*30) + rnorm(100, 0, 0.25))
y <- abs(cos(t*30))
# Convert to matrix[40x10] or time series
X <- matrix(x, nrow = 40)
Y <- matrix(y, nrow = 40)
# Plot noisy waves
plot(as.vector(X), col='blue', type='l', ylab = "X,Y",
 main = "Noisy Waves")
lines(as.vector(Y), col = "red")
legend("topright", c("X", "Y"), col = c("blue","red"),
 lty = c(1,1), lwd = c(1,1))
# Transpose
X <- t(X); Y <- t(Y)
# Training and testing index
train<-1:8; test <- train # total 10 sequences
model <- trainr(Y = Y[train,], X = X[train,],
 learningrate = 0.05, hidden_dim = 16,
 network_type = "rnn", numepochs = 3000)
# Predicted values
Yp <- predictr(model, X)
# Plot predicted vs actual. Testing set only.
plot(as.vector(t(Y[test,])), col = 'red', type='l',
 main = "Actual vs predicted: testing set", ylab = "Y,Yp")
lines(as.vector(t(Yp[test,])), type = 'l', col = 'blue')
legend("topright", c("Predicted", "Real"),
 col = c("blue","red"), lty = c(1,1), lwd = c(1,1))
```

The plot of predicted versus real values is shown in Figure 6.11.

6.3.3   Long Short-Term Memory Networks

The RNN suffers from a short-term memory problem. If the sequence of events, such as an English text, is long enough, the common RNN will have a hard time carrying information from earlier time steps to later ones. *Long Short-Term Memory Networks* (LSTMs) are explicitly designed to deal with long-term dependency problems.

The cell state is represented by the horizontal line on the top of the diagram (Figure 6.12). The cell state can be changed by adding or removing information through gates. A gate is a plain vanilla ANN (two-layer perceptron).

Actual vs predicted: testing set

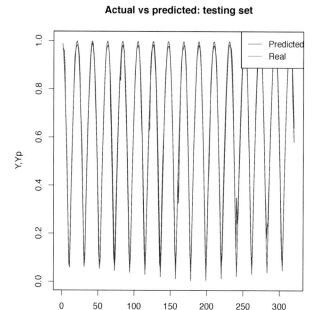

FIGURE 6.11
Real versus Predicted Wave Series of Test Data

Different gates regulate information flow differently. Commonly used gates are sigmoid and tanh gates. A *sigmoid* gate outputs numbers between 0 (nothing goes through) and 1 (everything goes through). A tanh gate outputs a value between -1 and 1. A typical LSTM unit (three such units in Figure 6.12) has three *sigmoid* (σ) gates and two *tanh* gates to control the information flow.

The first σ-gate (Figure 6.12), the so-called forget gate, determines what information is not important and forget's it. At time t, the cell with initial state C_{t-1} carries information from the previous state at time t–1, receives new information $\mathbf{x}_t$, and outputs a value between 0 and 1 through the function $\mathbf{f}_t = \sigma\left(\mathbf{W}_f[\mathbf{h}_{t-1}, \mathbf{x}_t] + \mathbf{b}_f\right)$. Here, $\mathbf{h}_{t-1}$ is the hidden state or output from the previous state, and $\mathbf{b}_f$ is the bias. The vector $\mathbf{f}_t$ will determine the fractions of C_{t-1} from the previous state to keep.

To determine the amount of information to add to the cell state, which is expressed as the *remember vector* $\mathbf{i}_t \otimes \tilde{\mathbf{C}}_t$, where candidate cell state $\tilde{\mathbf{C}}_t = \tanh\left(\mathbf{W}_C[\mathbf{h}_{t-1}, \mathbf{x}_t] + \mathbf{b}_C\right)$ is the total information added into the cell, and $\mathbf{i}_t = \sigma\left(\mathbf{W}_i[\mathbf{h}_{t-1}, \mathbf{x}_t] + \mathbf{b}_i\right)$ is the fraction to be added. The operator $\otimes$ denotes the *Hadamard product* (an element-wise product).

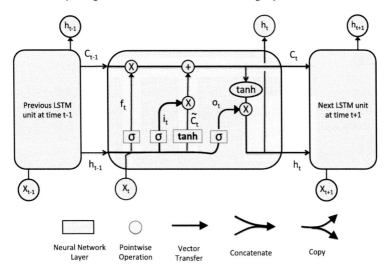

FIGURE 6.12
Diagram of a Long Short-Term Memory Network

Now we can update the cell's state to $\mathbf{C}_t$ from $\mathbf{C}_{t-1}$:

$$\mathbf{C}_t = \mathbf{f}_t \otimes \mathbf{C}_{t-1} + \mathbf{i}_t \otimes \tilde{\mathbf{C}}_t.$$

Lastly, the output (called the *focus vector*) of a LSTM unit will depend on the current cell state, that is,

$$\mathbf{o}_t = \sigma \left(\mathbf{W}_o \left[\mathbf{h}_{t-1}, \mathbf{x}_t \right] + \mathbf{b}_o \right),$$

where $\mathbf{b}_o$ is the bias. Then, we put the cell state through tanh (to push the values to be between -1 and 1) and multiply it by the output of the sigmoid gate, so that the hidden output is

$$\mathbf{h}_t = \mathbf{o}_t \otimes \tanh \left(\mathbf{C}_t \right).$$

In summary, the LSTM is a RNN architecture in deep learning. It can not only process single data points (such as images), but also entire sequences of data (such as speech or video). Arguably, RNN is the most commercial AI achievement, used for everything from predicting diseases to composing music. A LSTM consists of a sequence of STM units that is usually composed of a cell, an input gate, an output gate and a *forget gate*. The cell state acts as long-term memory, while hidden states act as short-term or working memory. The cell remembers values over arbitrary time intervals, and the three gates regulate the flow of information into and out of the cell. Relative insensitivity to gap length is an advantage of LSTM hidden Markov models and other sequence learning methods in numerous applications.

Mathematically we can rewrite the state function and (hidden) output function as

$$
\left\{
\begin{array}{c}
\mathbf{C}_t = \sigma\left(\mathbf{W}_f \cdot [\mathbf{h}_{t-1}, \mathbf{x}_t]\right) \otimes \mathbf{C}_{t-1} \\
+\sigma\left(\mathbf{W}_i \cdot [\mathbf{h}_{t-1}, \mathbf{x}_t]\right) \otimes \tanh\left(\mathbf{W}_C \cdot [\mathbf{h}_{t-1}, \mathbf{x}_t]\right) \\
\mathbf{h}_t = \sigma\left(\mathbf{W}_o [\mathbf{h}_{t-1}, \mathbf{x}_t]\right) \otimes \tanh\left(\mathbf{C}_t\right),
\end{array}
\right.
$$

where bias terms are included in the weights and $\mathbf{x}_t$ to simplify things.

The backpropagation algorithm is still applicable for LSTMs. There are many variants of the LSTM, such as peehole LSTMs (Gers & Schmidhuber, 2000) and the Gated Recurrent Unit, or GRU (Cho et al., 2014).

LSTM networks were developed by Hochreiter and Schmidhuber in 1997. In 2009, a Connectionist Temporal Classification (CTC)-trained LSTM network was the first RNN to win pattern recognition contests for its successes in handwriting recognition. In 2014, the Chinese search giant Baidu used CTC-trained RNNs to break the Switchboard Hub 5'00 speech recognition benchmark. Google uses LSTM for speech recognition on the smartphone for the smart assistant Allo and Google Translate. Apple uses LSTM for the Quicktype function on the iPhone and for Siri. Amazon uses a LSTM for Amazon Alexa. In 2017, Facebook performed some 4.5 billion automatic translations every day using LSTMs. Using LSTMs Microsoft reported in 2017 reaching 95.1% recognition accuracy on the Switchboard corpus. In 2018, bots developed by OpenAI were able to beat humans in the game of Dota (Rodriguez, 2018). The bots have a 1024-unit LSTM that sees the current game state and emits actions through several possible action heads. In 2019, DeepMind's program AlphaStar used a deep LSTM core to excel at the complex video game Starcraft. This was viewed as significant progress towards Artificial General Intelligence (Standford, 2019).

Leveraging large historical data in electronic health records (EHRs), Choi, E. et al. (2016) developed Doctor AI, a generic predictive model that covers observed medical conditions and medication use (https://github.com/mp2893/doctorai). Doctor AI, a temporal model using a RNN, was developed and applied to longitudinal time-stamped EHR data from 260 K patients over eight years. Encounter records (e.g., diagnosis codes, medication codes or procedure codes) were input to the RNN to predict the diagnosis and medication categories for a subsequent visit. Doctor AI assesses the history of patients to make multilabel predictions (one label for each diagnosis or medication category). According to their report, based on separate blind test set evaluation, Doctor AI can perform differential diagnosis with a significantly higher success rate than several baselines.

6.3.4   Sentiment Analysis Using LSTMs in R

Sentiment analysis (also known as opinion mining or emotion AI) refers to the use of natural language processing (NLP) and biometrics to identify, extract, and quantify information. Sentiment analysis is widely applied to reviews and survey responses, online and social media, and healthcare materials for applications that range from marketing to customer service to medicine and health services.

As an example, we will use the featured Code Competition (Price $25,000) for Quora Insincere Questions Classification to detect toxic content to improve online conversations at https://www.kaggle.com/c/quora-insincere-questions-classification. According to the description of the competition, an existential problem for any major website today is how to handle toxic and divisive content. Quora wants to tackle this problem head-on to keep their platform a place where users can feel safe sharing their knowledge with the world.

Here are a few examples of insincere questions from the dataset: How well are you adapting to the Trump era? Why is San Francisco so liberal that they can't even give Kate Steinle justice? We can see why these questions are considered "insincere": they're not after a genuine answer, but tend to instead either state the questioner's beliefs as fact or try to be deliberately provocative. This appears challenging since it requires social context beyond the immediate question.

Note: Quora is a platform that empowers people to learn from each other. On Quora, people can ask questions and connect with others who contribute unique insights and quality answers. A key challenge is to weed out insincere questions – those founded upon false premises, or that intend to make a statement rather than look for helpful answers.

In this competition, Kagglers will develop models that identify and flag insincere questions. To date, Quora has employed both machine learning and manual review to address this problem. Be aware that this is being run as a Kernels Only Competition, requiring that all submissions be made via a Kernel output. Please read the Kernels FAQ and the data page very carefully to fully understand how this is designed. The datasets (train.csv, test.csv) and descriptions can be found at https://www.kaggle.com/c/quora-insincere-questions-classification/data.

Datafields: *qid*—unique question identifier, *question_text*—Quora question text, *target*—a question labeled "insincere" has a value of 1, otherwise its value is 0.

The semantic modeling requires mapping words to numbers or word embedding. *Word embedding* is the collective name for language modeling and feature-learning techniques in NLP where words or phrases from the vocabulary are mapped to vectors of real numbers. Conceptually, it involves a mathematical embedding from a space with many dimensions per word to a continuous vector space with a much lower dimension. Word and phrase embedding, when used as the underlying input representation, have been shown to boost the performance in NLP tasks such as *syntactic parsing* and *distributional semantics*.

Word embedding for *n-grams* in biological sequences (e.g., DNA, RNA, and proteins) for bioinformatics applications have been proposed by Asgari and Mofrad (2015). Named bio-vectors (BioVec) to refer to biological sequences in general with protein-vectors (ProtVec) for proteins (amino-acid sequences) and gene-vectors (GeneVec) for gene sequences, this representation can be widely used in applications of deep learning in proteomics and genomics. The results presented by Asgari and Mofrad (2015) suggest that BioVectors can characterize biological sequences in terms of biochemical and biophysical interpretations of the underlying patterns.

Word embeddings along with the dataset that can be used in the models are as follows:

1. GoogleNews-vectors-negative300 - https://code.google.com/archive/p/word2vec/

2. glove.840B.300d - https://nlp.stanford.edu/projects/glove/

3. paragram_300_sl999 - https://cogcomp.org/page/resource_view/106

4. wiki-news-300d-1M - https://fasttext.cc/docs/en/english-vectors.html

As a way of learning how the RNN works, we use a LSTM to model the problem in *R*, crediting Aindow (2019). Running the code step by step and checking the annotations along with the code should help you understand the LSTM better.

```
# LSTM
# https://www.kaggle.com/taindow/simple-lstm-with-r
setwd("/Users/markchang/Desktop/PharmaAI")
getwd()
library(tidyverse) # importing, cleaning, visualising
library(tidytext) # working with text
library(keras) # deep learning with keras
library(data.table) # fast csv reading
options(scipen=999) # turn off scientific display
# Import data
train <- fread('train.csv', data.table = FALSE)
test <- fread('test.csv', data.table = FALSE)
test %>% head()
# Initial look
train %>% filter(target == 1) %>% sample_n(5)
# Tokenizing
# Setup some parameters
max_words <- 4096 # 15000 Maximum number of words to consider as features
maxlen <- 64 # 64 Text cutoff after n words
# Prepare to tokenize the text
full <- rbind(train %>% select(question_text), test %>% select(question_text))
texts <- full$question_text
tokenizer <- text_tokenizer(num_words = max_words) %>%
 fit_text_tokenizer(texts)
# Tokenize - i.e. convert text into a sequence of integers
```

```
sequences <- texts_to_sequences(tokenizer, texts)
word_index <- tokenizer$word_index
# Pad out texts so everything is the same length
data = pad_sequences(sequences, maxlen = maxlen)
# Datasplit
# Split back into train and test
train_matrix = data[1:nrow(train),]
test_matrix = data[(nrow(train)+1):nrow(data),]
# Prepare training labels
labels = train$target
# Prepare a validation set
set.seed(1337)
training_samples = nrow(train_matrix)*0.90
validation_samples = nrow(train_matrix)*0.10
indices = sample(1:nrow(train_matrix))
training_indices = indices[1:training_samples]
validation_indices = indices[(training_samples + 1): (training_samples +
validation_samples)]
x_train = train_matrix[training_indices,]
y_train = labels[training_indices]
x_val = train_matrix[validation_indices,]
y_val = labels[validation_indices]
# Training dimensions
dim(x_train)
table(y_train) # We see pretty major imbalance here, we will need
to address this later
# Embeddings
# Our first model will be based off one of the provided word
embeddings. Since we're limited to 2 hours of kernel time
# (6 hours if we use CPU), let's start with the smaller embedding
files. I'll go with the fasttext wiki-news-300d embeddings.
lines <- readLines("/Users/markchang/Desktop/PharmaAI/embeddings/
wiki-news-300d-1M/wiki-news-300d-1M.vec")
fastwiki_embeddings_index = new.env(hash = TRUE, parent = emptyenv())
lines <- lines[2:length(lines)]
pb <- txtProgressBar(min = 0, max = length(lines), style = 3)
for (i in 1:length(lines)){
 line <- lines[[i]]
 values <- strsplit(line, " ")[[1]]
 word<- values[[1]]
 fastwiki_embeddings_index[[word]] = as.double(values[-1])
 setTxtProgressBar(pb, i)
}
# Create our embedding matrix
```

```
fastwiki_embedding_dim = 300
fastwiki_embedding_matrix = array(0, c(max_words, fastwiki_embedding_dim))
for (word in names(word_index)){
 index <- word_index[[word]]
 if (index < max_words){
 fastwiki_embedding_vector = fastwiki_embeddings_index[[word]]
 if (!is.null(fastwiki_embedding_vector))
 fastwiki_embedding_matrix[index+1,] <- fastwiki_embedding_vector
 # Words without an embedding are all zeros
 }
}
gc()
# Model Architecture
# We start with a simple LSTM topped with a single dense layer for prediction
# Setup input
input <- layer_input(
 shape = list(NULL),
 dtype = "int32",
 name = "input"
)
# Model layers
embedding <- input %>%
 layer_embedding(input_dim = max_words, output_dim = fastwiki_
embedding_dim, name = "embedding")
lstm <- embedding %>%
 layer_lstm(units = maxlen,dropout = 0.25, recurrent_dropout =
0.25, return_sequences = FALSE, name = "lstm")
dense <- lstm %>%
 layer_dense(units = 128, activation = "relu", name = "dense")
predictions <- dense %>%
 layer_dense(units = 1, activation = "sigmoid", name = "predictions")
# Bring model together
model <- keras_model(input, predictions)
# Freeze the embedding weights initially to prevent updates propgating back
through and ruining our embedding
get_layer(model, name = "embedding") %>%
 set_weights(list(fastwiki_embedding_matrix)) %>%
 freeze_weights()
# Compile
model %>% compile(
 optimizer = optimizer_adam(),
 loss = "binary_crossentropy",
 metrics = "binary_accuracy"
)
```

```
# Print architecture (plot_model isn't implemented in the R package yet)
print(model)
# Model Training
# The embedding weights will be frozen to keep training fast for an
initial benchmark model.
# Train model
history <- model %>% fit(
  x_train,
  y_train,
  batch_size = 2048,
  validation_data = list(x_val, y_val),
  epochs = 15,
  view_metrics = FALSE,
  verbose = 0
)
# Look at training results
print(history)
plot(history)
# This model could easily be improved with some careful fine
tuning: just unfreeze the embedding
# layer and train the model for a few more epochs, being careful not to overfit.
# Submission
# Produce and save submission
predictions <- predict(model, test_matrix)
predictions <- ifelse(predictions >= 0.5, 1, 0)
submission = data.frame(cbind(test$qid, predictions))
names(submission) = c("qid", "prediction")
write_csv(submission, "submission.csv")
```

The line "print(history)" outputs show:
Trained on 1,175,509 samples (batch_size=2,048, epochs=15)
Final epoch (plot to see history):

loss:	0.1137
binary_accuracy:	0.9561
val_loss:	0.1149
val_binary_accuracy:	0.9554

The output plot for the loss function and accuracy is shown in Figure 6.13.

6.3.5 Applications of LSTMs in Molecular Design

LSTMs can be used in many ways for natural language processing (NLP), including:

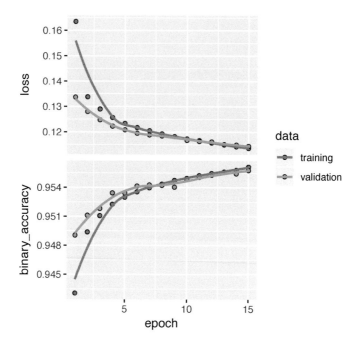

FIGURE 6.13
LSTM Accuracy

1. Text Classification, or sentiment analysis where class labels are used to represent the emotional tone of the text, usually as "positive" or "negative," filtering spam or classifying email text as spam, language identification, or classifying the language of the source text, genre classification, or classifying the genre of a fictional story.

2. Language Modeling, for predicting the probabilistic relationships between words, enabling one to predict the next word.

3. Speech Recognition, for understanding speech, to either generate text readable by humans or issue commands. Examples include transcribing a speech and creating text captions for a movie or TV show.

4. Caption Generation, to describe the contents of a digital image or video. This language model can be strategic as it allows one to create searchable text for search engines.

5. Machine Translation, for translating source text from one particular language into another language.

6. Document Summarization, to create a short description about a

document, such as creating a heading/abstract for a document or summarizing a news article.

7. Question Answering, to take a question posed in a natural language and provide an answer.

In NLP we essentially deal with sequences of words. Similarly, in drug discovery, we deal with gene sequences or molecular structures represented by a sequence of substructures, a protein with various folding structures, and structure-activity relationships (SARs). Therefore, LSTMs can be used for compound screening and molecular design. Let's outline the basic ideas on how to accomplish that.

For a sequence S of symbols $s_t \in S$ at steps $t \in T$, the key in the language model is to obtain the probability

$$P_\theta(S) = P_\theta(s_1) \prod_{t=2}^{T} P_\theta(s_t|s_{t-1}, ..., s_1), \tag{6.1}$$

where the parameter vector $\boldsymbol{\theta}$ is learned from the training dataset, for example, using maximum likelihood estimation. The conditional probability $P_\theta(s_t|s_{t-1}, ..., s_1)$ of the next symbol given the already seen sequence is usually a multinomial model and can be estimated using a binary output vector $\mathbf{y}_t = (y_{t1}, ..., y_{tK})$ of the RNN at time step t via

$$P_\theta(k|s_{t-1}, ..., s_1) = \frac{\exp(y_{tk})}{\sum_{m=1}^{K} \exp(y_{tm})}. \tag{6.2}$$

For training the RNN model, we can use ChEMBL22 (www.ebi.ac.uk/chembl) with 677,044 SMILES strings with annotated nanomolar activities. After cleaning, Gupta et al. (2018) trained their RNN on 541,555 SMILES strings, with lengths from 34 to 74 SMILES characters (tokens). RNN models can be used to generate sequences one token at a time, as these models can output a probability distribution over all possible tokens at each time step. Typically, the RNN aims to predict the next token of a given input. Sampling from this distribution would then allow generating novel molecular structures. Specifically, after sampling a SMILES symol s_{t+1} for the next time step, $t+1$, we can construct a new input vector $\mathbf{x}_{t+1}$, which is fed into the model, and via y_{t+1} and Eq. (6.2) yeilds $P_\theta(k|s_{t-1}, ..., s_1)$. Using a similar process we can further generate s_{t+2}, s_{t+3},... for new molecules (Figure 6.14).

6.4 Deep Belief Networks

There are two major challenges in current high-throughput screening drug design: (1) the large number of descriptors, which may also have autocorrelations, and (2) proper parameter initialization in model prediction to avoid

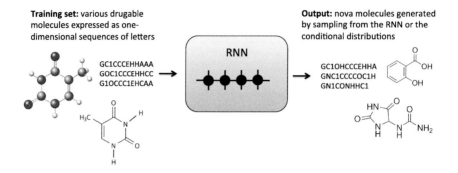

RNN trained on data (left) and Nova Molecule
Generated by sampling (Right)

FIGURE 6.14
Sampling Nova Molecules from Trained RNN

an over-fitting problem. Deep architecture structures have been recommended
to predict a compound's biological activity. Performance of deep neural networks is not always acceptable in quantitative structure–activity relationship
(QSAR) studies (Ghasemi et al., 2018a).

A deep belief neural network (DBN) consists of a sequence of restricted
Boltzmann machines. The output of a hidden layer is used as input of the
next layer. Each DBN layer is trained independently during the unsupervised
portion and thus can be trained concurrently. After the unsupervised portion
is complete, the output from the layers is refined with supervised logistic regression. The top logistic regression layer predicts probabilistically the class
to which the input belongs. The purpose of the unsupervised training is to
select better features. Supervised learning is used for the purpose of classification. Therefore, a DBN combines unsupervised and supervised learning for
the purpose of efficient learning.

6.4.1   Restricted Boltzmann machine

A *restricted Boltzmann machine* (RBM) is an algorithm useful for dimensionality reduction, classification, regression, collaborative filtering, feature

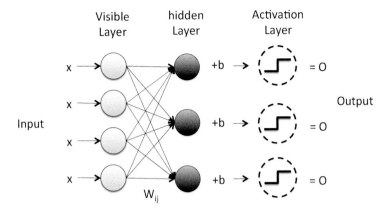

FIGURE 6.15
A Restricted Boltzmann Machine

learning, and topic modeling (Figure 6.15). RBMs are shallow, two-layer neural nets that constitute the building blocks of deep-belief networks. The first layer of the RBM is called the visible, or input layer, and the second is the hidden layer. Unlike Boltzmann machines, in restricted Boltzmann machines there are only connections (dependencies) between hidden and visible units, and none between units of the same type (no hidden-hidden or visible-visible connections).

The standard RBM consists of m visible and n hidden units with weight w_{ij} between visible unit i and hidden unit j; binary (Boolean) value v_i associates with visible unit i, and binary value h_j associates with unit j, as well as bias weights (offsets) a_i for the visible units and b_j for the hidden units. The energy of a configuration determined by the paired vectors $(\mathbf{v}, \mathbf{h})$ is defined as

$$E\left(\mathbf{v}, \mathbf{h}\right) = -\left(\sum_i^m a_i v_i - \sum_j^n b_j h_j - \sum_i^m \sum_j^n v_i w_{ij} h_j\right). \qquad (6.3)$$

Inspired by ideas from statistical mechanics, probability distributions over hidden and visible vectors are defined in terms of the energy function:

$$P\left(\mathbf{v}, \mathbf{h}\right) = \frac{1}{Z} e^{-E(\mathbf{v}, \mathbf{h})}, \qquad (6.4)$$

where the partition function or normalization factor Z is defined as the sum of $e^{-E(v,h)}$ over all possible configurations; that is, Z is a normalizing constant to ensure the probability distribution adds to 1.

$$P\left(\mathbf{v}, h_1, ..., h_m\right) = \left(\prod_{i=1}^{m-2} P\left(h_i | h_{i+1}\right)\right) P\left(h_{m-1} | h_m\right). \qquad (6.5)$$

Similarly, the (marginal) probability of a visible (input) vector of Booleans is the sum over all possible hidden layer configurations:

$$P\left(\mathbf{v}\right) = \frac{1}{Z} \sum_{h} e^{-E(\mathbf{v},\mathbf{h})}, \tag{6.6}$$

Since the RBM structure has the shape of a bipartite graph, with no intra-layer connections, the hidden unit activations are mutually independent given the visible unit activations, and conversely, the visible unit activations are mutually independent given the hidden unit activations. That is, for m visible units and n hidden units, the conditional probability of a configuration of the visible units $\mathbf{v}$, given a configuration of the hidden units $\mathbf{h}$, is

$$P\left(\mathbf{v}|\mathbf{h}\right) = \prod_{i=1}^{m} P\left(v_i|\mathbf{h}\right). \tag{6.7}$$

Conversely, the conditional probability of $\mathbf{h}$ given $\mathbf{v}$ is

$$P\left(\mathbf{h}|\mathbf{v}\right) = \prod_{j=1}^{m} P\left(h_j|\mathbf{v}\right). \tag{6.8}$$

The individual activation probabilities are given by

$$P\left(h_j = 1|\mathbf{v}\right) = \sigma\left(b_j + \sum_{i=1}^{m} w_{ij}v_i\right) \tag{6.9}$$

and

$$P\left(v_i = 1|\mathbf{h}\right) = \sigma\left(a_i + \sum_{i=1}^{m} w_{ij}h_j\right), \tag{6.10}$$

where σ denotes the logistic function.

The unsupervised learning is to train each RBM, that is, to find weigts W that maximize the likelihood via training set $\mathbf{V}$ (a matrix, each row of which is treated as a visible vector $\mathbf{v}$),

$$\max \prod_{\mathbf{v}\in\mathbf{V}} P\left(\mathbf{v}\right) \text{ or } \max_{\mathbf{W}} E\left[\ln P\left(\mathbf{v}\right)\right]. \tag{6.11}$$

After the feature vector is determined through the unsupervised train-ing, the supervised learning of DBN is through the same backpropagation approach.

6.4.2   Application of Deep Belief Networks

Jaekwon et al. (2017) compared DBNs with other methods in cardiovascu-lar risk prediction. They proposed a cardiovascular disease prediction model using the sixth Korea National Health and Nutrition Examination Survey

(KNHANES-VI) 2013 dataset to analyze cardiovascular-related health data. First, a statistical analysis was performed to find variables related to cardiovascular disease using health data related to cardiovascular disease. Then, a model of cardiovascular risk prediction by learning based on the deep belief network (DBN) was developed. They show that statistical DBN-based prediction model has an accuracy and a ROC curve of 83.9% and 0.790, respectively. There are eight variables/features in the DBN and only six variables (age, SBP, DBP, HDL, diabetes, smoking) with statistical significance. Total cholesterol and gender are excluded. The nonparametric Mann-Whitney U-test and chi-square test for univariate analysis are used.

Ghasemi et al. (2018a) used a deep belief network to evaluate the DBN's performance using Kaggle datasets with 15 targets containing more than 70 k molecules. The results revealed that an optimization in parameter initialization will improve the ability of deep neural networks to provide high quality model predictions. The mean and variance of squared correlation for the proposed model and deep neural network (multiple-layer perceptron) are $0.618 \pm 0.407 \times 10^{-4}$ and $0.485 \pm 4.82 \times 10^{-4}$, respectively. The DBN outperformed a DNN with MLP.

6.5 Generative Adversarial Networks

Generative adversarial networks (GANs) are deep neural net architectures comprised of two nets, pitting one adversarially against the other. *Generative algorithms* work differently from *discriminative algorithms*. Discriminative algorithms are used to classify input data; that is, given the features of a data instance, they predict a label or category to which that data belongs. For example, given all the words in an email, a discriminative algorithm could predict whether the message is "spam" or "not spam." Spam is one of the labels, y, and the bag of words gathered from the email are the features that constitute the input data x. The probability that an email is spam given the words it contains can be expressed as $p(y|x)$. In contrast, generative algorithms predict features given a certain label. In the spam email example, the question a generative algorithm tries to answer is: Assuming this email is spam, how likely are these features? That is, the probability of x given y, $p(x|y)$ or the probability of features given a class. That said, generative algorithms can also be used as classifiers. It just so happens that they can do more than classification. Discriminative models learn the boundary between classes, while generative models model the distribution of individual classes.

GAN can be viewed as the combination of a counterfeiter and a cop, where the counterfeiter is learning to pass false notes, and the cop is learning to detect them. Both are dynamic in the zero-sum game, each side comes to learn the other's methods in a constant escalation. As the discriminator changes its

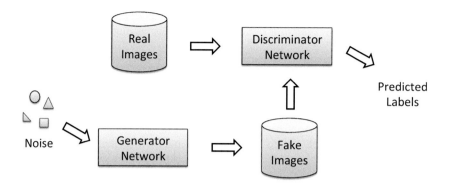

FIGURE 6.16
Generative Adversarial Networks as a Binary Classifier

behavior, so does the generator, and vice versa. Their losses push against each other.

The discriminator network can be a standard convolutional network that can categorize the images fed to it, a binomial classifier labeling images as real or fake, while the generator is an inverse convolutional network in a sense.

In the big picture, here is how a GAN works (Figure 6.16):

1. The generator takes in random numbers and returns an image.

2. This generated image is fed into the discriminator along with a stream of images taken from the actual dataset.

3. The discriminator takes in both real and fake images and returns probabilities of authenticity.

Imaging markers can be used for monitoring disease progression with or without treatment. Models are typically based on large amounts of data with annotated examples of known markers aiming at automating detection. Christian Doppler et al. (2017) developed a deep convolutional generative adversarial network that can learn a manifold of normal anatomical variability, accompanied by a novel anomaly scoring scheme based on the mapping from the image space to a latent space. Applied to new data such as images containing retinal fluid or hyperreflective foci, the model labels anomalies and scores image patches indicating their fit into the learned distribution.

Deep generative adversarial networks are the emerging technology in drug discovery and biomarker development. Kadurin et al. (2017) demonstrated a proof-of-concept in implementing a deep GAN to identify new molecular fingerprints with predefined anticancer properties. They also developed a new GAN model for molecular feature extraction problems, and showed that the model significantly enhances the capacity and efficiency of development of the new molecules with specific anticancer properties using the deep generative models.

Yahi et al. (2017) propose a framework for exploring the value of GANs in the context of continuous laboratory time series data. The authors devise an unsupervised evaluation method that measures the predictive power of synthetic laboratory test time series and show that when it comes to predicting the impact of drug exposure on laboratory test data, incorporating representation learning of the training cohorts prior to training the GAN models is beneficial.

Putin et al. (2018) proposed a *Reinforced Adversarial Neural Computer* (RANC) for the de novo design of novel small-molecule organic structures based on the GAN paradigm and reinforcement learning. The study shows RANCs can be reasonably regarded as a promising starting point to develop novel molecules with activity against different biological targets or pathways. This approach allows scientists to save time and covers a broad chemical space populated with novel and diverse compounds.

Example of a GAN in *R* can be found at https://skymind.ai/wiki/generative-adversarial-network-gan.

6.6 Autoencoders

An autoencoder (autoassociative network) is a type of artificial neural network used to learn efficient data codings in an unsupervised manner. Autoencoders encode input data as vectors. They create a hidden, or compressed, representation of the raw data (Figure 6.17). Such networks are useful in dimensionality reduction; that is, the vector serving as a hidden representation compresses the raw data into a smaller number of salient dimensions. Autoencoders can be paired with a so-called decoder, which allows one to reconstruct input data based on its hidden representation, the encoder can be an autoassociative neural network.

An autoencoder learns to compress data from the input layer into a short code, and then decompresses that code into something that closely matches the original data. A simple autoassociative network can be a multiple-layer perceptron, where the output is identical to the input and the middle hidden layer is smaller. This means we can use the compressed middle layer to generate the original image.

Kadurin et al. (2017) presented the first application of generative adversarial autoencoders (AAE) for generating novel molecular fingerprints with a defined set of parameters. In their model of a seven-layer AAE architecture with the latent middle layer serving as a discriminator; the input and output use a vector of binary fingerprints and concentration of the molecule. They

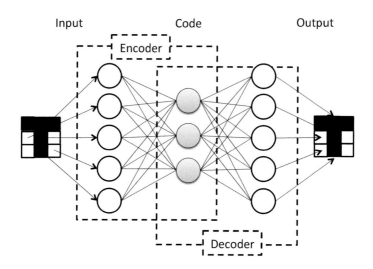

FIGURE 6.17
Autoassociative Network for Data Compression

introduce a neuron responsible for growth inhibition percentage to model the
reduction in the number of tumor cells after the treatment. To train the AAE,
the NCI-60 cell line assay data for 6252 compounds, profiled on the MCF-7
cell line, are used. The output of the AAE was used to screen 72 million compounds in PubChem and select candidate molecules with potential anti-cancer
properties.

An autoencoder example in R can be found at https://statslab.eighty20.co.
za/posts/autoencoders_keras_r/.

6.7   Summary

Deep learning (DL) is a class of machine-learning algorithms that use multiple layers to progressively extract higher-level features from raw input. The
four major deep learning neural network architectures are (1) the Feedforward Neural Network (FNN), (2) the Convolution Neural Network (CNN),
the Recurrent Neural Network (RNN), and the Deep Belief Network (DBN)
for Disease Diagnosis and Prognosis.

FNNs are discussed in a previous chapter. The main idea in CNNs is at
the convolution layers, where different filters are used. Each filter identifies
or filters out particular features or image elements and uses overlapped local

features to formulate the "overall picture." A CNN can be used not just for images, but also to categorize other types of data as long the data can be transformed to look like image data in matrix form. However, a CNN is not the most efficient tool to deal with sequential data. The rule of thumb is: if the data is just as useful after swapping any of your columns with each other, then you can't use a CNN. Location invariance and local compositionality are two key ideas behind CNNs. They make sense for computer vision applications but do not make much sense in the case of natural language processing or time-series events.

A recurrent neural network (RNN) is a class of artificial neural network for modeling temporal dynamic behavior. Unlike feedforward neural networks, RNNs can use their internal state as memory to process sequences of inputs. The common RNN suffers from a short-term memory problem. If the sequence of events, such as an English text, is long enough, the common RNN will have a hard time carrying information from earlier time steps to later ones. Long Short-Term Memory Networks (LSTMs) are explicitly designed to deal with long-term dependency problems. LSTMs are used for LNP, stock market prediction, voice recognition, motion picture captioning, and poem and music generation.

There are two major challenges in quantitative structure–activity relationship (QSAR) studies and drug design: (1) the large number of descriptors, which may also have autocorrelations, and (2) proper parameter initialization in model prediction to avoid an over-fitting problem. A deep belief neural network (DBN) combines unsupervised and supervised learning for the purpose of efficient learning. Each DBN layer is trained independently during the unsupervised portion and thus can be trained concurrently. The unsupervised learning is used for dimension reduction or selecting subsets of descriptors. After the unsupervised portion is complete, the output from the layers is refined with supervised logistic regression. DBNs combine unsupervised and supervised learning to produce efficient learning.

Generative adversarial networks (GANs) are deep neural net architectures comprised of two nets, pitting one adversarially against the other. Generative algorithms work differently from discriminative algorithms. Discriminative algorithms are used to classify input data; that is, given the features of a data instance, they predict a label or category to which that data belongs. A GAN can be viewed as the combination of a counterfeiter and a cop, where the counterfeiter is learning to pass false notes, and the cop is learning to detect them. As the discriminator changes its behavior, so does the generator, and vice versa. Their losses push against each other.

An *autoencoder* (autoassociative network) is used to learn efficient data codings in an unsupervised manner. Autoencoders encode input data as vectors. They create a compressed representation of the raw data in a middle layer. Autoencoders are paired with a decoder, allowing the reconstruction of input data based on its hidden representation. In other words, an autoencoder learns to compress data from the input layer into a short code, and then decompress that code into something that closely matches the original data. This means we can use the compressed middle layer to generate the original image.

6.8   Problems

6.1: How will the order of the input sequence change a perceptron learning?

6.2: What are the architectural differences among CNNs, RNNs, LSTMs, and DBNs?

6.3: What are the differences between hidden Markov Chain models and RNNs?

6.4: Discuss how LSTMs can be used in establishing *ngrams* and *skiptons* for NLP.

6.5: Try the examples in this chapter and modify the deep learning net architectures to see how the model characteristics will change.

6.6: What is the vanishing gradient problem in an ANN? Discuss solutions to the problem.

6.7: There are many datasets available at www.kaggle website. Download some and try out the deep learning network R code and modify it as needed.

7

Kernel Methods

7.1 Subject Representation Using Kernels

In classical statistical models for regression and classification, the form of the mapping $y(x, w)$ from input x to output y is governed by a vector w of adaptive parameters. During the learning phase, a set of training data is used either to obtain a point estimate of the parameter vector or to determine a posterior distribution over this vector. The training data is then discarded, and predictions for new inputs are based purely on the learned parameter vector w. This approach is also used in nonlinear parametric models such as neural networks, but SBML and kernel methods are memory-based approaches that involve storing an entire training set (dimension reduction is possible with modifications) in order to make future predictions. These methods are generally fast to "train" but slow at making predictions for test data points.

Kernels are the basic ingredients shared by all kernel methods. A kernel is used as an alternative representation of an object. An object is usually identified or represented by a set of selected attributes, e.g., age, size, weight, color, or race. A kernel's representation of an object emerges through the object's relationships to other objects (or data points). For instance, if Alice, Bob, and Cindy are $x_1 = 2$, $x_2 = 3$, and $x_3 = 5$ years old, respectively, and we define the kernel as

$$k(x_i, x_j) = x_i x_j, \tag{7.1}$$

then the kernel representations for the group of three people can be a matrix whose elements are the product of paired ages:

	Alice	Bob	Cindy
Alice	4	6	10
Bob	6	9	15
Cindy	10	15	25

At first look, it seems very odd that there are nine elements used to represent three persons, as due to symmetry there are only six unique elements. But the kernel dual representations, though redundant, do provide a huge algorithmic advantage, called a kernel trick.

Kernel trick (Scholkopf, etc., 2005): Any algorithm for vectorial data that can be expressed only in terms of dot products between vectors can be performed implicitly in the feature space associated with any kernel, by replacing each dot product by a kernel evaluation.

The kernel trick has huge practical implications. It is a very convenient way of transforming linear methods, such as linear discriminant analysis (Hastie et al., 2001) or principal component analysis (PCA; Jolliffe, 1986), into nonlinear methods by simply replacing the classic dot product with a more general kernel, such as the Gaussian RBF kernel. Nonlinearity via the new kernel is then obtained at no extra computational cost, as the algorithm remains exactly the same. Second, the combination of the kernel trick with kernels defined on nonvectorial data permits the application of many classic algorithms on vectors to virtually any type of data, as long as a positive-definite kernel (see below) can be defined (Scholkopf, Tsuda, Vert, 2005). Kernel methods have the following advantages:

1. The representation as a square matrix does not depend on the nature of the objects (images, persons, DNA sequences, molecules, protein sequences, languages) to be analyzed. Therefore, an algorithm developed for molecules can be used for image or language processing. This suggests a full modularity of analysis algorithms to cover various problems, while algorithm design and data processing can proceed independently.

2. The size of the kernel matrix used to represent a dataset of n objects is always $n \times n$, whatever the nature or the complexity or the number of attributes of the objects.

3. There are many cases where comparing objects is an easier task than finding an explicit representation for each object that a given algorithm can process. As an example, there is no obvious way to represent protein sequences as vectors in a biologically relevant way, even though meaningful pairwise sequence comparison methods exist.

In order to use the kernel trick, a kernel has to meet the following positive-definite criterion. A real function $k\left(\mathbf{x}_i, \mathbf{x}_j\right)$ of two objects $\mathbf{x}_i$ and $\mathbf{x}_j$ is called a positive-definite kernel if and only if it is symmetric, that is $k\left(\mathbf{x}_i, \mathbf{x}_j\right) = k\left(\mathbf{x}_j, \mathbf{x}_i\right)$, and positive definite:

$$\sum_{i=1}^{n} \sum_{j=1}^{n} c_i c_j k\left(\mathbf{x}_i, \mathbf{x}_j\right) \geq 0, \qquad (7.2)$$

for any integer $n > 0$, any choice of n objects, $\mathbf{x}_1, ... \mathbf{x}_n$, and any real $c_1, ..., c_n$.

Kernel methods can only be applied to vector data. This requires the objects are first represented by a vector $\phi\left(\mathbf{x}\right) \in \mathbb{R}^p$, and then a kernel is defined by the inner product,

$$k\left(\mathbf{x}, \mathbf{x}'\right) = \phi\left(\mathbf{x}\right)^T \phi\left(\mathbf{x}'\right) \qquad (7.3)$$

Fortunately such a function ϕ (mapping to a Hilbert space) always exists (Theorem 2.2, Scholkopf, Tsuda, Vert, 2005). The Hilbert space that consists of all possible vectors for a given $\phi(\mathbf{x})$ is called a feature space.

It is noteworthy to mention that the positive-definite properties of a kernel is a requirement mainly for mathematical convenience. More importantly, the kernel should be meaningful scientifically, e.g., it should measure the similarity between two points or subjects.

7.2 Prediction as Weighted Kernels

Once the kernel $k(\mathbf{x}, \mathbf{x}')$ is determined, the predicted outcome for the ith subject can be expressed as a linear combination of kernels

$$\hat{y}_i = \sum_{j=1}^{n} w_{ij} k(\mathbf{x}_i, \mathbf{x}_j), \tag{7.4}$$

where the weights w_{ij} are nonnegative constants. The loss function can be defined as the L_2 error:

$$E = \sum_{i=1}^{n} (y_i - \hat{y}_i)^2 = \sum_{i=1}^{n} \left(y_i - \sum_{j=1}^{n} w_{ij} k(\mathbf{x}_i, \mathbf{x}_j) \right)^2. \tag{7.5}$$

Taking the partial derivatives of E with respect to w_{ik},

$$\frac{\partial E}{\partial w_{ik}} = 2 \sum_{i=1}^{n} \left(y_i - \sum_{j=1}^{n} w_{ij} k(\mathbf{x}_i, \mathbf{x}_j) \right) k(\mathbf{x}_i, \mathbf{x}_k), k = 1, ..., n, \tag{7.6}$$

where the kronecker-delta function is define as $\delta_{jk} = 1$ if $j = k$ and 0, otherwise.

Setting the derivatives (7.6) equal to zero, we can find the solutions for $w_k, k = 1, ..., n$. However, because the number of weight parameters is the same as the number of data points, the model based on such solutions w_k will unavoidably lead to overfitting, since a smaller training error often leads to a larger test error. To deal with the issue, different methods such as regularization are proposed.

As a side note, there are similarities between *SBML* and *kernel methods*. However, a kernel method has n weights to determine, while SBML has only K (usually less than n) scaling factors to determine.

Example 5.1: Take the three children, Alice, Bob, and Cindy, as an example again, and suppose that in addition to their ages we also know their heights: $y_1 = 34$, $y_2 = 37$, and $y_3 = 43$ inches. Use the kernel $k(x, x_j) = xx_j$, kernel model (7.4), and the data from the three children as the training set. We can learn the child's height at age x as:

$$\hat{y}(x) = 2xw_1 + 3xw_2 + 5xw_3.$$

To train the model, i.e., determine the weight w_i, we use the training set $(x, y) = (\text{age, height}) = \{(2, 34), (3, 37), (5, 43)\}$ and minimize either the error or a defined loss function as we have discussed with other methods.

The weights should die off smoothly with distance from the target point. An example would be the so-called Nadaraya–Watson kernel-weighted for curve smooth averaging (confusingly but understandably, we also often call $k(\mathbf{x}_i, \mathbf{x}_j)$ a weight function):

$$\hat{y}_i = \frac{\sum_{i=1}^{n} k(\mathbf{x}_i, \mathbf{x}_j) y_j}{\sum_{m=1}^{n} k(\mathbf{x}_m, \mathbf{x}_m)} \tag{7.7}$$

The kernel function can also be the Epanechnikov quadratic kernel,

$$k_\lambda(\mathbf{x}_i, \mathbf{x}_j) = \frac{3}{4}\left(1 - \frac{||\mathbf{x}_i - \mathbf{x}_j||^2}{\lambda^2}\right) I\left(||\mathbf{x}_i - \mathbf{x}_j|| \leq 1\right). \tag{7.8}$$

The smoothing parameter λ, which determines the width of the local neighborhood, has to be determined. A large λ usually implies lower variance but higher bias. The parameter λ can be obtained through cross-validation.

Kernel methods are similar to SBML but do not have the attribute-scaling factors. Therefore, when mixed types of attributes are involved, the kernel as a similarity function can be completely inappropriate.

7.3 Support Vector Machine

7.3.1 Hard-Margin Model

Linear discriminant analysis (LDA) is based on the construction of the hyperplane that minimizes the misclassification error. Similarly, a support vector machine (SVM), developed in the mid-1960s, is a generalization of LDA used to construct a hypersurface (multiple hyperplanes) that minimizes the misclassification or regression error. SVM searches for a linearly separable hyperplane, or a decision boundary that separates members of one class from the other. In case that such a hyperplane does not exist, SVM uses a nonlinear mapping to transform the training data into a higher dimension before seeking the linear optimal separating hyperplane. With an appropriate nonlinear mapping to a sufficiently high dimension, data from two classes can always be separated by a hyperplane. SVM has successfully been applied to handwritten digit recognition, text classification, speaker identification, etc., and it is less

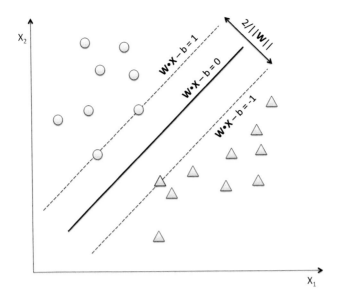

FIGURE 7.1
Support Vector Machine in Action

prone to overfitting. In the medical field, SVM has been used in breast cancer diagnosis (Akay, 2009).

In classification, a training dataset contains features $\mathbf{x}_i$ and labels y_i (either 1 or -1), $i = 1, 2, ..., n$. Each $\mathbf{x}_i$ is a K-dimensional real vector. The *support vector machine* algorithm is to find a *hyperplane* that most effectively divides the points $\mathbf{x}_i$ with label $y_i = 1$ from the points with label $y_i = -1$. Here "effectively" can mean that the distance between the hyperplane and the nearest point x_i from either group is maximized. Such a hyperplane is called a *maximum-margin hyperplane*, though other definitions are possible (e.g., a plane that maximizes the distance between the centers of the two groups).

We know that any hyperplane can be written as the set of points x satisfying $\mathbf{w} \cdot \mathbf{x} - b = 0$, where $\mathbf{w}$ is usually a vector perpendicular to the hyperplane. The quantity $(b/(||\mathbf{w}||))$ determines the offset of the hyperplane from the origin along the normal vector w.

If the training data is standardized with standard deviation of 1 and linearly separable, we can select two parallel hyperplanes (Figure 7.1)

$$\mathbf{w} \cdot \mathbf{x} - b = 1 \text{ and } \mathbf{w} \cdot \mathbf{x} - b = -1$$

to separate the two classes of data, so that the distance (margin) between them is as large as possible.

Since the distance (margin) between these two hyperplanes is $\frac{2}{||\mathbf{w}||}$, maximizing the distance between the planes is equivalent to the following *hard-margin* optimization problem:

$$\text{minimize} \frac{1}{2}||\mathbf{w}||^2$$

$$\text{subject to } y_i\left(\mathbf{w}\cdot\mathbf{x}_i - b\right) \geq 1 \text{ for } i = 1, ..., n. \tag{7.9}$$

The condition $y_i\left(\mathbf{w}\cdot\mathbf{x}_i - b\right) \geq 1$ for $i = 1, 2, ..., n$ means correct classifications for all datapoints and that they are separated by the margin (Figure 7.1).

The solution $\mathbf{w}$ and b for the optimization problem (7.9) will construct the classifier:

$$\hat{y}\left(\mathbf{x}\right) = sign(\mathbf{w}\cdot\mathbf{x} - b). \tag{7.10}$$

Optimization (7.9) can be rewritten as an unconstrained optimization problem using *Lagrange multipliers* $\boldsymbol{\alpha} = \left(\alpha_1, ... \alpha_n\right)^T$:

$$Q\left(\mathbf{w}, \boldsymbol{\alpha}, b\right) = \frac{1}{2}\mathbf{w}\cdot\mathbf{w} - \sum_{i=1}^{n}\alpha_i\left\{\mathbf{y}_i\left(\mathbf{w}\cdot\mathbf{x}_i - b\right) - 1\right\}. \tag{7.11}$$

Setting the derivatives $\frac{\partial Q}{\partial \mathbf{w}} = \mathbf{0}$ and $\frac{\partial Q}{\partial b} = 0$, we obtain, respectively,

$$\mathbf{w} = \sum_{i=1}^{n}\alpha_i y_i \mathbf{x}_i \tag{7.12}$$

and

$$\sum_{i=1}^{n}\alpha_i y_i = 0. \tag{7.13}$$

Substituting (7.12) and (7.13) into (7.11), we obtain the following optimization problem:

$$\text{maximize } Q\left(\boldsymbol{\alpha}\right) = \sum_{i=1}^{n}\alpha_i - \frac{1}{2}\sum_{i=1}^{n}\alpha_i\alpha_j y_i y_j \mathbf{x}_i \cdot \mathbf{x}_j$$

$$\text{subject to } \sum_{i=1}^{n}\alpha_i y_i = 0, \alpha_i \geq 0 \text{ for } i = 1, ..., n. \tag{7.14}$$

The classifier now can be rewritten as

$$\hat{y}\left(\mathbf{x}\right) = sign(\sum_{i=1}^{n}\alpha_i y_i \mathbf{x}_i \cdot \mathbf{x} - b). \tag{7.15}$$

If the datapoints are linearly separable, the global optimal solution $\boldsymbol{\alpha}$ exists. To solve the optimization problem (7.14), *quadratic programming* can be used.

7.3.2 Soft-Margin Model

The question is that in the optimization (7.14), the max-margin hyperplane and classifier are fully determined by a few datapoints $\mathbf{x}_s$ that lie nearest to the hyperplane. These $\mathbf{x}_s$ are called support vectors. To take all the datapoints into consideration in a classifier, we can emmploy the soft-margin method, which is used when the data are not linearly separable. Let's introduce the hinge loss function, $\max(0, 1 - \mathbf{y}_i(\mathbf{w} \cdot \mathbf{x}_i - b))$. This function is zero if the ith datapoint lies on the correct side of the margin. For data on the wrong side of the margin, the function's value is proportional to the distance from the margin. We also introduce a penalty term $\lambda||\mathbf{w}||^2$ in the loss function for minimization:

$$\frac{1}{n}\sum_{i=1}^{n}\max(0, 1 - \mathbf{y}_i(\mathbf{w} \cdot \mathbf{x}_i - b)) + \lambda||\mathbf{w}||^2, \qquad (7.16)$$

where the parameter λ determines the trade-off between increasing the margin size and ensuring that the $\mathbf{x}_i$ lie on the correct side of the margin. Following the similar derivations for the hard-margin model, we can obtain the same optimization problem as (7.14), but have a slightly modified constraint on α_i, that is, $0 \le \alpha_i \le \frac{1}{2\lambda n}$.

In general SVMs, we can replace $\mathbf{x}_i \cdot \mathbf{x}_j$ with a general kernel (dot product) $k(\mathbf{x}_i, \mathbf{x}_j)$. The classifier is

$$\hat{y} = sign\left(\sum_{i=1}^{n} y_i\alpha_i k(\mathbf{x}_i, \mathbf{x}_j) + b\right). \qquad (7.17)$$

Classifiers (7.16) and (7.17) are the kernel representations of support vector machine models.

7.3.3 R Program for Support Vector Machine

There are a number of R packages with implementations for the support vector machine and other *kernel methods*, including *e1071*, *kernlab*, *klaR*, and *svmPath*. The most comprehensive of these is the kernlab package. See R-documents, Platt (2000) and Osuna et al. (1997) for more information. We use the *e1071* library in R to demonstrate *SVM* for classification. The *svm()* function can be used to fit a *svm()* support vector classifier when the argument kernel="linear" is used. A cost argument is used for specifying the penalty for a soft margin. When the cost argument is small, then the margins will be wide. When the cost argument is large, then the margins will be narrow and the result will be close to that from *SVM* with hard margins. The *tune()* function in e1071 can be used to perform *crosstune()* validation. By default, tune() performs ten-fold cross-validation on a set of models of interest. The following is the example *R*-code for the Alzheimer Disease dataset.

Example 7.1: Cognitive Impairment Prediction

Alzheimer's disease (AD) is a cognitive impairment disorder characterized by memory loss and a decrease in functional abilities above and beyond what is typical at a given age. It is the most common cause of dementia in the elderly. Biologically, Alzheimer's disease is associated with amyloid-β ($A\beta$) brain plaques as well as brain tangles associated with a form of the tau protein. Diagnosis of AD focuses on clinical indicators that, once manifested, indicate that the progression of the disease is severe and difficult to reverse. Early diagnosis of Alzheimer's disease could lead to a significant improvement in patient care. As such, there is an interest in identifying biomarkers (Kuhn and Johnson, 2013).

Clinicopathological studies suggest that Alzheimer's disease (AD) pathology begins 10 to 15 years before the resulting cognitive impairment elicits medical attention. Biomarkers that can detect AD pathology in its early stages and predict dementia onset would, therefore, be invaluable for patient care and efficient clinical trial design (Craig-Schapiro et al. 2011). There are known non-imaging biomarkers: protein levels of particular forms of the $A\beta$ and tau proteins and the Apolipoprotein E genotype. For the latter, there are three main variants: $E2$, $E3$, and $E4$. $E4$ is the allele most associated with AD (Kim et al. 2009; Bu 2009).

Craig-Schapiro et al. (2011) describe a clinical study of 333 patients, including some with mild (but well-characterized) cognitive impairment as well as healthy individuals. CSF samples were taken from all subjects. The goal of the study was to determine if subjects in the early states of impairment could be differentiated from cognitively healthy individuals (Kuhn and Johnson, 2013). Data collected on each subject included:

- Demographic characteristics such as age and gender;

- Apolipoprotein E genotype;

- Protein measurements of $A\beta$, tau, and a phosphorylated version of Tau (called pTau);

- Protein measurements of 124 exploratory biomarkers; and

- Clinical dementia scores.

Data from the 333 patients are included in the R package *AppliedPredictiveModel*, where the cognitive outcomes are converted to scores forming two classes: impaired and healthy. The interesting question is to predict the outcome using the demographic and assay data to predict which patients have early stages of disease. We use 130 biomarkers (predictors) to predict the cognitive outcomes via the support vector machine. We will use function *svm* in the *ISLR* package after the data preparation.

```
# Support Vecor Machine
library (ISLR)
library(e1071) # Lib for support vector machine
library(AppliedPredictiveModeling)
# Prepare training and test datasets
data("AlzheimerDisease", package = "AppliedPredictiveModeling")
summary(diagnosis) # labels: either "Impaired" or "Control".
summary(predictors) # predictors
ALZ <- cbind(predictors, diagnosis)
n = nrow(ALZ)
trainIndex = sample(1:n, size = round(0.7*n), replace=FALSE)
ALZtrain=ALZ[trainIndex, ]
ALZtest=ALZ[-trainIndex, ]
# Train SVM
ALZ.svm <- svm(diagnosis~., data=ALZtrain, kernel="linear", cost=10)
summary(ALZ.svm)
pred.test=predict (ALZ.svm , newdata =ALZtest ) # prediction for new data
table(pred.test, ALZtest$diagnosis)
# Study the effect of cost
set.seed (1)
ALZtune=tune.svm(diagnosis~., data=ALZtrain, kernel="linear",
cost=c(0.001, 0.1, 1, 10, 100) )
summary (ALZtune)
```

The output confusion matrix is

pred.te	Impaired	Control
Impaired	14	11
Control	9	66

From the output confusion table, we can further calculate other quantities of interest: sensitivity (true positive rate) $=14/(14+9) = 0.61$, specificity (true negative rate) $= 66/(11+66) = 0.86$, positive predictive value $= 14/(14+11) = 0.56$, and negative predictive value $= 66/(9 + 66) = 0.88$.

7.4   Feature and Kernel Selections

In kernel methods, a key is to select an appropriate set of features. Different types of objects such as sequencing, trees (e.g., chemical compounds), networks (graphs), images, text/strings, and sound data require examining different sets of features to characterize them. Since a kernel is the dot product of feature vectors, the selection of appropriate features is essential for constructing a good kernel, and thus requires field knowledge. We will first discuss some basic methods that do not involve great deal of field knowledge.

Point data can be continuous (e.g., age and weight), binary (e.g., gender), ordinal (e.g., pain scale), or nominal (e.g., race). For more complicated data types such as natural language, DNA sequences and sound sequences, vector, or time sequences can be used to reflect the order of the tokens or elements.

In application of kernel methods, we need a kernel for gene sequences of length $50 \sim 1000$ over a 20-letter alphabet A (the amino acids) (Vert, 2004). For example, for fixed-length sequences of nucleotide sequences with alphabet (A, C, G, T), the codes for A, C, G, and T would, respectively, be $(1, 0, 0, 0)$, $(0, 1, 0, 0)$, $(0, 0, 1, 0)$, and $(0, 0, 0, 1)$ and the code for the sequence of ATG would be $(1, 0, 0, 0, 0, 0, 0, 1, 0, 0, 1, 0)$. The reason we use vectors of length 4 is that there are 4 possible different letters in nucleotide sequences.

For kernels with variable-length sequences we can replace each letter by one or several numerical features, such as physico-chemical properties of amino-acids, and then extract features from the resulting variable-length numerical time series using classical signal processing techniques such as Fourier transforms (Wang et al., 2004) or autocorrelation analysis (Zhang et al., 2003a). Let $h_1, ..., h_n$ denote n numerical features associated with the successive letters of a sequence of length n. Then the autocorrelation function r_j for a given $j > 0$ is defined by

$$r_j = \frac{1}{n-j} \sum_{i=1}^{n-j} h_i h_{i+j}.$$

The analysis of biological systems can be carried out by investigating interacting molecules through biological networks. For example, the prediction of interacting proteins to reconstruct the interaction network can be posed as a binary classification problem: given a pair of proteins, do they interact or not? Binary kernel discrimination has thus far mostly been used for distinguishing between active and inactive compounds in the context of virtual screening (Wilton et al. 2006, Harper 2001).

A kernel that is a dot product of vectors has many nice properties that are helpful in constructing complicated kernels from simple ones. For example, a linear combination of kernels is a kernel. A polynomial function with positive coefficients of kernels is a kernel. Here is a list of properties that can be repeatedly/recursively used to construct complicated kernels that meet our needs. If $k_1(x, x')$, $k_2(x, x')$ and $k_3(x, x')$ are kernels, the following functions are all kernels: $c_1 k_1(x, x') + c_2 k_2(x, x')$, $g(k_1(x, x'))$, $f(x) k_1(x, x') f(x')$, $\exp(k_1(x, x'))$, $k_1(x, x') k_2(x, x')$, $k_3(\phi(x), \phi(x'))$, $x^T A x\prime$, where $c_1, c_2 \geq 0$ are a constants, $f(\cdot)$ is any function, $g(\cdot)$ is a polynomial with positive coefficients, and $\phi(x)$ is a function from x to $\mathbb{R}^M$, We require that the kernel $k(x, x')$ be symmetric and positive semidefinite and that it expresses the appropriate form of similarity between x and x' according to the intended application. Some popular kernels as measures of similarity and dot products of features are:

Fisher kernel: $k(x, x') = U_x^T I^{-1} T_{x'}$, where $I = $ *Fisher information matrix* and $U_x = \nabla_\theta \ln P(x|\theta)$, with θ being vector of parameters.

Sigmoid kernel: $S(x) = \frac{e^{x \cdot x'}}{1 + e^{x \cdot x'}}$.

Polynomial kernel: $k(x, x')^\rho$, $\rho \geq 0$. When $\rho = 1$, it is called linear kernel.

Gaussian radial kernel (RBF): $k(x, x') = \exp\left(-\frac{\|x - x'\|^2}{2\sigma^2}\right)$, where σ is a parameter.

7.5 Application of Kernel Methods

Kernel methods have been broadly used in bioinformatics. In their edited book *Kernel Methods in Computational Biology*, Scholkopf, Tsuda, and Vert (2004) collect different applications of kernel methods (and support vector machines):

1. Inexact Matching String Kernels for Protein Classification by Leslie, Kuang, and Eskin

2. Fast Kernels for String and Tree Matching by Vishwanathan and Smola

3. Local Alignment Kernels for Biological Sequences by Vert, Saigo, and Akutsu

4. Kernels for Graphs by Kashima, Tsuda, and Inokuchi

5. Diffusion Kernels by Kondor and Vert

6. A Kernel for Protein Secondary Structure Prediction by Guermeur, Lifchitz, and Vert

7. Heterogeneous Data Comparison and Gene Selection with Kernel Canonical Correlation Analysis

8. Kernel-Based Integration of Genomic Data Using Semidefinite Programming

9. Protein Classification via Kernel Matrix Completion by Kin, Kato, and Tsuda

10. Accurate Splice Site Detection for Caenorhabditis elegans by Rätsch and Sonnenburg

11. Gene Expression Analysis: Joint Feature Selection and Classifier Design by Krishnapuram, Carin, and Hartemink

12. Gene Selection for Microarray Data by Hochreiter and Obermayer

7.6   Dual Representations

Many linear models for regression and classification such as support vector machine can be reformulated in terms of a dual representation in which the kernel function arises naturally. Following Bishop's presentation (2005), consider a linear regression model

$$\hat{y}(x) = \sum_{i=1}^{N} w_i \phi_i(x), \tag{7.18}$$

where the parameters w_i are determined by minimizing a regularized sum-of-squares error function given by

$$L(w) = \frac{1}{2} \sum_{n=1}^{N} \left\{ \mathbf{w}^T \phi(x_n) - y_n \right\}^2 + \frac{\lambda}{2} \mathbf{w}^T \mathbf{w}, \tag{7.19}$$

where $\lambda \geq 0$.

If we set the gradient of $L(w)$ with respect to w equal to zero, we obtain

$$\mathbf{w} = -\frac{1}{\lambda} \sum_{n=1}^{N} \left\{ \mathbf{w}^T \phi(\mathbf{x}_n) - y_n \right\} \phi(\mathbf{x}_n) = \sum_{n=1}^{N} a_n \phi(\mathbf{x}_n) = \mathbf{\Phi}^T \mathbf{a}, \tag{7.20}$$

where $\mathbf{\Phi}$ is the design matrix, whose nth row is given by $\phi(x_n)^T$. Here the vector $a = (a_1, ..., a_N)^T$, and

$$a_n = -\frac{1}{\lambda} \left\{ \mathbf{w}^T \phi(\mathbf{x}_n) - y_n \right\}. \tag{7.21}$$

Instead of working with the parameter vector $\mathbf{w}$, we can now reformulate the least-squares algorithm in terms of the parameter vector a by substituting $\mathbf{w} = \mathbf{\Phi}^T$ in (7.19), and we obtain

$$L(\mathbf{a}) = \frac{1}{2} \mathbf{a}^T \mathbf{\Phi} \mathbf{\Phi}^T \mathbf{\Phi} \mathbf{\Phi}^T \mathbf{a} - \mathbf{a}^T \mathbf{\Phi} \mathbf{\Phi}^T \mathbf{y} + \frac{1}{2} \mathbf{y}^T \mathbf{y} + \frac{\lambda}{2} \mathbf{a}^T \mathbf{\Phi} \mathbf{\Phi}^T \mathbf{a}, \tag{7.22}$$

where $\mathbf{y} = (y_1, ..., y_N)^T$. We define the *Gram matrix* $\mathbf{K} = \mathbf{\Phi} \mathbf{\Phi}^T$, which is an $N \times N$ symmetric matrix with elements

$$k_{nm} = k(x_n, x_m) = \phi(x_n)^T \phi(x_m). \tag{7.23}$$

Now we can rewrite (7.22) as

$$L(\mathbf{a}) = \frac{1}{2} \mathbf{a}^T \mathbf{K} \mathbf{K} \mathbf{a} - \mathbf{a}^T \mathbf{K} \mathbf{y} + \frac{1}{2} \mathbf{y}^T \mathbf{y} + \frac{\lambda}{2} \mathbf{a}^T \mathbf{K} \mathbf{a}. \tag{7.24}$$

Setting the gradient of $L(\mathbf{a})$ with respect to $\mathbf{a}$ to zero, we obtain the following solution:

$$\mathbf{a} = (\mathbf{K} + \lambda \mathbf{I}_N)^{-1} \mathbf{y}. \tag{7.25}$$

If we substitute **a** into the linear regression model (7.18), we obtain the following prediction for a new input **x**

$$\hat{y}(\mathbf{x}) = \mathbf{w}^T \phi(x) = \mathbf{a}^T \mathbf{\Phi} \phi(x) = \mathbf{k}(\mathbf{x})^T (\mathbf{K} + \lambda \mathbf{I}_N)^{-1} \mathbf{t}. \tag{7.26}$$

7.7 Summary

In regression modeling, each object is represented by a set of attributes. For an example, a patient may be represented by their age, gender, race, weight, and baseline characteristics. We then model the outcome as a function of this set of attributes. A kernel is used as an alternative representation of an object. A kernel method is the process of modeling the outcome as a function of the kernel. Kernel $k(\mathbf{x}_i, \mathbf{x}_j)$ defines an object via the relationship to each of the objects under consideration. Such a relationship can be a similarity between them, often expressed mathematically as a dot product of two vectors.

A kernel method is an instance-based method, that is, the model includes all the data points in the training dataset and no parameters are involved. In contrast, a regression model is a parametric model, in which training data points are disregarded after the parameters are determined. SBML has both scaling parameters and training data points, but the dimension reduction method can be easily applied as we discussed in Chapter 4. The scaling factors are important – without them it is difficult to objectively measure the relative importance of different attributes.

Kernel methods can be viewed as a special case of SBML when the scaling parameters are preset to the same value for all features. A support vector machine can be expressed as a kernel method; penalty terms can be included in a soft margin model to reduce the variance of the prediction. SVM is one of the popular ML methods that we expect to find useful for many applications in drug discovery and other fields.

7.8 Problems

7.1: Use support vector machine to analyze the infertility in the *infert* dataset in the *R datasets* package.

7.2: Can we derive a dual presentation using L_1 or *elastic net* loss function instead of L_2 loss function?

8

Decision Tree and Ensemble Methods

8.1 Classification Tree

Tree-based methods, sometimes known as tree methods, are among the most popular methods in statistical machine learning. They are intuitive, as well as easy to use and interpret. As an example, a physician may wish to use a decision tree like the one shown in Figure 8.1 (adopted from Jessie Li, Penn State University) for classifying whether a patient is at low or high risk of death in 30 days based on initial 24-hour data of a medical event or exam.

The decision rules might be something like this: the minimum systolic blood pressure within the initial 24 hours is checked, and if it's 90 or lower the patient is classified as high-risk. Otherwise, check patient's age; if they are no more than 60 years old, classify the patient as low risk. If age is more than 60 years old, then further check for sinus tachycardia; if it is present, classify patient as a high-risk patient. Otherwise, he is a low-risk patient. In this example, given that the low and high risks are defined, one of the key questions is how to determine the threshold for each risk factor to minimize the error or the loss function.

In a decision tree, the terminal nodes or leaves of the tree are denoted by $R_m (m = 1, M)$, each of which corresponds to a region of the variable. There are two types of trees based on the outcome: classification and regression trees (CART). In a regression tree, y is treated as a continuous variable, while in classification trees, y is treated as discrete variable.

In a classification tree, each predictor $x_i (i = 1, ..., K)$ is divided into categories in terms of value (e.g., age divided into age groups) sequentially. A binary tree is a simple example, in which each x_i is divided into two categories. At the end of the tree are the leaves, denoted by R_j $(i = 1, 2, ..., m)$. R_j represents a classification of x. The goal of any tree method is to classify a subject based on input variables $x = (x_1, ..., x_N)$. The construction of a classification tree involves the following three tasks (Figure 8.2):

1. Select the splits: one variable a time, $x_i < c_i$ versus $x_i \geq c_i$, where constant c_i is to be optimized based on the Gini index (CART), entropy (ID3, C4.5), or misclassification error).

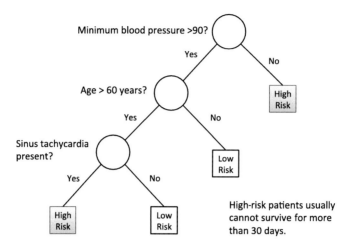

FIGURE 8.1
Decision Tree for Hypertension Patient Classification

2. Decide when to declare a node terminal without further splitting. Stopping criteria tend to be myopic; instead, we can grow to a full tree and then prune it using cross-validation to prevent overfitting.

3. Assign each terminal node to a class for classification trees or associated value for regression trees.

The proportion of class k observations in node m is

$$\hat{p}_{mk} = \frac{1}{n_m} \sum_{x_i \in R_m} I(y_i = k),$$

where $I(y_i = k)$ is an indicator function, which is equal to 1 if $y_i = k$ and 0 otherwise; n_m is the number of observations at node R_m. We can use the majority rule to classify the observations in node m, that is, to class $k(m) = \arg\max_k \hat{p}_{mk}$.

The common impurity measures are:
Misclassification error:

$$ME = \frac{1}{n_m} \sum_{x_i \in R_m} I(y_i \neq k) = 1 - \hat{p}_{mk}. \tag{8.1}$$

Gini index:

$$GI = \sum_{k=1}^{K} \hat{p}_{mk} (1 - \hat{p}_{mk}). \tag{8.2}$$

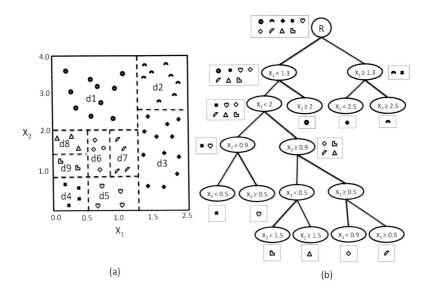

(a)   (b)

FIGURE 8.2
Decision Tree Method

Gross-entropy or deviance:

$$GE = -\sum_{k=1}^{K} \hat{p}_{mk} \ln \hat{p}_{mk}. \tag{8.3}$$

Note that the number of leaves in the binary tree is much less than the number of unique categories (2^p) formulated by the n input variables.

For two classes, if p is the proportion in the second class, $ME = 1 - \max(p, 1-p)$, $GI = 2p(1-p)$ and $GE = -p \ln p - (1-p) \ln(1-p)$. All three measures are similar, but cross-entropy and the Gini index are more sensitive to changes in the node probabilities than the misclassification rate. For this reason, either the Gini index or cross-entropy should be used when growing the tree. To guide cost-complexity pruning, any of the three measures can be used, but typically it is the misclassification rate. Interestingly, whether we classify observations to the majority class in the node or classify them to class k with probability p_{mk}, the expected training error rate of this rule in the node is equal to the Gini index (Hastie, Tibshirani, Friedman, 2016).

The iterative dichotomiser 3 (ID3) Algorithm can be outlined as follows:

1. Calculate the entropy of every attribute x of the dataset S, $H(S) = -\sum_{x \in X} p(x) \log_2 p(x)$, where X is the set of classes in S, and $p(x)$

is the proportion of the number of elements in class x to the number of elements in set S. When $H(S) = 0$, the set S is perfectly classified (i.e., all elements in S are of the same class).

2. Split the set S into subsets using the attribute for which the resulting entropy after splitting is minimized (maximum information gain).

3. Make a decision tree node containing that attribute.

4. Recur on subsets using remaining attributes until the entropy cannot be further minimized or no more information gain is possible.

ID3 does not guarantee an optimal solution. It can converge upon local optima. ID3 algorithm is easy to interpret, handles mixed discrete and continuous inputs, and performs automatic variable selection. ID3 is robust to outliers and insensitive to monotone transformations of the inputs because the split points are based on ranking the data points. ID3 scales well to large data sets and can be modified to handle missing features. However, it does not predict very accurately compared to other kinds of models due to the greedy nature of the tree construction algorithm. A single big tree is not stable because a single error in classification can propagate to the leaves. Different methods have been proposed to remedy this, such as bagging, boosting, and random forests. The boosted trees can be used for regression-type and classification-type problems.

There are several obvious questions with the tree methods, e.g., how many classes should the tree have? Or how do we determine the tree depth? How do we split the data for training and validation? A large training set will usually make the model better. The question is: do you want a better model, without knowing how good it is (if all the data available are used for training only), or a worse model with information on the quality of the model?

Why do we use binary splits? We could classify a variable into more than two categories. While this can sometimes be useful, it is not a good general strategy because using three or more categories split the data too quickly, leaving insufficient data at the next level down. We can rather replace multi-category splits with a series of binary splits.

Threshold Determination and Tree Pruning

There are two competing factors in determining an optimal tree model: the accuracy of the tree method and computational efficacy.

For a given tree depth D, a commonly used approach to obtaining an optimal tree (minimize ME, GI, or GE) is the greedy algorithm: for each parameter, try different thresholds (there are only limited for the given input dataset). We can let the tree grow larger than what we need at the final state, then prune it.

Tree size is a tuning parameter governing the model's complexity and should be determined based on the data. An obvious idea is to split tree nodes only when the decrease in sum-of-squares due to the split exceeds some threshold. This strategy is too short-sighted, however, since a seemingly worthless split might lead to a very good split below it. The preferred strategy is to grow a large tree, stopping the splitting process only when some minimum node size or tree depth is reached. Then this large tree is pruned using cost-complexity pruning. The cost-complexity function can be defined as

$$C_\alpha = ME + \alpha D. \tag{8.4}$$

The idea is to find, for each α, the subtree to minimize C_α. The tuning parameter $\alpha \geq 0$ determines the trade-off between tree size and its goodness of fit to the data. Large values of α will result in smaller trees. Setting the parameter $\alpha = 0$ will result in the full tree.

Example 8.1: Classification Tree

The *R* package rpart provides functions for decision tree methods. Let's take the breast cancer dataset as an example.

```
library(rpart) # for rpart function
library(rattle)
library(mlbench)
library(rpart.plot) # for fancyRpartPlot function
library(RColorBrewer) # for fancyRpartPlot function
## Breast Cancer Example
## data(BreastCancer)
n=nrow(BreastCancer)
trainIndex = sample(1:n, size = round(0.7*n), replace=FALSE)
BCtrain=BreastCancer[trainIndex, ]
BCtest=BreastCancer[-trainIndex, ]
mytree <- rpart(Class ~Cl.thickness + Cell.size + Marg.adhesion,
        data=BCtrain, method = "class")
mytree
fancyRpartPlot(mytree, caption = NULL) # plot mytree
```

The output decision tree for breast cancer is shown in Figure 8.3.

We can plot *mytree* by loading the *rattle* package and using the *fancyRpartPlot()* function. By default, rpart uses *Gini impurity* to select splits when performing classification. You can use information gain instead by specifying it in the parms parameter.

```
mytree <- rpart(Class ~Cl.thickness + Cell.size + Marg.adhesion,
        data=BCtrain, method = "class",
        parms = list(split = 'information'))
```

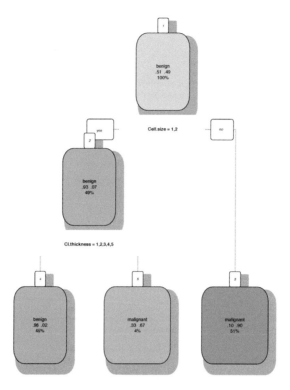

FIGURE 8.3
Breast Cancer Decision Tree

The complexity measure is a combination of the size of a tree and the ability of the tree to separate the classes of the target variable. If the next best split in growing a tree does not reduce the tree's overall complexity by a certain amount specified by cp (e.g., 0.2), rpart will terminate the growing process. If $cp = -1$ the tree will be fully grown.

```
mytree <- rpart(Class ~Cl.thickness + Cell.size + Marg.adhesion,
        data=BCtrain, method = "class", cp = -1)
fancyRpartPlot(mytree, caption = NULL)
```

A fully-grown tree often leads to over-fitted tree, we can prune it back using *cp*:

```
mytree <- prune(mytree, cp = 0.2)
fancyRpartPlot(mytree)
```

When rpart grows a tree it performs 10-fold cross-validation on the data. We can use *printcp()* to see the cross-validation results.

```
printcp(mytree)
```

We can also restrict a tree depth by setting the *maxdepth* argument:

```
mytree <- rpart(Class ~., data=BCtrain, method = "class", maxdepth = 1)
mytree <- prune(mytree, cp = 0.2)
fancyRpartPlot(mytree)
```

To use the trained tree model to predict the class and the probability on the test dataset, we can use the *predict()* function.

```
# Predict the outcome and the possible outcome probabilities
PredClass <- predict(mytree, newdata = BCtest, type = "class")
PredProb <- predict(mytree, newdata = BCtest, type = "prob")
```

Readers can try this code and compare the results.

8.2 Regression Tree

We now turn to regression tree. The data consists of K inputs and a response, $(\mathbf{x}_i, y_i)$ for the ith subject, $i = 1, 2, ..., N$, *with* $\mathbf{x}_i = (x_{i1}, x_{i2}, ..., x_{iK})$. At leaf j, a value r_j is attached. The prediction with the tree method is

$$\hat{y} = \sum_{j=1}^{N_m} r_j I \left(x \in R_j\right), \qquad (8.5)$$

where r_j is a constant; the terminal node (leaf) R_i can be considered a range of values, and $I\left(x \in R_j\right)$ is an indicator function, which is equal to 1 if $x \in R_j$ and 0 otherwise.

Here are two questions that tree methods try to answer: (1) for a given tree structure, how should the value r_i be determined so that the error can be minimized? (2) How should the tree be constructed so that the error can be minimized?

To determine the constant r_j, we can minimize the error:

$$MSE_i = \frac{1}{N_i} \sum_{i=1}^{N_j} (y_{ij} - \hat{y}_i)^2, \qquad (8.6)$$

which leads to

$$r_i = ave\left(y_j | x_j \in R_i\right). \tag{8.7}$$

Here we have assumed there is no missing data. We can see that the classification tree is a special case of a regression tree, where r_i is either 0 or 1.

Now our goal is to find $R_1, ..., R_K$ that minimize the MSE given by

$$MSE = \frac{1}{N} \sum_{j=1}^{K} \sum_{i \in R_j} (y_i - r_i)^2.$$

To reduce overfitting, we can use a penalty on tree complexity (size) $|T|$ and build a tree based on minimization of loss function:

$$LF = \frac{1}{N} \sum_{j=1}^{K} \sum_{i \in R_j} (y_i - r_i)^2 + \alpha |T|.$$

The main steps in building a regression tree are:

1. Use recursive binary splitting to grow a large or full tree on the training data.

2. For a given α, apply cost complexity pruning to the large tree in order to obtain a sequence of best subtrees.

3. Use M-fold cross-validation to choose α.

4. Return the subtree from Step 2 that corresponds to the chosen value of α.

5. Stop when the loss function stops reducing.

Example 8.2: Regression Tree R

We use the Ames Housing data that has been included in the *AmesHousing* package to demonstrate how to use the *rpart* package in *R* for regression tree analysis.

```
library(rpart) # performing regression trees
library(rpart.plot) # plotting regression trees
library(ipred) # bagging
library(caret) # bagging
set.seed(453)
ames <- make_ames()
n=nrow(ames)
trainIndex = sample(1:n, size = round(0.7*n), replace=FALSE)
ames_train=ames[trainIndex, ]
```

```
ames_test=ames[-trainIndex, ]
RTree <- rpart(
 formula = Sale_Price ~.,
 data = ames_train,
 method = "anova",
 control = list(minsplit = 11, maxdepth = 8, cp = 0.01),
 )
pred <- predict(RTree, newdata = ames_test)
RMSE(pred = pred, obs = ames_test$Sale_Price)
rpart.plot(RTree)  # Plot tree
plotcp(RTree)      # Plot Error
```

The results from the code execution include the error 42590.29, the tree plot and error plot.

The parameter *minsplit* specifies the minimum number of data points required for further split. The default is 20. The *maxdepth:* the maximum number of internal nodes between the root node and the terminal nodes. The default is 30, which is quite liberal and allows for fairly large trees to be built. When the control statement is not used the default 10-fold cross-validation is performed to find the tuning parameter α.

8.3   Bagging and Boosting

Bagging

A single big tree is not stable because a single error in classification can propagate to the leaves. The commonly used remedies are bagging, boosting, and random forests.

Bagging is simply forming an average of many different trees. We first illustrate why bagging might be a good solution to the problem of error propagation. Suppose that A, B, C, D, and E are the five members of a trial jury. Guilt or innocence for the defendant is determined by simple majority rule. There is a 5% chance that A gives the wrong verdict; for B, C, and D it is 10%, and E is mistaken with a probability of 20%. When the five jurors vote independently, the probability of bringing the wrong verdict is about 1%. Paradoxically, this probability increases to 1.5% if E (who is most probably mistaken) abandons his own judgment and always votes the same as A (who is most rarely mistaken). Even more surprisingly, if the four jurors B, C, D, and E all follow A's vote, then the probability of delivering the wrong verdict is 5%, five times more than that when they vote independently (Székely, 1986; Chang, 2012, 2014). From this example, we are naturally lead to conclude that a committee decision can be better than individual decisions. Applying this idea to the decision tree method leads us to the tree-averaging method known as bagging (bootstrap aggregating), a variance reduction method that averages predictions from multiple tree models. The tree models are generated by multiple (N_b) sampling from training sets with replacement.

$$\hat{y}_{bag} = \frac{1}{N_b} \sum_{i=1}^{N_b} \hat{y}_i. \tag{8.8}$$

Example 8.3: Breast Cancer Classification with Bagging Tree

```
library(mlbench) # for BreastCancer dataset
library(ipred) # for the bagging function
data(BreastCancer)
# Test set error bagging (nbagg = 50): 3.7% (Breiman, 1998, Table 5)
myBag <- bagging(Class ~Cl.thickness + Cell.size
      + Cell.shape + Marg.adhesion
      + Epith.c.size + Bare.nuclei
      + Bl.cromatin + Normal.nucleoli
      + Mitoses, data=BreastCancer, coob=TRUE)
print(myBag)
```

The output shows the out-of-bag estimate of misclassification error: 0.0454.

Boosting

Similar to bagging is boosting. Weak classifiers $G_i(x)$ with values of either -1 or 1 from multiple sampling, $i = 1, 2, ..., n_t$, are those whose misclassification error rates are only slightly better than random guessing. The predictions from all of them are then combined through a weighted majority vote to produce the final prediction:

$$G(\boldsymbol{x}) = sgn\left(\sum_{i=1}^{n_t} \alpha_i G_i(\boldsymbol{x})\right), \tag{8.9}$$

where $\alpha_1, \alpha_2, ..., \alpha_{n_t}$ are computed by the boosting algorithm in such a way that more accurate classifiers in the sequence will get larger weights.

The data modifications at each boosting step consist of applying weights $w_1, ..., w_{n_t}$ to each of the training observations (x_i, y_i), $i = 1, 2, ..., n_t$, so that the weight shifts more to misclassified observations. In the commonly used *AdaBoost* (adaptive boosting) the weights are determined based on the errors of the previous classifiers made on the training set (Hastie et al., 2001). The algorithm is outlined in the following.

AdaBoost Algorithm

1. Initialize weights $w_i = 1/N, i = 1, 2, ..., N$.

2. From $m = 1$ to M:

    (a) Fit a learner $G_m(x)$ to the training set using w_i.

    (b) Compute the error, $\epsilon_m = \frac{\sum_{i=1}^{N} w_i I(y_i \neq G_m(x_i))}{\sum_{i=1}^{N} w_i}$.

    (c) Compute $\alpha_m = \ln\left(\frac{1-\epsilon_m}{\epsilon_m}\right)$.

    (d) Update weights $w_i \leftarrow w_i \exp\left(\alpha_m I(y_i \neq G_m(x_i))\right), i = 1, 2, ...N$.

3. To classify x, output $G(x) = sign\left[\sum_{m=1}^{M} \alpha_m G_m(x)\right]$.

Here N is the number of observations in the training set, and M is the number of weak learners/classifiers to be used.

Example 8.4: Breast Cancer Classification with Boosting Tree

```
library(adabag)
library(mlbench) # for BreastCancer dataset
library(rpart)
data(BreastCancer)
trainIndex = sample(1:n, size = round(0.7*n), replace=FALSE)
BCtrain=BreastCancer[trainIndex, ]
BCtest=BreastCancer[-trainIndex, ]
myBoost <- boosting(Class ~Cl.thickness + Cell.size
        + Cell.shape + Marg.adhesion
      + Epith.c.size + Bare.nuclei
      + Bl.cromatin + Normal.nucleoli
      + Mitoses, data=BCtrain,mfinal=3, coeflearn="Zhu",
         control=rpart.control(maxdepth=5))
myBoost.pred <- predict.boosting(myBoost,newdata=BCtest)
myBoost.pred$confusion # Confusion Table
myBoost.pred$error
#comparing error evolution in training and test set
evol.train <- errorevol(myBoost ,newdata=BCtrain)
evol.test <- errorevol(myBoost ,newdata=BCtest)
plot.errorevol(evol.test,evol.train)
```

The outputs indicate the frequencies in predicted versus observed classes and an error of 0.04. The confusion matrix is:

	Observed Class	
Predicted Class	benign	malignant
benign	319	10
malignant	9	128

The predicted error is:

```
> myBoost.pred$error
[1] 0.04077253
```

8.4 Random Forests

Random forest (or random forests) is an *ensemble classifier* that consists of many decision trees and outputs the class that is the mode of the class's output by individual trees. The method was induced by Ho and Breiman (Ho, 1998; Breiman, 2001). The method combines Breiman's "bagging" idea and the random selection of features. There are many versions of random forest algorithms. Here is the basic one. Each tree is constructed using the following algorithm:

1. Choose p for the number of variables in the classifier and n_t for the number of training cases.

2. Choose $m \ll p$ input variables for determining the decision at a node of the tree.

3. Choose a training set for this tree by choosing n times with replacement from all available training cases.

4. For each node of the tree, randomly choose m variables, based on which the decision at that node is made. Calculate the best split based on these m variables in the training set.

Example 8.5: Breast Cancer Classification with Random Forest

```
# Random forest
library(mlbench) # for BreastCancer dataset
library(rpart)
library(randomForest)
data(BreastCancer)
trainIndex = sample(1:n, size = round(0.7*n), replace=FALSE)
```

```
BCtrain <- BreastCancer[trainIndex, ]
BCtest <- BreastCancer[-trainIndex, ]
# Create a Random Forest model with default parameters
myFT1 <- randomForest(Class ~Cl.thickness + Cell.size + Marg.adhesion,
 data=BCtrain, importance = TRUE)
myFT1
```

The outputs include the confusion matrix:

	benign	malignant	class.error
benign	120	9	0.06976744
malignant	9	95	0.08653846

We now tune parameters of the random forest model:

```
myFT2 <- randomForest(Class ~Cl.thickness + Cell.size + Marg.adhesion,
 data=BCtrain, ntree = 100, mtry = 3, importance = TRUE)
myFT2
```

The output confusion matrix is:

	benign	malignant	class.error
benign	119	10	0.06976744
malignant	7	97	0.08653846

8.5 Summary

A tree method (TM) is a commonly used supervised learning method. Tree methods include classification trees for binary outcomes and regression trees for continuous outcomes. In a TM, all features are categorized or dichotomized for a binary tree. A tree model can be optimized based on Gini index (CART), entropy (ID3, C4.5), or misclassification error. For larger trees, an early misclassification error can propagate downstream, eventually leading to poor predictions. To overcome this problem, different methods have been proposed, such as tree pruning, bagging, boosting, and random forests.

8.6   Problems

8.1: Analyze Alzheimer's disease (dataset in Example 7.1) using simple
decision tree, bagging tree, boosting, and random forests.

8.2: Use the breast cancer dataset in Example 8.2 to use the effect of tree
pruning on the accuracy of a tree method.

9

Bayesian Learning Approach

9.1 Bayesian Paradigms

The term *Bayesian* refers to the 18th-century theologian and mathematician Thomas Bayes. Bayes conceived of and applied broadly a method of estimating an unknown probability on the basis of other, related known probabilities. Human beings ordinarily acquire knowledge through a sequence of learning events and experiences. We hold perceptions and understandings of certain things based on our prior experiences or prior knowledge. When new facts are observed, we update our perception or knowledge accordingly. No matter whether the newly observed facts are multiple or solitary, it is this progressive, incremental learning mechanism that is the central idea of the Bayesian approach. Bayes' rule (or theorem) therefore enunciates important and fundamental relationships among prior knowledge, new evidence, and updated knowledge (posterior probability). It simply reflects the ordinary human learning mechanism, and is part of everyone's personal and professional life (Chang and Boral, 2008).

Suppose our question is whether an investigational drug has efficacy. So far we have finished experiments with the drug on animals, and we are going to design and conduct a small phase 1 clinical trial. Bayesian statisticians would consider the drug effect, θ, is a random variable with a distribution. Such a distribution when estimated based on previous experiences or experiments, in this case, animal studies, is called the prior distribution of θ, or simply the prior. This prior knowledge or distribution of the drug effect will be integrated with current experimental data (the clinical trial data), x, to produce the so-called posterior distribution of the drug effect, θ. Generally speaking, the more prior data we have and the more relevant they are, the narrower the prior distribution will be.

As illustrated, the prior distribution is widely spread, indicating that we have no accurate and precise estimate of the drug effect when there are only animal data available. On the other hand, the posterior distribution is much narrower, indicating that the combination of the animal and human data can provide a much more accurate and precise estimate of the drug effect. Although our knowledge has improved as a result of doing the clinical trial, we still do not know exactly the effectiveness of the drug.

While both frequentist and Bayesian statisticians use prior information, the former use a prior in the design of experiments only, whereas the latter uses priors for both experimental designs and statistical inferences (estimations) on the parameter or the drug effect. The frequentist approach considers a parameter to be simply a fixed value, while Bayesians view that same parameter as a variable with a probability distribution and recognize that the variable is liable to be updated at some time in the future, when yet more information becomes available. One trial's posterior probability is only the next trial's prior probability! Having said that, there is another approach, called empirical Bayesianism, in which the prior is calculated from the current data.

Another main difference between the Bayesian and frequentist approaches is that Bayesianism emphasizes the importance of information synergy from different sources: the model $f(x|\theta)$, a *prior distribution* of the parameters, $\pi(\theta)$, and current data, x. *Bayes' Theorem* places causes (observations) and effects (parameters) on the same conceptual level, since both have probability distributions. However, the data x are usually observable, while the parameter θ is usually latent.

Bayes' Rule plays a fundamental role in Bayesian reasoning. If A and B are two random events, then the joint probability of A and B can, due to symmetry, be written as $\Pr(B|A)\Pr(A) = \Pr(A|B)\Pr(B)$, where $\Pr(X|Y)$ denotes the conditional probability of X given Y. From this formulation, we can obtain Bayes' rule:

$$\Pr(B|A) = \frac{\Pr(A|B)\Pr(B)}{\Pr(A)}. \tag{9.1}$$

Bayes' rule is useful when $\Pr(A|B)$ is known and $\Pr(B|A)$ is unknown. Here we often call $\Pr(B)$ the prior probability and $\Pr(B|A)$ the *posterior probability*. The data in the form of conditional probability $\Pr(A|B)$ is the likelihood when it is viewed as a function of the model parameter.

In general, the three commonly used hypothesis testing $(H_1$ versus $H_2)$ methods in the Bayesian paradigm are:

1. Posterior Probability: Reject H_1 if $\Pr(H_1|D) \leq \alpha_B$, where α_B is a predetermined constant.

2. Bayes' Factor: Reject H_1 if $BF = \frac{\Pr(D|H_1)}{\Pr(D|H_2)} \leq k_0$, where k_0 is a small value, e.g., 0.1. A small value of BF implies strong evidence in favor of H_2 or against H_1.

3. Posterior Odds Ratio: Reject H_1 if $\Pr(H_1|D) / \Pr(H_2|D) \leq r_B$, where r_B is a predetermined constant.

Keep in mind that the notions of the Bayes' factor and the likelihood ratio are essentially the same in hypothesis testing, even though there are several versions of likelihood tests. Practically, we often write Bayes' law in the following form:

$$\Pr\left(H_1|D\right) = \frac{\Pr\left(D|H_1\right)\Pr\left(H_1\right)}{\Pr\left(D|H_1\right)\Pr\left(H_1\right) + \Pr\left(D|H_2\right)\Pr\left(H_2\right)}, \tag{9.2}$$

where H_2 is the negation of H_1.

Let's use Sally Clark's court case in Section 1.2 to illustrate the application of Bayes' rule. A Bayesian analysis would have shown that the children probably died of SIDS as McGrayne showed (McGrayne, 2012):

Let H_1 be the hypothesis to be updated: the children died of SIDS, and H_2 is the opposite hypothesis of H_1, i.e., the children are alive or were murdered. Let data D be that both children died suddenly and unexpectedly.

The chance of one random infant dying from SIDS was about 1 in 1300 during this period in Britain. The estimated odds of a second SIDS death in the same family was much larger, perhaps 1 in 100, because family members can share a common environmental and/or genetic propensity for SIDS. Therefore, the probability $P(H_1) = \frac{1}{1300} \times \frac{1}{100} = 0.0000077$. Thus, the probability $P(H_2) = 1 - P(H_1) = 0.9999923$. Next, only about 30 children in England, Scotland, and Wales out of 650,000 births annually were known to have been murdered by their mothers. The number of double murders must be much lower, estimated at 10 times less. Therefore, given a pair of siblings, neither of whom died of SIDS, the probability that both die a sudden and unexpected death or are murdered is approximately $P(D|H_2) = \frac{30}{650000} \times \frac{1}{10} = 0.0000046$. And the probability of dying suddenly and expectedly given they die of SIDS is $P(D|H_1) = 1$.

The goal is to estimate $P(H_1|D)$, the probability that the cause of death was SIDS, given their sudden and unexpected deaths. Bayes' rule provides the formula

$$P(H_1|D) = \frac{1 \times 0.0000077}{1 \times 0.0000077 + 0.9999923 \times 0.0000046} = 0.626$$

Thus, in this Bayesian assessment, the probability that the infants died of SIDS is 62.6%, much larger than 1 in 73 million chance.

There are differences regarding the concept of probability, Bayesian probability versus frequentist probability. As we discussed earlier, frequentists would consider data X is random and the parameter θ (such as the effect of a drug on a given patient population) is fixed, while Bayesians consider both X and parameter Θ to be random. Here, the Bayesian model parameter Θ is the knowledge of frequentist parameter θ. In other words, the effect θ is fixed but unknown to us. But our knowledge of the effect of the drug is changing as the data accumulates.

9.2 Bayesian Networks

9.2.1 Bayesian Network for Molecular Similarity Search

A Bayesian network (BN) is a simple and popular way of doing probabilistic inference, which formalizes the way to synchronize the evidence using Bayes' rule:

$$P(H|E) = P(E|H)P(H)/P(E). \qquad (9.3)$$

A Bayesian network is a directed acyclic graph and has the following properties:

- Nodes of the BN represent random variables; the parents of a node are those judged to be direct causes for it.

- The roots of the network are the nodes without parents.

- The arcs represent causal relationships between these variables, and the strengths of these causal influences are expressed by conditional probabilities.

- Let $X = \{x_1, ..., x_n\}$ be a set of parents of y (a child node), where $x_i \cap x_j = \phi$ (empty) for $i \neq j$ and x_i is a direct cause of y. The influence of X on y can be quantified with the conditional probability $P(y|X)$.

To use a BN for molecular similarity search, the following steps are involved:

1. Network model generation, i.e., representing the system using a suitable network form,

2. Representation of importance of descriptors, i.e., determining the weighting schemes,

3. Probability estimation for the network model, and

4. Calculation of the similarity scores.

Figure 9.1 represents a BN with four layers: compound nodes (c_j), feature nodes (f_i), query nodes (q_k), and target nodes (A_m). The arcs or edges indicate the direct causality relationships. Abdo and Salim (2008) described the method as follows.

Random variable f_i is a binary variable indicating the presence (1) or absence (0) of the feature. $\{f_1, ..., f_n\}$ is an n-dimensional vector equivalent to fingerprinting with length n. Similarly, c_j is a binary variable with 0 indicating absence of the molecule and 1 indicating the existence of the molecule.

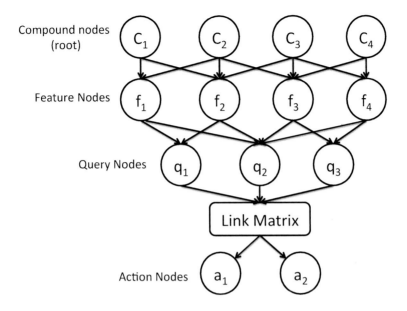

FIGURE 9.1
Bayesian Network of Similarity Search in Drug Discovery

The retrieval of an active compound compared to a given target structure is obtained by means of an inference process through a network of dependences using conditional probability. Note that in BN, each variable is conditionally independent of all its non-descendants in the graph given the value of all its parents, i.e.,

$$P(X_1, ..., X_n) = \Pi_{i=1}^{n} P(X_i | \text{parents}(X_i)). \qquad (9.4)$$

(1) If you conduct a random selection of molecules from a virtual molecular library, the prior probability associated with compound c_j can be estimated using the following flat distribution:

$$P(c_j) = \frac{1}{m}, \qquad (9.5)$$

where m is the collection size, i.e., the number of different molecules in the virtual library.

(2) The conditional probability of feature nodes given compound c_j is given by

$$P(f_i | c_j) = \frac{P(c_j | f_i) P(f_i)}{P(c_j)}. \qquad (9.6)$$

Practically, the following formulation can be used for the conditional probability:

$$P\left(f_i|c_j\right) = \frac{\hat{f}_{ij}}{\hat{f}_{ij} + 0.5 + 1.5\left(\frac{L_j}{\bar{L}}\right)} \frac{\ln\left(\frac{m+0.5}{f_i^*}\right)}{\ln\left(m+1\right)}, \tag{9.7}$$

where $\hat{f}_{ij}$ = the frequency of the ith feature within the jth compound, f^* is the inverse compound frequency of the ith feature in the collection, L_j is the size of jth compound, and $\bar{L}$ is the average length of the molecules in the collection,i.e., $\bar{L} = 1/m\sum_{j=1}^{m} L_j$.

(3) The conditional probabilities in the third layer are given for the query nodes and are defined as

$$P\left(q_k|f_i\right) = \frac{\tilde{f}_{i,q_k}}{\sum_{i=1}^{n_f} \tilde{f}_{i,q_k}}, \tag{9.8}$$

where n_f is the number of feature nodes, and $\tilde{f}_{i,q_k}$ is the frequency of the query node q_k in the feature node f_i.

(4) The conditional probability $P\left(a_m|q_k\right)$ for the fourth layer (activity-need node) given a query node can be only defined as empirically or based on beliefs.

We now can calculate the probability of activity a_m using the following equation:

$$P\left(a_m\right) = \sum_k \sum_i \sum_j P\left(A_m|q_k\right) P\left(q_k|f_i\right) P\left(f_i|c_j\right) P\left(c_j\right). \tag{9.9}$$

To select a list of the best compounds $\{c_j\}$, we can use the conditional probability such that

$$P\left(a_m|c_j\right) = \sum_k \sum_i \sum_j P\left(A_m|q_k\right) P\left(q_k|f_i\right) P\left(f_i|c_j\right) > \delta, \tag{9.10}$$

where δ is a predetermined threshold (Chang, 2010; Chang, 2011).

9.2.2 Coronary Heart Disease with Bayesian Network

Package bnlearn (Version 4.4.1) is developed for Bayesian Network Structure Learning, Parameter Learning and Inference. This package implements constraint-based, pairwise, score-based, and hybrid structure learning algorithms for discrete, Gaussian and conditional Gaussian networks, along with many score functions and conditional independence tests. The naive Bayes and the Tree-Augmented Naive Bayes (TAN) classifiers are also implemented. In addition, some utility functions and support for parameter estimation (maximum likelihood and Bayesian) and inference, conditional probability queries, and cross-validation are included. We will use the coronary heart disease data as a simple example of using the package. The dataset include 1830 observations of the following variables:

- Smoking (smoking): a two-level factor with levels no and yes.

- M. Work (strenuous mental work): a two-level factor with levels no and yes.

- P. Work (strenuous physical work): a two-level factor with levels no and yes.

- Pressure (systolic blood pressure): a two-level factor with levels <140 and >140.

- Proteins (ratio of beta and alpha lipoproteins): a two-level factor with levels < 3 and > 3.

- Family (family anamnesis of coronary heart disease): a two-level factor with levels neg and pos.

The first step in implementing a BN is to create the network. There several algorithms in deriving an optimal BN structure and some of them exist in bnlearn. The line, hc(coronary), is used to construct a BN using the score-based structure learning algorithms. The causality between some nodes is intuitive; however, some relations extracted from data do not seem to be correct, for example, Family as a variable condition on M. Work (Figure 9.2). Therefore, remove the link between M. Work and Family using the following statement.
res$arcs <- res$arcs[-which((res$arcs[,'from'] == "M..Work" & res$arcs[,'to'] == "Family")),]
To train the network we use the bn.fit function. The bn.fit function runs the *EM algorithm* to learn CPTs (conditional probability tables) for different nodes in the BN. After learning the structure, we need to determine the CPTs at each node. Here is the *R* implementation:

```
# R Code for Bayesian Network
library(bnlearn)
graphics.off()
par(mar=rep(2,4))
data(coronary)
print(coronary$'M. Work')
res = hc(coronary)  # Score-based structure learning algorithms
plot(res)
# make no sense to have Family condition on M. Work.
# Remove the link between M. Work and Family.
```

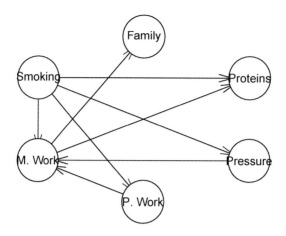

FIGURE 9.2
Bayesian Network for Coronary Disease

```
res$arcs <- res$arcs[-which(((res$arcs[,'from'] == "M. Work"
    & res$arcs[,'to'] == "Family")),]
plot(res)
fitted = bn.fit(res, coronary)
coefficients(fitted)
print(fitted$Proteins)
print(fitted$Pressure)
```

The partial outputs are shown below:

```
> print(fitted$Proteins)
Parameters of node Proteins (multinomial distribution)
Conditional probability table:
, , M. Work = no
            Smoking
Proteins   no          yes
<3         0.6685824   0.6167763
>3         0.3314176   0.3832237

, , M. Work = yes
            Smoking
Proteins   no          yes
<3         0.5671982   0.3235294
>3         0.4328018   0.6764706
```

```
> print(fitted$Pressure)
Parameters of node Pressure (multinomial distribution)
Conditional probability table:
          Smoking
Pressure  no          yes
 <140     0.5359001   0.6125000
 >140     0.4640999   0.3875000
```

Now, the BN is ready and we can start inferring from the network.

cpquery (fitted, event = (Pressure==">140"), evidence = (Proteins=="<3"))
which results in 0.4190842.

9.3 Bayesian Inference

9.3.1 Basic Formulations

There are different statistical paradigms or theoretical frameworks, which reflect different philosophies or beliefs. These differences have provoked quite a few controversies. However, within each paradigm or axiom system, consistency and completeness are expected. Our discussions in this chapter will focus on the Bayesian paradigm.

A Bayesian statistical model is made of a parametric statistical model, $f(x|\theta)$, and a prior distribution of the parameters, $\pi(\theta)$. Bayes' theorem places causes (observations) and effects (parameters) on the same conceptual level, since both have probability distributions. It is considered a major step from the notion of an unknown parameter to the notion of a random parameter (Robert, 1997). However, it is important to distinguish x and θ: x is usually observable, but θ is usually latent.

Denote the prior distribution by $\pi(\theta)$, and the sample distribution by $f(x|\theta)$. The following are four basic elements of the Bayesian approach:

(a) the joint distribution of (θ, x) given by

$$\varphi(\theta, x) = f(x|\theta) \pi(\theta), \tag{9.11}$$

(b) the marginal distribution of x given by

$$m(x) = \int \varphi(\theta, x) \, d\theta = \int f(x|\theta) \pi(\theta) \, d\theta, \tag{9.12}$$

(c) the posterior distribution of θ given by Bayes' formula

$$\pi(\theta|x) = \frac{f(x|\theta) \pi(\theta)}{m(x)}, \text{ and} \tag{9.13}$$

(d) the predictive probability distribution given by

$$P(y|x) = \int P(x|y,\theta)\,\pi(\theta|x)\,d\theta. \tag{9.14}$$

Example 9.1 Beta Posterior Distribution

Assume that $X \sim Bin(n,p)$ and $p \sim Beta(\alpha,\beta)$.

The sample distribution is given by

$$f(x|p) = \binom{n}{x} p^x (1-p)^{n-x}, \quad x = 0,1,...,n. \tag{9.15}$$

The prior about the parameter p is given by

$$\pi(p) = \frac{1}{B(\alpha,\beta)} p^{\alpha-1}(1-p)^{\beta-1}, \quad 0 \le p \le 1, \tag{9.16}$$

for the beta function $B(\alpha,\beta) = \frac{\Gamma(\alpha)\Gamma(\beta)}{\Gamma(\alpha+\beta)}$. The joint distribution then is given by

$$\varphi(p,x) = \frac{\binom{n}{x}}{B(\alpha,\beta)} p^{\alpha+x-1}(1-p)^{n-x+\beta-1}, \tag{9.17}$$

and the marginal distribution is

$$m(x) = \frac{\binom{n}{x}}{B(\alpha,\beta)} B(\alpha+x, n-x+\beta). \tag{9.18}$$

Therefore the posterior distribution is given by

$$\pi(p|x) = \frac{p^{\alpha+x-1}(1-p)^{n-x+\beta-1}}{B(\alpha+x, \beta+n-x)} = Beta(\alpha+x, \beta+n-x). \tag{9.19}$$

Example 9.2 Normal Posterior Distribution

Assume that X has a normal distribution: $X \sim N(\theta, \sigma^2/n)$ and $\theta \sim N(\mu, \sigma^2/n_0)$. The posterior distribution can be written as

$$\pi(\theta|X) \propto f(X|\theta)\,\pi(\theta) \tag{9.20}$$

or

$$\pi(\theta|X) = Ce^{-\frac{(X-\theta)^2 n}{2\sigma^2}} e^{-\frac{(\theta-\mu)^2 n_0}{2\sigma^2}}, \tag{9.21}$$

where C is a constant. We immediately recognize that (9.21) is the normal distribution $N\left(\frac{n_0\mu+nX}{n_0+n}, \frac{\sigma^2}{n_0+n}\right)$.

We now wish to make predictions concerning future values of X, taking into account our uncertainty about its mean θ. We may write $X = (X - \theta) + \theta$, so that we can consider X the sum of two independent quantities: $(X - \theta) \sim N\left(0, \sigma^2/n\right)$, and $\theta \sim N\left(\mu, \sigma^2/n_0\right)$. The predictive probability distribution is given by

$$X \sim N\left(\mu, \sigma^2\left(\frac{1}{n} + \frac{1}{n_0}\right)\right). \tag{9.22}$$

If we have already observed the mean of the first n_1 observations $\bar{x}_{n_1}$, the predictive distribution is given by

$$X|x_{n_1} \sim N\left(\frac{n_0\mu + n_1\bar{x}_{n_1}}{n_0 + n_1}, \sigma^2\left(\frac{1}{n_0 + n_1} + \frac{1}{n}\right)\right). \tag{9.23}$$

Example 9.3: Drug Label Extension Study with Binomial Endpoint

A drug shows a clinical improvement in 30% patients with a one-sided 95% confidence interval larger than 10%. The pharmaceutical company want to expand its use to a broad patient population. This information will be used in their phase-4 clinical trial design.

Prior information: We use a prior β-distribution with parameter a and b. To determine the parameters, we use the fact, a clinical improvement in 30% patients, so that

$$\int_0^{0.3} \frac{\Gamma(a+b)}{\Gamma(a)\Gamma(b)} \theta^{a-1}(1-\theta)^{b-1} d\theta = 0.5.$$

and the fact, 1-sided 95% confidence interval larger than 10%, so that

$$\int_0^{0.1} \frac{\Gamma(a+b)}{\Gamma(a)\Gamma(b)} \theta^{a-1}(1-\theta)^{b-1} d\theta = 0.05.$$

We can solve for the parameters a and b numerically using numerical solver *BBsolve* in *R*:

```
## install the BBsolve package
# install.packages("BB", repos="http://cran.r-project.org")
library(BB)
fn = function(x){qbeta(c(0.05,0.5), x[1], x[2]) - c(0.1, 0.3)}
BBsolve(c(1,1), fn)
```

The output parameters are $a = 2.88$ and $b = 6.29$.

The solver requires two arguments: (1) a function which must take arguments as vectors for the unknown values of a and b (coded $x[1]$ and $x[2]$) and (2) a set of initial values as your best guess of a and b (1 and 1 in this example). The solver sets the function to zero and then finds the best solutions for a and b (we won't go into the details of how it solves for the values in this class).

Data from a single clinical trial

Suppose we observed 12 responders from 60 patients in the trial and use a binomial model for the response X:

$$\text{Response } X|\theta \sim \text{Binomial}(n, \theta)$$
$$\theta \sim Beta\,(2.8, 6.29)$$
$$\theta|x \sim Beta\,(x + 2.8, 60 - x + 6.29)$$
$$\theta|17 \sim Beta\,(14.8, 54.29)$$

We can plot the likelihood, prior, and posterior in R. The command seq below creates a sequence of 500 numbers between 0 and 1 for the plots. The command line allows us to add more plots onto the original plot. From Figure 9.3, we can see that the posterior is a mixture of the likelihood and the prior.

```
th = seq(0,1,length=500)
a = 2.8; b = 6.29; n = 60; x = 12
prior = dbeta(th,a,b)
like = dbeta(th,x+1,n-x+1)
post = dbeta(th,x+a,n-x+b)
plot(th,post,type="l",ylab="Density",lty=2,lwd=3,xlab = expression(theta))
lines(th, like,lty=1,lwd=3)
lines(th, prior,lty=3,lwd=3)
legend(0.5,8,c("Prior","Likelihood","Posterior"), lty=c(3,1,2),lwd=c(3,3,3))
dev.off()
```

Point estimate

The Bayesian point estimate is the posterior expectation of the parameter, given by

$$E\,(\theta) = \int \pi\,(\theta|x)\,\theta d\theta. \tag{9.24}$$

Bayesian credible interval

Parallel to the frequentist confidence interval, the Bayesian credible interval (BCI) describes the variability of the parameter θ. For $0 < \alpha < 1$, a $100(1 - \alpha)\%$ credible set for θ is a subset $C \in \Theta$ such that

$$P\,\{C|X = x\} = 1 - \alpha. \tag{9.25}$$

The BCI is usually not unique. The commonly used equal-tailed credible interval does not necessarily have the smallest size. The smallest size credible interval is defined by the *highest posterior density* (HPD).

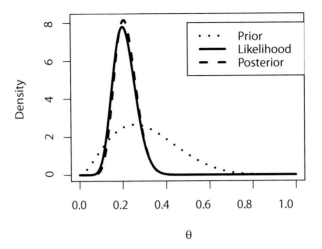

FIGURE 9.3
Likelihood, Prior, Posterior Distributions

Suppose the posterior density for θ is unimodal. Then the HPD interval for θ is the interval

$$C = \{\theta : \pi\left(\theta|X = x\right) \geq k\},\qquad(9.26)$$

where k is chosen such that

$$P\left(C|X = x\right) = 1 - \alpha.\qquad(9.27)$$

Hypothesis Testing

For the hypothesis test:

$$H_o : \theta \in \Theta_0 \text{ versus } H_a : \theta \in \Theta_a,\qquad(9.28)$$

we would use the posterior odds ratio (POR) $\frac{P(\Theta_0|x)}{P(\Theta_a|x)}$; a small value of POR is strong evidence against the null hypothesis H_o.

We can separate POR into the prior component and current evidence from the current data

$$\frac{P\left(\Theta_0|x\right)}{P\left(\Theta_a|x\right)} = \frac{P\left(\theta \in \Theta_0\right)}{P\left(\theta \in \Theta_a\right)}BF,\qquad(9.29)$$

where the Bayes' factor (BF) is defined as

$$BF = \frac{\int_{\Theta_0} f\left(x|\theta\right)\pi_0\left(\theta\right)d\theta}{\int_{\Theta_a} f\left(x|\theta\right)\pi_a\left(\theta\right)d\theta}.\qquad(9.30)$$

BF is independent of the prior and equal to POR when $P(\theta \in \Theta_0) = P(\theta \in \Theta_a)$. BF may be practically more useful than POR because the prior is subjective and varies from individual to individual, and it is difficult to convince everyone to use a particular individual prior. In general, we reject H_o if $BF \leq k_0$, where k_0 is a small value, e.g., 0.1. A small value of BF implies strong evidence in favor of H_a or against H_o.

9.3.2   Preclinical Study of Fluoxetine on Time Immobile

We are going to study the example discussed by Bruno Contrino and Stanley E. Lazic (2017), but use different approaches and obtain slightly different conclusions. The data come from an antidepression study, where 20 rats were randomized to one of four doses of the antidepressant fluoxetine given in the animals' drinking water. The time that the rats spent immobile in the *Forced Swim Test* (FST) was recorded. The FST is a standard behavioral test of "depression" in rodents. The experimental data are in the labstats package. We will use the *BayesCombo* package to perform the calculations.

Step 1: Specify Priors for Hypotheses and Prior for Effect Size

When the experimental data are small, the posterior results will be sensitive to the choice of the prior. Often three different (noninformative, pessimistic, optimistic) priors will be included in the study to see the robustness of the results.

Suppose we are interested in three hypotheses: fluoxetine decreases ($H < 0$), has no effect on ($H = 0$), and increases immobility time ($H > 0$). From the prior distribution of effect size, the priors for the three hypotheses can be calculated. The priors are easily determined: $\pi(H < 0) = 0.5$, $\pi(H = 0) = 0.0$, and $\pi(H > 0) = 0.5$.

In this example, the noninformative prior is the default, which is a normal prior, centered at zero, with the width calculated so that the 99% confidence interval (CI) of the prior matches the 99% CI of the data distribution. The default prior is suitable for most situations where one has no prior information to include, and the results are insensitive to small changes near the default value. Importantly, the prior for effect size and priors for effect size distribution must be consistent. For example, if the effect size has a continuous prior distribution, than the prior for particular values is infinitely small or zero. In this example, because the prior is a normal distribution centered at 0, it is consistent with the prior $\pi(H = 0) = 0.0$.

Step 2: Calculate Effect Size and Standard Error from Experimental Data

For the normal prior distribution of the effect size of immobility time, it is sufficient to calculate the mean (or the least-square mean) and standard

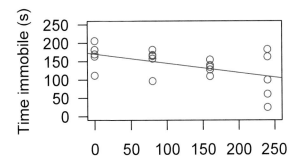

FIGURE 9.4
Linear Regression of Time Immobile

error. We use linear regression with a factor of dose. It is accomplished by the following *R* code.

```
library(labstats)
library(BayesCombo)
par(las=1)
plot(time.immob ~dose, data=fluoxetine, col="royalblue",
   ylab="Time immobile (s)", ylim=c(0, 250), xlim=c(0, 250),
   xlab="Dose of fluoxetine (mg/L in drinking water)")
   abline(lm(time.immob ~dose, data=fluoxetine), col="royalblue") # add
reg. line
summary(lm(time.immob ~dose, data=fluoxetine))$coef
```

From the above output (Figure 9.4) we see that the estimated slope is -0.252, with a standard error of 0.099, and a *p*-value of 0.020. We now have all the information needed to calculate the probability that fluoxetine increases, decreases, or has no effect on immobility time.

Step 3: Calculate the Posterior Probability of Each Hypothesis

The posterior probability of a hypotheses can be conveniently calculated using the *pph()* function in the BayesCombo package:

pph(beta, se.beta, beta0 = 0, se0 = NULL, ci = 99, se.mult = 1, H0 = c(0, 0), H.priors = rep(1/3, 3)),

where *beta* = effect size, *se.beta* = standard error for the effect, *beta0* = a prior value for the effect size, *se0* = a prior standard error for the effect size (the default is calculated automatically), *ci* = percentile used to calculate the prior standard error, and *se.mult* = standard error multiplier used to increase or decrease the prior SE and used in conjunction with ci when se0 = NULL. When combining diverse evidence across experiments (e.g., prior and experiment involve different animals), the se.mult option is often needed. *H0* = a vector of length two that defines the null hypothesis. If the values are identical (e.g. H0 = c(0,0)) a point null is used, otherwise the null is defined as the range between the lower and upper values. Lastly, *H.priors* specifies the prior hypothesis probabilities; the default is an equal probability of 1/3, and they are specified in the following order: $H < 0, H = 0, H > 0$.

Returning to the fluoxetine example, we can calculate the probability that the slope is negative, positive, or zero. Below, we specify the slope (*beta* = -0.252) and its standard error (*se.beta* = 0.099) that we obtained previously from the output of the lm() function. The priors for the hypotheses are specified in *H.priors* = *c(0.5, 0.0, 0.5)*. The default settings are used for all other options.

```
x=pph(beta = -0.252, se.beta = 0.099, H.priors = c(0.5, 0.0, 0.5))
x

par(las=1)
plot(x, leg.loc = "topright", xlab="Effect size (slope of regression line)")
```

The results show the priors 0.5, 0.0, and 0.5 for $H < 0$, $H = 0$, and $H > 0$, respectively. The three corresponding posterior probabilities are 0.989, 0.0, and 0.0115, respectively. Therefore we can conclude that fluoxetine decreases the immobility time with a probability of 0.989. The last two lines of code also produce the plots for the prior, likelihood, and posterior distributions (Figure 9.5).

Step 4: Sensitivity Study

As stated earlier, we need to study the effect of different priors. First, we view the uncertainty about the prior as large since there is no data to derive from. For this reason, it occurs to us that the spread of the effect size should be bigger than the standard error from the experimental data, and so we use se.mult = 2 in the option.

```
pph(beta = -0.252, se.beta = 0.099, se.mult=2, H.priors = c(0.5, 0.0, 0.5))
```

The resulting posteriors for the three hypotheses are similar: 0.993, 0.0, and 0.007. Furthermore, we chose fluoxetine (not a randomly picked compound) to study the effect on immobility time, so we must have some prior opinions on how fluoxetine will affect the immobility time directionally. More likely this will be a negative effect, say, with a 0.55 probability. In the following, I also specify a very small probability (0.01) for hypothesis H_0 to see how the outcomes will change. Because it is a null zero value, the *H0* option in the

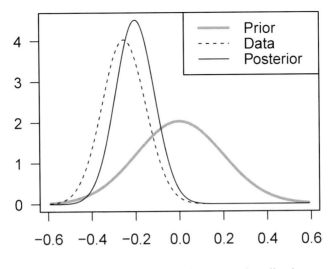

FIGURE 9.5
Likelihood, Prior, and Posterior Distributions

pph function needs to have different values, e.g., *H0 = c(-0.05, 0.05)*. Putting these together, we have the following code:

```
pph(beta = -0.252, se.beta = 0.099, beta0=0, se0=1.2,
 H0 = c(-0.05, 0.05), H.priors=c(0.55, 0.01, 0.44))
```

With these options, the probability that fluoxetine decreases immobility time is now 0.999. The outputs also include the Bayes' factors, 3.191, 0.052, and 0.004 for $H < 0$, $H = 0$, and $H > 0$, respectively. A value larger than 1 will support the hypothesis. Therefore, the BF of 3.191 strongly supports the hypothesis that fluoxetine decreases immobility time.

9.4 Model Selection

Suppose i.i.d. samples x are drawn from a density $f(x|\theta)$, with an unknown parameter θ. In this scenario, we face a model selection problem:

$$\begin{cases} M_0 : X \sim f(x|\theta) \text{ where } \theta \in \Theta_0, \\ M_1 : X \sim f(x|\theta) \text{ where } \theta \in \Theta_1. \end{cases} \quad (9.31)$$

Let $\pi_i(\theta)$ be the prior density conditional on M_i being the true model. We can use the Bayes' factor BF_{01} for model selection:

$$BF_{01} = \frac{\int_{\Theta_0} f(\boldsymbol{x}|\theta) \pi_0(\theta) d\theta}{\int_{\Theta_1} f(\boldsymbol{x}|\theta) \pi_1(\theta) d\theta}. \qquad (9.32)$$

We know that the Bayes' factor is the ratio of the posterior odds ratio of the hypotheses to the corresponding prior odds ratio. Thus, if the prior probability $P(M_0) = P(\Theta_0)$ and $P(M_1) = P(\Theta_1) = 1 - P(\Theta_0)$, then

$$P(M_0|\boldsymbol{x}) = \left\{ 1 + \frac{1 - P(\Theta_0)}{P(\Theta_0)} \frac{1}{BF_{01}(\boldsymbol{x})} \right\}^{-1}. \qquad (9.33)$$

Therefore, if conditional prior densities π_0 and π_1 can be specified, we should simply use the Bayes' factor BF_{01} for model selection. If $P(\Theta_0)$ is also specified, the posterior odds ratio of M_0 to M_1 can be used.

However, the computation of BF may not always be easy. In such cases, we can use the Bayesian information criterion (BIC). Using the Taylor series expansion, we have the following approximation (Schwarz 1978, Ghosh et al. 2006):

$$2\ln(BF_{01}) = 2\ln\left(\frac{f\left(\boldsymbol{x}|\hat{\theta}_0\right)}{f\left(\boldsymbol{x}|\hat{\theta}_1\right)} \right) - (\kappa_0 - \kappa_1)\ln n + O(1), \qquad (9.34)$$

where κ_0 and κ_1 are the number of parameters in models M_0 and M_1, respectively, and n is the number of data points in $\boldsymbol{x}$. Here the term $(p_0 - p_1)\ln n$ can be viewed as a penalty for using a more complex model. Alternatively, the Akaike information criterion (AIC) is often used:

$$\text{AIC} = 2\ln f\left(\boldsymbol{x}|\hat{\theta}\right) - 2\kappa, \qquad (9.35)$$

where κ is the number of parameters in the model. AIC has a smaller penalty for more complex models than BIC does.

9.5   Hierarchical Model

For modeling it is often convenient and appropriate to use Bayesian statistical models hierarchically with several levels of conditional prior distributions. A hierarchical Bayes model is a Bayesian statistical model, $(f(x|\theta), \pi(\theta))$, where the prior distribution $\pi(\theta)$ is decomposed into conditional distributions

$$\pi_1(\theta|\theta_1), \pi_2(\theta_1|\theta_2), \cdots, \pi_n(\theta_{n-1}|\theta_n) \qquad (9.36)$$

and a marginal distribution $\pi_{n+1}(\theta_n)$ such that

$$\pi(\theta) = \int_{\Theta_1 \times \cdots \times \Theta_n} \pi_1(\theta|\theta_1) \pi_2(\theta_1|\theta_2) \cdots \pi_n(\theta_{n-1}|\theta_n) \pi_{n+1}(\theta_n) d\theta_1 \cdots d\theta_n.$$

(9.37)

The parameters θ_i are called hyperparameters of level i.

Exchangeability

The parameters $(\theta_1, ..., \theta_n)$ are exchangeable in their joint distribution if the density $f(\theta_1, ..., \theta_n)$ is invariant with respect to permutations of the indexes $(1, ..., n)$.

The simplest form of an exchangeable distribution has each of the parameters θ_j as an independent sample from a prior (or population) distribution governed by some unknown parameter vector ϕ; thus,

$$f(\theta|\phi) = \prod_{j=1}^{n} f(\theta_j|\phi).$$

(9.38)

In general, ϕ is unknown, so our distribution for θ must average over our uncertainty in ϕ:

$$f(\theta) = \int \prod_{j=1}^{n} f(\theta_j|\phi) \pi(\phi) d\phi.$$

(9.39)

Elimination of nuisance parameters

In a multivariate or hierarchical model, our interest may not be in the full parameter vector $\boldsymbol{\theta} = (\theta_1, ..., \theta_n)$, but only certain components of $\boldsymbol{\theta}$. In the frequentist paradigm, there are three methods that can be used (Ghosh et al. 2006, p. 51): (1) constructing a conditional test, (2) using an invariance argument, and (3) using the profile likelihood defined by

$$L_p(\theta_1) = \sup_{\theta_2} f(x|\theta_1, \boldsymbol{\theta}_{1+}) = f\left(x|\theta_1, \hat{\boldsymbol{\theta}}_{1+}(\theta_1)\right),$$

(9.40)

where θ_1 is the parameter of interest, $\boldsymbol{\theta}_{1+} = (\theta_2, \theta_3, ..., \theta_n)$, and $\hat{\boldsymbol{\theta}}_{1+}(\theta_1)$ is the MLE of $\boldsymbol{\theta}_{1+}$ given the value of θ_1.

In the Bayesian paradigm, to eliminate the nuisance parameters we just integrate them out in the joint posterior. This integrate out approach can also be used in other quantities, e.g., analogous to the profile likelihood,

$$L(\theta_1) = \int f(x|\theta_1, \boldsymbol{\theta}_{1+}) \pi(\boldsymbol{\theta}_{1+}|\theta_1) d\boldsymbol{\theta}_{1+}.$$

(9.41)

9.6 Bayesian Decision-Making

Statistical analyses and predictions are motivated by one's objectives. When choosing between models, we evaluate their consequences or impacts. The impact is characterized by a loss function in Bayesian decision theory. The challenge is that different people have different perspectives on the loss, hence the different loss functions, including the error rate in the frequentist paradigm. Loss functions are often vague and not explicitly defined, especially when we make decisions in our daily lives. Decision theory makes this loss explicit and deals with it with mathematical rigor.

In decision theory, statistical models involve three spaces: the observation space X, the parameter space Θ, and the action space A. Actions are guided by a decision rule $\delta(\mathbf{x})$. An action $\alpha \in A$ always has an associated consequence characterized by the loss function $L(\theta, a)$. In hypothesis testing, the action space is $A = \{\text{accept, reject}\}$.

Because it is usually impossible to uniformly minimize the loss $L(\theta, a)$, in the frequentist paradigm, the decision rule δ is determined to minimize the following average loss:

$$
\begin{aligned}
R(\theta, \delta) &= E^{X|\theta}\left(L\left(\theta, \delta\left(\boldsymbol{x}\right)\right)\right) \\
&= \int_X L\left(\theta, \delta\left(\boldsymbol{x}\right)\right) f\left(\boldsymbol{x}|\theta\right) d\boldsymbol{x}.
\end{aligned} \tag{9.42}
$$

The rule $a = \delta(\mathbf{x})$ is often called an estimator in estimation problems. Commonly used loss functions are *squared error loss* (SEL) $L(\theta, a) = (\theta - a)^2$, *absolute loss*, $L(\theta, a) = |\theta - \alpha|$, the *0-1 loss*, $L(\theta, a) = I(|\alpha - \theta|)$, the indicator function, etc.

Definition 1 *Bayesian expected loss is the expectation of the loss function with respect to posterior measure, i.e.,*

$$
\rho\left(\delta\left(\boldsymbol{x}\right), \pi\right) = E^{\theta|X} L\left(\delta\left(\boldsymbol{x}\right), \theta\right) = \int_\Theta L\left(\theta, \delta\left(\boldsymbol{x}\right)\right) \pi\left(\theta|\boldsymbol{x}\right) d\theta \tag{9.43}
$$

An action $a^* = \delta^*(\mathbf{x})$ that minimizes the posterior expected loss is called a Bayes' action.

By averaging (10.36) over a range of θ for a given prior $\pi(\theta)$, we can obtain:

$$
\begin{aligned}
r(\pi, \delta) &= E^\pi\left(R\left(\theta, \delta\right)\right) \\
&= \int_\Theta \int_X L\left(\theta, \delta\left(\boldsymbol{x}\right)\right) f\left(\boldsymbol{x}|\theta\right) \pi\left(\theta\right) d\boldsymbol{x}\, d\theta.
\end{aligned} \tag{9.44}
$$

The two notions in (10.37) and (10.38) are equivalent in the sense that they lead to the same decision.

An estimator minimizing the integrated risk $r(\pi, \delta)$ can be obtained by selecting, for every $x \in X$, the value $\delta(x)$, which minimizes the posterior expected loss $\rho(a, \pi)$, because

$$r(\pi, \delta) = \int_X \rho(a, \delta(\boldsymbol{x}) | \boldsymbol{x}) m(\boldsymbol{x}) d\boldsymbol{x}. \tag{9.45}$$

The Bayes' estimators use uniform representations of loss functions.

Bayesian approaches have been broadly used in adaptive clinical trials, including adaptive dose escalation trials with continual reassessment methods (CRM) and Bayesian design for trials with efficacy-toxicity trade-off and drug combinations (Chang, 2014). Adaptive trial design is new innovative clinical trial designs that were recently broadly adopted by the pharmaceutical industry.

9.7 Summary and Discussion

Bayesian method as a ML tool emphasizes the importance of information synergy from different sources: the model $f(x|\theta)$, a *prior distribution* of the parameters, $\pi(\theta)$, and current data, x. Although both x and θ are considered random, the data x are observable, while the parameter θ is latent.

Frequentist and Bayesian statistics both use the symmetry of the events S and D in the joint probability

$$\Pr(S, D) = \Pr(S|D) \Pr(D) = \Pr(D|S) \Pr(S).$$

Thus

$$\Pr(S|D) = \frac{\Pr(D|S) \Pr(S)}{\Pr(D)}.$$

However, the frequentist and Bayesian approaches use this equation in different ways. The former emphasizes mathematical accuracy and requires the probabilities in the formulation to be defined at the same calendar time. The Bayesian approach emphasizes the process of learning or knowledge updating, thus the probabilities in the formation do not have to be defined at the same moment; that is, probabilities $\Pr(D|S)$ and $\Pr(S)$ can be defined at different timepoints, $\Pr(S)$ being prior (our knowledge at earlier time) while $\Pr(D|S)$ and $\Pr(S|D)$ are from a later time. In short, using the term "probability" without clarifying whether it is meant in the frequentist or Bayesian sense can lead to confusion!

Bayes' factor, posterior distribution, and credible intervals are commonly used in Bayesian inference. A Bayesian network is a simple and popular way of doing probabilistic inference, which formalizes the way to synchronize the evidence using Bayes' rule. Bayesian inference can be used in model selections. Bayesian decision-making is a probabilistic optimization in which minimization of the defined loss function is the goal.

9.8   Problems

9.1: In Section 3.1.4. we discussion Simpson's paradox with the following example:

Tumor Responses to a Drug in Males and Females

	Drug A	Drug B
Males	200/500	380/1000
Females	300/1000	140/500
Total	500/1500	520/1500

(a) If the entire data in the table is from one clinical trial, calculate the posterior of the cancer response rate for the women (you need to use your own prior).

(b) If the data for males were from an early clinical trial and the data for females are from the current trial, how do you calculate the posterior distribution and Bayes' factor for female?

(c) If the data for females were from an early clinical trial and the data for males are from the current trial, how do you calculate the posterior distribution and Bayes' factor for females?

9.2: Do you use different model when you face different priors as in the scenarios (a), (b), and (c) in problem 9.1/

10
Unsupervised Learning

10.1   Needs of Unsupervised Learning

Unlike supervised learning, for unsupervised learning, there are no correct answers. The goal of unsupervised learning is to identify or simplify data structure. Unsupervised learning is of growing importance in a number of fields; examples are seen when grouping breast cancer patients by their genetic markers, shoppers by their browsing and purchase histories, and movie viewers by the ratings assigned by movie viewers. We may want to organize documents into different mutually exclusive or overlapping categories, or we just might want to visualize the data. It is often easier to obtain unlabeled data than labeled data, which often require human intervention.

Unsupervised learning problems can be further divided into *clustering*, *association*, and *anomaly detection*. A clustering problem occurs when we want to discover the inherent groupings in the data, such as grouping customer by purchasing behavior. An association rule learning problem is one where we want to discover rules that describe connections in large portions of our data, e.g., people who buy product A may also tend to buy product B. The third type of problem, *anomaly detection* or *outlier detection*, involves identifying items, events, or observations that do not conform to an expected pattern, such as instances of bank fraud, structural defects, medical problems, or errors in a text. Anomalies are also referred to as outliers, novelties, noise, deviations, and exceptions. In particular, in the context of abuse of computer networks and network intrusion detection, the interesting objects are often not rare objects, but unexpected bursts in activity. This pattern does not adhere to the common statistical definition of an outlier as a rare object, and for this reason there are many outlier detection methods (Zimek, 2017).

10.2   Association or Link Analysis

In many situations, finding causality relationships is the goal. When there are a larger number of variables, the task is not trivial. However, association is a necessary condition for a causal relationship. Finding a set of events that correlate many others is often the focus point and springboard for further research. Link-analysis provides a way to find the event set with high density. Finding sale items that are highly related (or frequently purchased together) can be very helpful for stocking shelves, cross-marketing in sales promotions, catalog design, and consumer segmentation based on buying patterns. A commonly used algorithm for such density problems is the so-called *Apriori* (Motoda and Ohara, 2009, in Wu and Kumar, 2009, p. 61).

In network theory, link analysis is a data-analysis technique used to evaluate relationships (connections) between nodes. Relationships may be identified among various types of nodes (objects), including organizations, people, and transactions. Link analysis has been used in the investigation of criminal activity (fraud detection, counterterrorism, and intelligence), computer security analysis, search engine optimization, market research, medical research, and even in understanding works of art.

Apriori, proposed by Agrawal and Srikant (1994) is an algorithm for frequent items set over transactional databases. The algorithm proceeds by identifying the frequent individual items in the database and extending these to larger and larger sets of items as long as those item sets appear sufficiently often in the database. Apriori uses a bottom-up approach, where frequent subsets are extended one item at a time, and groups of candidates are tested against the data. The algorithm terminates when no further successful extensions are found.

Apriori uses breadth-first search (level-wise complete search) and a hash tree structure to count candidate item sets efficiently. It generates candidate item sets of length k from item sets of length $k - 1$, and then prunes the candidates, which have an infrequent sub-pattern. According to the *downward closure property* (if an item is not frequent, any of its supersets that contain the item are not frequent), the candidate set contains all frequent k-length item sets.

Kuo et al. (2009) study the suitability of the Apriori association analysis algorithm for the detection of adverse drug reactions (ADR) in health care data. The Apriori algorithm is used to perform association analysis on the characteristics of patients, the drugs they are taking, their primary diagnosis, co-morbid conditions, and the ADRs or adverse events (AE) they experience. The analysis produces association rules that indicate what combinations of medications and patient characteristics lead to ADRs

10.3   Principal Components Analysis

Principal component analysis (PCA) is an important and useful unsupervised learning tool for dimension reduction in drug design and discovery. Giuliani (2017) believes the reason that PCA is broadly used in the pharmaceutical industry is that it is a tool creating a statistical mechanics framework for biological systems modeling without the need for strong a priori theoretical assumptions. This makes PCA of the utmost importance, as it enables drug discovery from a systemic perspective, overcoming other too-narrow reductionist approaches.

PCA can be used as a tool for data pre-processing before supervised techniques are applied. PCA produces a low-dimensional representation of a dataset. It finds sequentially a set of linear combinations of the predictors that have maximal variance and are orthogonal with each other.

The first principal component of a set of features $\mathbf{X}_1, \mathbf{X}_2, ..., \mathbf{X}_K$ is the linear combination of the features or predictors $\mathbf{X}$,

$$\mathbf{Z}_1 = \boldsymbol{\phi}_1^T \mathbf{X} = \phi_{11}\mathbf{X}_1 + \phi_{21}\mathbf{X}_2 + ... + \phi_{K1}\mathbf{X}_K \tag{10.1}$$

that has the largest variance, where ϕ_{j1} satisfy

$$\sum_{j=1}^{K} \phi_{j1}^2 = 1. \tag{10.2}$$

The ϕ_{ji} are called loadings and can be found using a singular-value decomposition of the $n \times K$ matrix $\mathbf{X}$. The second principal component is the linear combination of $X_1, ..., X_K$ that has maximal variance among all linear combinations that are orthogonal with $\mathbf{Z}_1$.

$$\mathbf{Z}_2 = \boldsymbol{\phi}_2^T \mathbf{X} = \phi_{12}\mathbf{X}_1 + \phi_{21}\mathbf{X}_2 + ... + \phi_{K1}\mathbf{X}_K, \tag{10.3}$$

where $\boldsymbol{\phi}_2$ is the second principal component loading vector, satisfying $\boldsymbol{\phi}_1 \cdot \boldsymbol{\phi}_2 = 0$, that is, $\boldsymbol{\phi}_1 \perp \boldsymbol{\phi}_2$. This process continues until all $K^* = \min(n-1, K)$ principal components are found. Here the inner products $\boldsymbol{\phi}_i \cdot \boldsymbol{\phi}_j = 0$ and $\mathbf{Z}_i \cdot \mathbf{Z}_j = 0$ for all $i \neq j$.

The meaning of loadings can be interpreted as follows:

1. The loading vector $\boldsymbol{\phi}_1$ with elements $\phi_{11}, \phi_{21}, ..., \phi_{K1}$ defines a direction in the feature space along which the data vary the most.

2. The principal component directions $\boldsymbol{\phi}_1, \boldsymbol{\phi}_2, ...\boldsymbol{\phi}_K$ are the ordered sequence of right singular vectors of the matrix $\mathbf{X}$, and the variances of the components are $1/n$ times the squares of the singular values.

3. The loading vector $\boldsymbol{\phi}_1$ defines the line in K-dimensional space that is closest to the n observations in terms of average squared Euclidean distance.

Assuming that the variables have been centered to have mean zero, the total observed variance is

$$\sum_{j=1}^{K} Var(X_j) = \sum_{j=1}^{K} \frac{1}{n} \sum_{i=1}^{n} x_{ij}^2, \tag{10.4}$$

and the variance explained by the mth principal component is

$$Var\left(Z_m\right) = \frac{1}{n}\sum_{i=1}^{n} z_{im}^2. \tag{10.5}$$

Therefore, the proportional variance explained by the mth principal component is given by the positive quantity between 0 and 1,

$$\frac{\sum_{i=1}^{n} z_{im}^2}{\sum_{j=1}^{K}\sum_{i=1}^{n} x_{ij}^2}. \tag{10.6}$$

One of the disadvantages of using PCA is that the transformed features Z are not as intuitive as the original measures X.

PCA, in a typical quantitative structure–activity relationship (QSAR) study, analyzes an original data matrix in which molecules are described by several intercorrelated quantitative dependent variables (molecular descriptors). Although extensively applied, there is disparity in the literature with respect to the applications of PCA in QSAR studies. Yoo and Shahlaei (2017) investigate the different applications of PCA in QSAR studies using a dataset including CCR5 inhibitors. They conclude that PCA is a powerful technique for exploring complex datasets in QSAR studies for identification of outliers and can be easily applied to the pool of calculated structural descriptors.

A related method, *principal component regression* (PCR), is similar to a standard linear regression model, but uses PCA for estimating the unknown regression coefficients in the model.

Example 10.1: PCA for Alzheimer's Disease

Using the *prcomp()* function, we do not need to explicitly multiply the data by the principal component loading vectors to obtain the principal component score vectors. We use Alzheimer's dsease to illustrate the *procomp()* for PCA.

```
library(AppliedPredictiveModeling)
data("AlzheimerDisease", package = "AppliedPredictiveModeling")
# remove non numerical variable and covariance matrix
s = cor(predictors[,-130])
s.eigen <- eigen(s) # eigen values
# Proportions of total variance explained by the eigenvalues.
for (ev in s.eigen$values) {
 print(ev / sum(s.eigen$values))
}
# Do PCA with s using the prcomp() function, set scale = TRUE.
ALZ.pca.scaled <- prcomp(predictors[,-130], scale = TRUE)
```

10.4 *K*-Means Clustering

Clustering refers to a very broad set of techniques for finding subgroups, or clusters, in a data set. The goal of clustering is to find a partition of the data into distinct groups so that the observations within each group are quite similar to each other in some sense. Such a sense of similarity is often a domain-specific consideration that must be made based on knowledge of the data being studied. In Chapter 4, SBML, we discussed the fact that similarity is related to the purpose or outcome variable; therefore, here the similarity must be related to some vague outcome or possible multiple outcomes/purposes. In libraries, we organize the books by different categories and sub-categories, while such selections of categories and sub-categories are based on customers' needs that are often not clearly defined. At home, we organize our stuff into clothes, shoes, kitchen utilities, and others for convenience when we use them. Therefore, clustering must have some purposes that are difficult to characterize using a simple outcome measure. For this reason, supervised learning is not applicable to the problem.

A good clustering example in the commerce would be *clustering for market segmentation*. Suppose we have access to big data (e.g., median household income, occupation, distance from nearest urban area) from a large number of people who may or may not be our existing customers. Our goal is to identify subgroups of people who might be more receptive to a particular form of advertising, or to group them (in terms of data) according to the likelihood of purchasing a particular product.

Unlike PCA, which looks for a low-dimensional representation of the observations that explains a good fraction of the variance, clustering looks for homogeneous subgroups among the observations.

There are two commonly-used clustering methods: *K*-means clustering and hierarchical clustering. In *K*-means clustering, we seek to partition the observations into a pre-specified number of clusters, while in hierarchical clustering we do not know in advance how many clusters we want. Instead, hierarchical clustering will end up with a tree-like visual representation of the observations, called a dendrogram, that allows us to view at once the clustering obtained for each possible number of clusters (Figure 10.1).

The code to generate Figure 10.1 is presented in the following:

```
library(stats)
USArrestsHC<- scale(na.omit(USArrests))
head(USArrestsHC)
 # Dissimilarity matrix
 disimMatrix <- dist(USArrestsHC, method = "euclidean")
 # Hierarchical clustering using Complete Linkage
 hc1 <- hclust( disimMatrix, method = "complete" )
```

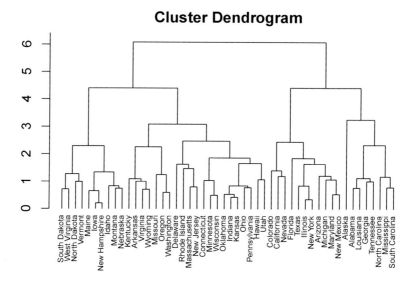

FIGURE 10.1
US Arrest Hierarchical Clustering

```
# Plot the dendrogram
plot(hc1, cex = 0.6, hang = -1)
```

K-Means Algorithm

The K-means algorithm is one of the simplest and most popular clustering methods. Let the data be $X = \{x_i; i = 1, ..., n\}$ and cluster $C = \{c_j; j = 1, ..., k\}$. For a given k, the goal of clustering is to find C, such that

$$\min_{c_j; j} \left(\sum_{i=1}^{n} ||x_i - c_j|| \right). \tag{10.7}$$

The K-means algorithm is described as follows:

1. Randomly choose k data points from X as the initial cluster C.

2. Reassign all $x_i \in X$ to the closest cluster mean c_j.

3. Update all $c_j \in C$ with the mean of their corresponding clusters.

4. Repeat Steps 2 and 3 until the cluster assignments don't change.

The convergence of the algorithm is guaranteed in a finite number of iterations. However, when the distance $\sum_{i=1}^{n} ||x_i - c_j||$ is a nonconvex function, the convergence can lead to a local optimum. The algorithm is also sensitive to outliers and could lead to some empty clusters. Here we have assumed that k is fixed. The clustering problem with variable k is very challenging, computation-wise.

k-means (k-medoids) clustering fits exactly k clusters as specified and the final clustering assignment depends on the chosen initial cluster centers

Haraty et al. (2015) studied an enhanced k-mean clustering algorithm for pattern discovery in health data. Yildirim et al. (2014) use a k-means algorithm to discover hidden knowledge in health records of children and data mining in bioinformatics. They conclude that medical professionals can investigate the clusters that our study revealed, thus gaining useful knowledge and insight into this data for their clinical studies. Wallner et al. (2010) studied correlation and cluster analysis of immunomodulatory drugs based on cytokine profiles. In the study, they determined cytokine profiles for 26 non-biological immunomodulatory drugs or drug candidates and used these profiles to cluster the compounds according to their effect in a preclinical *ex vivo* culture model of arthritis and to predict the functions and drug target of a novel drug candidate based on profiles. Thus their method provides a fingerprint-like identification of a drug as a tool to benchmark novel drugs and to improve descriptions of mode of action.

Example 10.2: K-means Clustering for Alzheimer's Disease

```
library(AppliedPredictiveModeling)
library("factoextra", lib.loc="/Library/Frameworks/R.framework/Versions/
3.5/Resources/library")
data("AlzheimerDisease", package = "AppliedPredictiveModeling")
AD.scaled = scale(predictors[,-130]) # Remove non-scale variable
# K-means clustering
km.AD <- kmeans(AD.scaled, 4, nstart = 10)
km.AD$cluster # cluster number for each subject
km.AD$size # Cluster size
km.AD$centers # Cluster means
# Visualize kmeans clustering
# use repel = TRUE to avoid overplotting
fviz_cluster(km.AD, predictors[,-130], ellipse.type = "norm")
# Change the color palette and theme
fviz_cluster(km.AD, predictors[,-130], palette = "Set2", ggtheme =
theme_minimal())
```

The output for clusters are presented in Figure 10.2. Since the final K-means clustering result is sensitive to the random starting assignments, we specify $nstart = 10$ (the default value is 1). This means that R will try 10 different random starting assignments and then select the best results corresponding to the one with the lowest intracluster variation

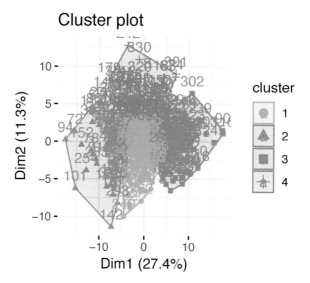

FIGURE 10.2
k-Means Clustering of Alzheimer's Disease Data

10.5   Hierarchical Clustering

Hierarchical clustering seeks to build a hierarchy of clusters. Strategies for hierarchical clustering can be either agglomerative or divisive. An *agglomerative strategy* is a bottom-up approach, that is, each observation starts in its own cluster, and pairs of clusters are merged as one moves up the hierarchy. A divisive strategy is a top-down approach, that is, all observations start in one cluster, and splits are performed recursively as one moves down the hierarchy.

Divisive clustering with an exhaustive search is $O(2^n)$ in complexity, but it is common to use faster heuristics to choose splits, such as k-means. The results of hierarchical clustering are usually presented in a dendrogram.

To decide which clusters should be combined (agglomerative strategy), or where a cluster should be split (divisive strategy), a measure of dissimilarity or similarity between sets of observations is required, such as distance or linkage between pairs of observations. The distance can be Euclidean distance, squared Euclidean distance, Manhattan distance, maximum distance, Mahalanobis distance, or something else. The linkage criterion can be *complete-linkage clustering*, $\max\{d(a,b)\,;a \in A, b \in B\}$, *single-linkage clustering*, $\min\{d(a,b)\,;a \in A, b \in B\}$, or *average-linkage clustering*, $\frac{1}{|A|\cdot|B|}\sum_{a\in A}\sum_{b\in B}d(a,b)$. Single linkage suffers from chaining because to merge two groups, we only need one pair of points to be close, irrespective of all others. Therefore clusters can be too spread out. In contrast, complete linkage avoids chaining, but suffers from crowding. Average linkage tries, using the average pairwise dissimilarity, to strike a balance.

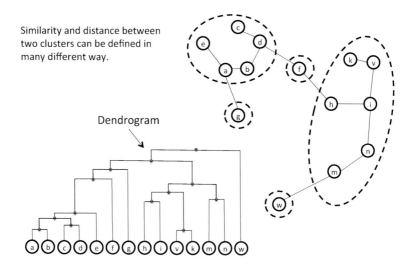

Similarity and distance between two clusters can be defined in many different way.

Dendrogram

FIGURE 10.3
Agglomerative Clustering Algorithm and Dendrogram

Agglomerative Algorithm (Figure 10.3):

1. Choose a dissimilarity measure between two subjects or clusters, e.g., the minimum *Euclidian distance* between subjects from two clusters.

2. In the set of n subjects, identify the most similar pair of subjects (with the minimum distance) and combine them into one cluster. Now there are $n-1$ clusters (a cluster can just have one subject).

3. Among the new set of $n-1$ clusters, identify the most similar pair of clusters with the smallest distance and combine them into one cluster.

4. Among the new set of $n-2$ clusters, identify the most similar pair of cluster based on the distance and combine them into one cluster.

5. This procedure continues until all n subjects have been combined into one cluster.

Divisive Algorithm:

1. Choose a dissimilarity measure.

2. Start with all points in one cluster.

3. Until all points are in their own cluster, repeatedly: split the group into two resulting in the largest dissimilarity.

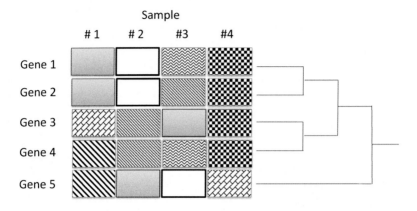

FIGURE 10.4

Gene Heatmap and Hierarchical Clustering

Dissimilarity scores between merged clusters only increases as we run the algorithm, meaning that we can draw a proper dendrogram, where the height of a parent is always higher than height of its children.

In drug discovery and bioinformatics, a typical use of hierarchical clustering is a *heatmap*, where one axis is for the gene and the other is the sample (subject). The color (the number on each cell) on the cell is an indication of the gene expression level (Figure 10.4).

Unlike K-means clustering, for each given pairwise dissimilarities d_{ij} between data points, hierarchical clustering produces a unique result without the need to choose initial starting positions (number of clusters). At one end, all points are in their own cluster; at the other end, all points are in one cluster.

In the latter 1990s, the United States National Cancer Institute conducted an anticancer drug discovery program in which approximately 10,000 compounds were screened in successive years in vitro against a panel of 60 human cancer cell lines from different organs (Shi LM et al., 1998). They tested approximately 62,000 compounds to collect information on activity patterns. Anticancer activity patterns of 112 ellipticine analogs were analyzed using a hierarchical clustering algorithm. A dramatic coherence between molecular structures and their activity patterns was discovered from the cluster tree: the first subgroup consisted principally of normal ellipticines, whereas the second subgroup consisted principally of N2-alkyl-substituted ellipticiniums. The ellipticiniums were more potent on average against p53 mutant cells than against p53 wild-type cells. This study, with its application of unsupervised learning, provided insights into the relationship between activity patterns of anticancer drugs and the molecular pharmacology of cancer.

The application of established drug compounds to new therapeutic indications, known as drug repositioning, offers several advantages over traditional drug development, including the reduction of both development time and costs. Sirota et al. (2011) used hierarchical clustering to predict novel therapeutic indications on the basis of comprehensive testing of molecular signatures in drug-disease pairs. Integrating gene expression measurements from 100 diseases and gene expression measurements on 164 drug compounds, the team rediscovered many known drug-disease relationships and predicted many new indications for these 164 drugs. They also experimentally validated some of the predictions.

Böcker et al. (2005) uses hierarchical clustering for large compound libraries. Zhang et al. (2017) applied hierarchical cluster analysis in clinical research with heterogeneous study population, focusing on visualization.

Example 10.3: Hierarchical Clustering for Breast Cancer Patients

The R function *hclust()* implements hierarchical clustering in the *stats* package. In order to show a clear plot for the hierarchical clustering, only 30 patients from the BreastCancer dataset are included in the following analysis.

```
library(stats) # for hclust() function
require(graphics)
library(AppliedPredictiveModeling)
BC=BreastCancer[1:30,-11] # Remove the non-value outconme
hc <- hclust(dist(BC), "ave")
plot(hc)
head(BC)
```

We may scale the attributes before performing the clustering or we can use a different distance definition. See Figure 10.5 for the dendrogram.

10.6 Self-Organizing Maps

A self-organizing map (SOM, Kohonen, 1980s) or self-organizing feature map (SOFM) is a type of artificial neural network (ANN) that is trained using unsupervised learning to produce a low-dimensional (here dimension is not the feature dimension, but the number of data points), discretized representation of the input space of the training samples, called a map.

As an example, suppose we have 100 subjects, each characterized by 3 attributes, gender, age, and weight, denoted by input data x_{ij} ($i = 1, 2, ..., 100$,

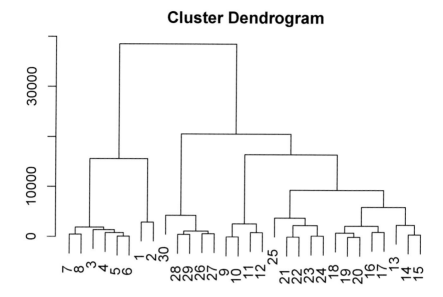

FIGURE 10.5
Hierarchical Clustering of 30 Breast Cancer Patients

$j = 1, 2, 3$). We want to artificially create 8 typical persons to represent the 100 subjects without significant information loss for the problem at hand. How to make up the 8 people? Maybe we should include male, female, young, old, light, and heavy, $2 \times 2 \times 2 = 8$ subjects. But how do we define the age and weight thresholds? The SOM is an AI method used to determine such virtual "typical subjects," called nodes or neurons (Figure 10.6). The number of typical subjects is predetermined; the thresholds for all the attributes are called the weights w_{kj} of the SOM. An SOM is essentially an iterative process of finding the optimal weights for the corresponding attributes at each node.

Like most artificial neural networks, SOMs operate in two modes: training and mapping. Training builds the map using input examples, while mapping automatically classifies a new input vector. The output of an SOM can be visualized in the map space, which consists of components called nodes or neurons. The number of nodes (equivalent to clusters) is defined beforehand, usually as a finite two-dimensional region where nodes are arranged in a regular hexagonal or rectangular grid. SOM is to convert n data point in K-feature space to a collection of nodes (neurons) organized in a 2-D space, called the map. Each node in the map has hidden K-features.

Figure 10.6 is a diagram to illustrate how a SOM works. The data include $n = 100$ observations of 3 features and 8 nodes. Before training, an initial weight vector is given to each node. When a training example is fed to the network, its Euclidean distance between input vector $x = (x_1, x_2, x_3)$ and

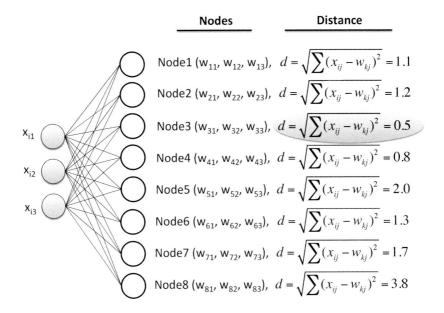

FIGURE 10.6
Best Matching Unit (BMU) is Highlighted with Distance 0.4.

weight vector $w_i = (w_{i1}, w_{i2}, w_{i3})$ at each neuron is computed. The neuron with smallest distance is called the best matching unit (BMU). During the learning process, the weights W_v of the BMU and neurons close to it in the SOM grid are adjusted towards the input vector. The magnitude of the change decreases with time and with the grid-distance from the BMU. The update formula for a neuron v with weight vector $W_v(s)$ is

$$W_v(s+1) = W_v(s) + \Theta(u, v, s) \cdot \alpha(s) \cdot (D(t) - W_v(s)),$$

where s is the step index, t is an index into the training sample, and u is the index of the BMU for $D(t)$. Here $\alpha(s)$ is a monotonically decreasing learning rate, which a Gaussian function is commonly chosen for. $D(t)$ is the input vector, and $\Theta(u, v, s)$ is the neighborhood function which gives the distance between the neuron u and the neuron v in step s.

The effect of weight updating is equivalent to dragging nodes within the defined neighborhood towards the BMU. The closer the node is to the BMU, the more dragging will be applied. Regardless of the functional form of the neighborhood function, the learning coefficient $\alpha(s)$ shrinks with time s during the training so that the weight updating will eventually converge. At the beginning when the neighborhood is broad, the self-organizing takes place on the global scale. When the neighborhood has shrunk to just a couple of neurons, the weights are converging to local estimates. In some implementations, the neighborhood function Θ also decreases steadily with increasing s. The final weights w_{kj} characterize the nodes or the virtual "typical subjects" produced by the SOM.

In applications, Schneider et al. (2009) presented a critical discussion on the advantages, limitations and potential future applications of SOMs along with case studies. Ihmaid, et al. (2016) discussed the SOM-based classification of inhibitors with different selectivity profiles using different structural molecular fingerprints. Daniel et al. (2014) used SOM-based prediction of drug equivalence relationships (SPiDER) in a study that merged the concepts of SOM, consensus scoring, and statistical analysis to successfully identify targets for both known drugs and computer-generated molecular scaffolds. They discovered a potential off-target liability of fenofibrate-related compounds, and in a comprehensive prospective application. Their results demonstrate that SPiDER can be used to identify innovative compounds in chemical biology and in the early stages of drug discovery and can help to investigate the potential side effects of drugs and their repurposing options.

Example 10.4: Creating Self-Organized Maps for Alzheimer's Disease

The *R* package *kohonen* facilitates the creation and visualization of SOMs, including *supersoms*. A supersom is an extension of SOMs to multiple data layers, possibly with different numbers and different types of variables. NAs are allowed. A weighted distance over all layers is calculated to determine the winning units during training. Functions *som* and *xyf* are simply wrappers for supersoms with one and two layers, respectively.

In the following the SOM example by Lynn (2019), we analyze the Alzheimer's disease dataset using *kohonen*.

```
# Self-organized Map
library(kohonen)
library(tempR)
library(AppliedPredictiveModeling)
data("AlzheimerDisease", package = "AppliedPredictiveModeling")
ALZ <- cbind(predictors, diagnosis)
length(diagnosis)
n = nrow(ALZ)
trainIndex = sample(1:n, size = round(0.9*n), replace=FALSE)
ALZtrain=ALZ[trainIndex, 1:120]
ALZtest=ALZ[-trainIndex, 1:120]
# Change the data frame with training data to a matrix
# Also center and scale all variables to give them equal importance
data_train_matrix <- as.matrix(scale(ALZtrain))
```

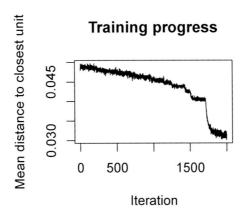

FIGURE 10.7
SOM Training Process

```
# Create the SOM Grid (hexagonal and circular ) to train the SOM.
som_grid <- somgrid(xdim=12, ydim=12, topo="hexagonal")
# Finally, train the SOM, options for the number of iterations (rlen),
# the learning rates (alpha), and others are available
som_model <- som(data_train_matrix, grid=som_grid, rlen=2000,
    alpha=c(0.05,0.01), keep.data = TRUE )
```

The *kohonen.plot* function can be used to visualize a user-generated SOM and to explore the relationships between the variables in a dataset. There are a number of different plot types available. Understanding them is key to exploring one's SOM and discovering relationships in one's data. As the SOM training iterations progress, the neighborhood function will reduce and eventually reach a minimum plateau. This can be visualized using a *plot* function (Figure 10.7).

```
# Visualize distance reduction in SOM training
plot(som_model, type = "changes")
```

The term, *node counts* is the number of samples that are mapped to each node on the map. This metric can be used as a measure of map quality. That is, ideally the sample distribution should be relatively uniform. Large values in some map areas suggests that a larger map would be beneficial, while empty nodes indicate that the map size is too big for the number of samples. We can use the following *R* code to visualize the node counts (Figure 10.8).

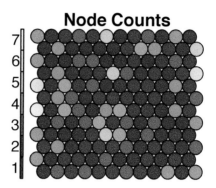

FIGURE 10.8
Node Counts of the Alzheimer's Disease SOM

```
#Node count plot
plot(som_model, type="count", main="Node Counts")
```

A SOM heatmap allows the visualization of the distribution of a single variable across the map. Figure 10.9 is the heatmap for the attribute adiponectin from the Alzheimer's disease dataset. Typically, SOM multiple heatmaps can be generated to compare and identify interesting areas on the map. Note that the individual sample positions do not move from one visualization to another; the map is simply colored by different variables.

The default *Kohonen heatmap* is created by using the type heatmap, and then providing one of the variables from the set of node weights. In this case we visualize the average education level on the SOM. Here is the code to create the heatmap for *Adiponectin* in the Alzheimer's disease dataset:

```
# Kohonen Heatmap creation
coolBlueHotRed <- function(n, alpha = 1) {rainbow(n, end=4/6,
alpha=alpha)[n:1]}
plot(som_model, type = "property", property = getCodes(som_model)[,4],
    main=colnames(getCodes(som_model))[4], palette.name=coolBlueHotRed)
```

Clustering can be performed on the SOM nodes to isolate groups of samples with similar metrics. The Kohonen package documentation shows how a map can be clustered using hierarchical clustering.

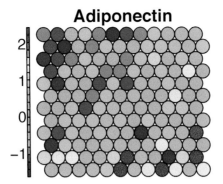

FIGURE 10.9
Heatmap of Alzheimer's Disease

10.7 Network Clustering and Modularity

A *network* is a collection of members (called nodes) that have relatively stable interrelationships (called links). Such stable interrelationships allow regular "information" exchanges among its members. Such information exchange can lead to changes in the member's behavior and eventually affect the topology and dynamics of the network. For example, each of us has friends who in sum form our friends network. Such a relatively stable group allows us to share "information," more often within the network than outside the network. The information exchanged can take many forms; the passing on of disease is but one unusual example. Different networks can be formed and used for different purposes; for instance, the internet, social, transportation, biological, communication, electronic, and electrical networks, just to list a few. A network can be artificial, such as an artificial neural network that mimics human neural networks and can be used to effectively solve real-world problems.

A *social network* is one of the most interesting networks since individual nodes benefit (or suffer) from direct and indirect relationships. Friends might provide access to favors from their friends. Just as Jackson (Jackson, 2008, p. 3) describes it: "Social networks permeate our social and economic lives. They play a central role in the transmission of information about job opportunities and are critical to the trade of many goods and services. They are the basis for the provision of mutual insurance in developing countries. Social networks are also important in determining how diseases spread, which products we buy, which languages we speak, how we vote, as well as whether we become criminals, how much education we obtain, and our likelihood of succeeding professionally. The countless ways in which network structures affect our well-being make it critical to understand (1) how social network structures affect behavior and (2) which network structures are likely to emerge in a society."

Biological networks (pathways) are interesting to many scientists. A biological pathway is a molecular interaction network in biological processes. The pathways can be classified into the fundamental categories: regulatory, metabolic, and signal transduction pathways. There are about 10,000 pathways in humans, including nearly 160 pathways involving 800 critical chemical interactions.

A *gene regulatory pathway or genetic regulatory pathway* is a collection of DNA segments in a cell which interact with each other and with other substances in the cell, thereby governing the rates at which genes in the network are transcribed into mRNA. In general, each mRNA molecule goes on to make a specific protein with particular structural properties. The protein can be an enzyme needed for the breakdown of a food source or toxin. By binding to the promoter region at the start of other genes, some proteins can turn the genes on, initiating or inhibiting the production of another protein.

A *metabolic pathway* is a series of chemical reactions occurring within a cell, catalyzed by enzymes, resulting in either the formulation of a metabolic product to be used or stored by the cell, or the inhibition of another metabolic pathway. Pathways are important to the maintenance of homeostasis within an organism.

A *signal transduction pathway* is a series of processes involving a group of molecules in a cell that work together to control one or more cell functions, such as cell division or cell death. Molecular signals are transmitted between cells by the secretion of hormones and other chemical factors, which are then picked up by different cells. After the first molecule in a pathway receives a signal, it activates another molecule, and then another, until the last molecule in the signal chain is activated. Abnormal activation of signaling pathways can lead to diseases such as cancer. Drugs are being developed to block these disease pathways.

Systems biology is a newly emerging, multi-disciplinary field that studies the mechanisms underlying complex biological processes by treating these processes as integrated systems of many interacting components (Materi and Wishart, 2007).

Networks often share certain common structures. Network topology is the study of the static properties or structures of a network. What follows is some basic terminology used in describing the topology of a network. A walk is a sequence of links connecting a sequence of nodes. A cycle is a walk that starts and ends at the same node, with all nodes appearing once except the starting node, which also appears as the ending node. A path is a walk in which any given node appears at most once in the sequence. A tree is a connected network that has no cycle. In a tree, there is unique path between any two nodes. A forest is a network with tree components. A component is any connected network.

How many friends (*links*) a person has often reflects, and is one way to measure, the importance of the person (their *centrality*) in a network. Thus, by the degree of a node we mean the number of direct links the node has. A node's *authority* is the number of nodes that point to it, assuming all links point in one direction or another We often need to know how fast information can travel from one person to another, which is described by *geodesics*, the shortest paths between two nodes. The centrality of a node can also be measured by its so-called *closeness*, which is defined as the inverse of average geodesics from this node to all the others in the network. *Closeness* measures how easily a node can reach other nodes, whereas the *diameter of a network*, i.e., the largest distance between any two nodes in the network, can be an indicator of the influential range of a person (node).

One of the mathematically simplest network models is the so-called *random graph*, in which each node is connected (or not) with an independent probability having a binomial distribution of degrees. However, most networks in the real world, e.g., the Internet, the worldwide web, biological networks, and some social networks, are highly right-skewed, meaning that a large majority of nodes have low degree but a small number, known as "hubs," have high degree. This means that only a relatively small number of nodes play the big roles. Similarly, the distribution of geodesics is a measure of the closeness of a network. Calculating the geodesics' distribution is straightforward. Denote a direct connection between nodes i and j by g_{ij}; $g_{ij} = 1$ if i and j are connected; otherwise $g_{ij} = 0$. Then matrix $\left[g^k\right]$ represents all possible paths with length k in the network. For example, $\left[g^2\right]_{ij} = \sum_k g_{ik}g_{kj} = 3$ implies there are 3 walks of length 2 between nodes i and j, while $\left[g^3\right]_{ij} = \sum_k \sum_m g_{ik}g_{km}g_{mj} = 4$ means there are 4 walks of length 3 between nodes i and j.

Some relationships in a network have the property of direction as well as other properties that are necessary to model real world problems. A Petri net (Petri, 1962) is a mathematical network model for simulating dynamic systems, such as electronic signal processes, traffic systems, biological pathways, etc. (Chang, 2010).

The study of game theory in the network setting is called network evolutionary game theory. Game players interact with each other through network connections. In such interactions, your friends' strategy (and hence their gain or loss) will affect your strategy. Thus, a "good guy" can become a "bad guy" or vice versa, with various probabilities. The dynamics of network games can be studied through Monte Carlo simulations.

We now discuss clustering and modularity. The calculation of the local clustering coefficient, $C(V)$, is presented in Figure 10.12. The minimum and maximum numbers of links among V's neighbors are 0 and $K_v (K_v - 1)$. Therefore, the clustering coefficient ranges from 0 to 1. The global clustering coefficient is the average of $C(V)$ over all nodes V in the network.

A *clique* in the social sciences, refers a group of individuals who interact with one another and share similar interests. Interacting with cliques is part of normative social development regardless of gender, ethnicity, or popularity. Cliques are most commonly studied during adolescence and middle childhood development. *Homophily* or birds of a feather flock together, describes the

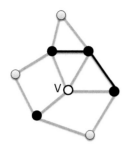

V = a node with K_v degrees
N_v = number of links between neighbors of V
CC(V) = clustering coefficient
  = Fraction of possible interconnections

$$K_v = 4, N_v = 2,$$

$$CC(V) = \frac{2N_v}{K_v(K_v - 1)} = \frac{2 * 2}{4 * 3} = \frac{1}{3}$$

FIGURE 10.10
Calculation of Clustering Coefficient

tendency that people link up with others based on sharing similar character-istics. The existence of homophily is also very prevalent in today's society. This concept can be seen as a possible main cause for clique formation.

The global clustering coefficient can be used as a metric for measuring network *modularity*. Modularity measures the strength of division of a net-work into modules (also called groups, clusters, or communities). Networks with high modularity have dense connections among the nodes within clusters but sparse connections between nodes in different clusters. Network topolog-ical properties characterize certain community structure that has substantial importance in building an understanding regarding the dynamics of the net-work. For instance, a closely connected social community will imply a faster rate of transmission of information or diseases among them than a loosely con-nected community. Biological networks, including animal brains, exhibit a high degree of modularity. Modularity is the fraction of the edges that fall within the given groups minus the expected fraction if edges were distributed at ran-dom. Modularity, however, suffers a resolution limit; that is, it is unable to detect small communities. Biological networks, including animal brains, ex-hibit a high degree of modularity.

There is a large amount research on unweighted networks, but very lim-ited study of weighted networks (Newman, 2001; Brandes, 2001; Opsahl and Panzarasa, 2009; Barrat et al., 2004; Zhang and Horvath, 2005) The *Louvain method* for community detection (clustering) is a method of extracting com-munities from large networks created by Blondel et al. (2008). In the Louvain method of community detection, first small communities are found by opti-mizing modularity locally on all nodes, then each small community is grouped into one node and the first step is repeated. The method is similar to an earlier method by Clauset, Newman, and Moore (2004) that connects communities whose amalgamation produces the largest increase in modularity.

The value to be optimized is modularity, defined as a value between -1 and 1 that measures the density of links inside communities as compared to links between communities. For a weighted graph, the modularity is defined as:

$$Q = \frac{1}{2m} \sum_{i,j} \left(w_{ij} - \frac{k_i k_j}{2m} \right) \delta\left(c_i, c_j\right), \tag{10.8}$$

where A_{ij} represents the edge weight between nodes i and j; k_i and k_j are the sum of the weights of the edges attached to nodes i and j, respectively, $2m$ is the sum of all of the edge weights in the network; c_i and c_j are the communities of the nodes, and δ is Kronecker-delta function.

10.8   Unsupervised to Supervised Learning

We are going to discuss an interesting method that converts the unsupervised problem into a supervised problem.

A clustering problem can be described as a density estimation problem: find a subset s_j of data (or support S_j), $j = 1, ..., p$, such that

$$P\left[\cap_{j=1}^{p} \left(X_j \in s_j\right)\right] \tag{10.9}$$

is high. To solve this unsupervised density problem, we can to convert it to a supervised problem (Chang, 2011; Hastie et al., 2001).

Let $x_1, x_2, ..., x_n$ be the i.i.d. random sample from an unknown p.d.f. $g(x)$. Let $x_{n+1}, x_{n+2}, ..., x_{n+n_0}$ be the i.i.d. random sample from a reference (e.g., uniform) p.d.f. $g_0(x)$. Assign weight $w = n_0/(n_0 + n)$ to each observation from $g(x)$ and $w_0 = n/(n_0 + n)$ to each observation from $g_0(x)$. Then the resulting pooled data with $n + n_0$ observations can be viewed as a sample drawn from the mixture density $(g(x) + g_0(x))/2$. Assign $Y = 1$ to each sample point from $g(x)$ and $Y = 0$ to each sample point from $g_0(x)$. Then

$$\mu(x) = E(Y|x) = \frac{g(x)}{g(x) + g_0(x)} \tag{10.10}$$

can be estimated by the supervised learning using the combined sample

$$(y_1, x_1), (y_2, x_2), ... (y_{n+n_0}, x_{n+n_0}) \tag{10.11}$$

as the training data set. The resulting estimate $\hat{\mu}(x)$ can be inverted to provide an estimate for $g(x)$:

$$\hat{g}(x) = g_0(x) \frac{\hat{\mu}(x)}{1 - \hat{\mu}(x)}. \tag{10.12}$$

Similarly, for logistic regression, since the log-odds is given by

$$f(x) = \ln \frac{g(x)}{g_0(x)}, \tag{10.13}$$

$g(x)$ can be estimated by

$$\hat{g}(x) = g_0(x) e^{\hat{f}(x)}. \qquad (10.14)$$

10.9 Summary

Unlike *supervised learning*, for *unsupervised learning* there are no clear, correct answers. The goal of *unsupervised learning* is to identify or simplify data structure. Unsupervised learning problems can be divided into problems of *clustering, association,* and *anomaly detection*. A *clustering problem* is where you want to discover the inherent groupings in the data, such as grouping customers by purchasing behavior. An *association rule learning* problem is where you want to discover rules that describe large portions of your data, such as people buying product A also tending to buy product B. The third type of problem, *anomaly detection* or *outlier detection*, seeks to identify items, events or observations that do not conform to an expected pattern, such as bank fraud, structural defects, medical problems, or text errors.

Clustering is used when the response is not available or there are potentially multiple purposes, or when we may not know the future purposes exactly, but based on the similarity principle, the subjects in the same cluster will likely have similar outcomes that are potentially of interest.

There are also AI problems sitting between supervised and unsupervised learning, which can be approached via *semi-supervised learning*. In semi-supervised learning problems, we have only have parts of the input data labeled, and the rest are unlabeled. An example would be a photo archive where only some of the images (e.g., dogs, cats, persons) are labeled, and the majority are unlabeled.

Finding the set of events that correlate many others is often the focal point and springboard for further causality research. *Link-analysis* also provides a way to find the event sets with high density. Finding sale items that are highly related (or frequently purchased together) can be very helpful when stocking shelves or doing cross-marketing in sales promotions, catalog design, and *consumer segmentation* based on buying patterns. A commonly used algorithm for such density problems is the so-called *Apriori*, which is an algorithm that identifies the frequent individual items in the database and extends them to larger and larger item sets as long as those item sets appear sufficiently often in the database.

Principal component analysis (PCA) can be used as a tool for data pre-processing before supervised techniques are applied. PCA produces a low-dimensional representation of a dataset. It finds, sequentially, a set of linear combinations of the predictors that have maximal variance and are orthogonal with each other.

Unlike PCA, which looks for a low-dimensional representation of the observations that explains a good fraction of the variance, clustering looks for homogeneous subgroups among the observations. There are two commonly used clustering methods: *K-means* clustering and *hierarchical clustering*. In *K*-means clustering we seek to partition the observations into a pre-specified number of clusters, while in hierarchical clustering we do not know in advance how many clusters we want. Instead, hierarchical clustering yields a tree-like visual representation (dendrogram) of the observations.

A *self-organizing map* (SOM) is a type of *artificial neural network* (ANN) used to produce a low-dimensional (data points), discretized representation of the input space of the training samples, called a map. Like most artificial neural networks, SOMs operate in two modes: training and mapping. Training builds the map using input examples, while mapping automatically classifies a new input vector. The output of SOM can be visualized in the map space, which consists of components called nodes or neurons. The number of nodes (equivalent to clusters) is defined beforehand, usually as a finite two-dimensional region where nodes are arranged in a regular hexagonal or rectangular grid.

Network clustering is a way of finding communities within a network in terms of some network properties, such as clustering coefficients.

Unsupervised learning is a critical foundation of supervised learning. Any application of *supervised learning* must involve some sort of unsupervised learning. We call this phenomenon entanglement of supervised and unsupervised learning. For instance, when we decide which features need to be collected for our supervised learning model, we have already used implicitly unsupervised learning or clustering. That is, we perform simple clustering based on one set of features, instead of any other features, and assume that as long as two objects have identical values for these features their outcomes will be the same, regardless of whether they might be different in other aspects. Another example, where clustering is implicitly performed before any supervised learning, is when we decide the number of digits or decimal digits to keep in a set of measurements: we have implicitly put objects with the same values into the same cluster, even though the rest of the decimal digits of their measurements might be different

10.10 Problems

10.1: Use the two different clustering algorithms to analyze the *infert* dataset in the *R* datasets package (hint: with and without considering the outcome variable for the clustering).

10.2: Explain the weight difference between ANN and SOM.

11

Reinforcement Learning

11.1   Introduction

Reinforcement learning (RL) emphasizes learning through interaction with (real or virtual) environments. Feedback from one's environment is essential for learning. RL can be used when the correct answer is difficult to define or there are too many steps for the agent to take to complete the task. Taking a driverless car as an example, we cannot define the road conditions manually. In such a case, RL can be helpful in building the overall architecture (Figure 11.1), though each component, such as a sensor, may still use other learning methods (e.g., CNN).

RL can be particularly useful in attacking problems that require strong interactions with different environments, such as self-cleaning vacuum cleaners and rescue robots. A rescue robot is a robot that has been designed for the purpose of rescuing people in mining accidents, urban disasters, hostage situations, and explosions. The benefits of rescue robots to these operations include reduced personnel requirements, reduced fatigue, and access to otherwise unreachable areas.

A well-known RL example is AlphaGo Zero. AlphaGo is a computer program that plays the board game Go. It was developed by Alphabet Inc.'s Google DeepMind in London. At the 2017 Future of Go Summit, its successor AlphaGo Master beat Ke Jie, the world No.1-ranked player at the time, in a three-game match. AlphaGo and its successors use a Monte Carlo tree search algorithm to find its moves based on knowledge learned by an ANN from both human and computer play.

AlphaGo Zero, a version created without using data from human games, is stronger than any previous version. By playing games against itself, AlphaGo Zero surpassed the strength of AlphaGo Lee in three days by winning 100 games to 0, reaching the level of AlphaGo Master in 21 days. The next version, AlphaZero, within 24 hours achieved a superhuman level of play in three games by defeating world-champion programs, Stockfish, Elmo, and the three-day version of AlphaGo Zero. In each case it made use of custom tensor processing units (TPUs) that the Google programs were optimized to use.

Reinforcement learning can embody a model-based approach, such as a Markov decision process. The methods used to solve the optimization problem include dynamic programming with either policy-based or value-based

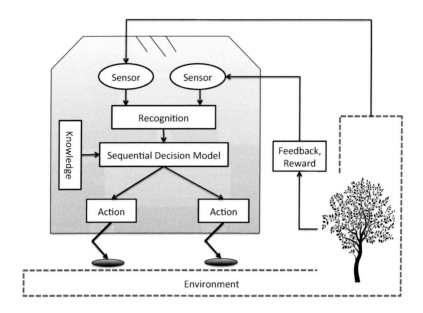

FIGURE 11.1
Agent with Reinforcement Learning Architecture

algorithms. Model-free RL approaches include Bayesian Q-learning. There are
also game–theory-based formulations. Chang discusses all these approaches f
in the field of drug development (Chang, 2010). Czibula et al. (2015) proposed
a RL model to solve the protein-folding problem, predicting the bidimensional
structure of proteins in the hydrophobic-polar model.

11.2   Sequential Decision-Making

11.2.1   Descriptive and Normative Decision-Making

Life is about decisions. You may not be aware that you are making decisions
but every second you are, be they big or small. There are two distinctive approaches to making decisions: descriptive and normative decision theories.
Descriptive decision-making considers a decision as a specific information
processing process; it employs a study of the cognitive processes that lead
to decisions. For example, the ways humans deal with conflicts and per-
ceive the values of the solutions are one form of descriptive decision-making.

Descriptive decision-making seeks explanations for the ways individuals or groups of individuals arrive at decisions so that methods can be developed for influencing and guiding the decision process. Normative decision-making considers a decision as a rational act of choice under the viable alternatives. It is a mathematical or statistical theory for modeling decision-making processes. Classical decision-making theory and Bayesian decision theory fall into this category. Normative decision-making strives to make the optimal decision, given the available information. Hence, it is an optimization approach.

Normative decision-making can be further divided into deterministic and probabilistic approaches. For the deterministic approach, information is considered as completely known, whereas for the probabilistic approach, information is associated with uncertainties or probabilities. In RL for artificial general intelligence (AGI), both descriptive and normative designs should be included. However, it is a narrow form of AI that we are dealing with in the book, and for this we will only focus on the probabilistic normative approach that can be combined with Monte Carlo simulations. We will use many examples from drug development to illustrate the learning process. However, we should be aware that drug development is a sequence of decision processes, and the processes are mixtures of descriptive and normative decisions.

11.2.2 Markov Chain

The simplest and widely studied stochastic process is *Markov chain*, named after Andrey Markov, in which the probability of transiting from one state to a next state does not dependent on how the agent g arrived at the current state. The transition probability from state i to state j is given by

$$P_{ij} = \prod_{k=i}^{j-1} \Pr(s_k|s_{k+1}).$$ (11.1)

A simple example is the board game Candyland. In one version of the game, two players roll dice in turns to determine the number of steps they should move ahead. The person who arrives at the destination first wins. Obviously, the probability of moving from one position to another position is fully determined by the current position and independent of how the player got to the current position.

Example 11.1: Markov Chain for Clinical Development Program

As we discussed early in Chapter 2, clinical trials are often conducted in sequences from phase 1 to phase 3. Sufficient positive results from a phase set off the next phase trial. Promising phase 3 results will trigger the company's submission of a nondisclosure agreement to the FDA for marketing approval.

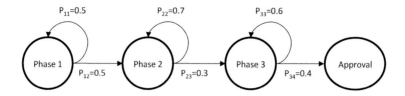

FIGURE 11.2
Markov Chain for Clinical Development Program

The results from earlier phases other than the immediate phase play only a minor role in the decision process. Such a transition process can be modeled by a Markov chain (Figure 11.2).

The average success rates are around 50%, 30%, 40% for phase 1, 2, and 3 trials, respectively (Chang, 2010; Chang et al., 2019). Based on these rates, the transition probabilities are 0.5 from phase 1 to phase 2 (the probability of a no-go after phase 1 is $P_{11} = 0.5$) and 0.3 from phase 2 to phase 3 (probability of a no-go is $P_{22} = 0.7$). About 40% of phase 3 trials finally receive approval for marketing (the probability of failure is $P_{33} = 0.6$).

The transition probability matrix for the clinical trial processes in Figure 11.2 can be written as

$$[T] = \begin{bmatrix} 0.5 & 0.5 & 0.15 & 0.06 \\ 0 & 0.7 & 0.3 & 0.12 \\ 0 & 0 & 0.6 & 0.4 \\ 0 & 0 & 0 & 0.6 \end{bmatrix}$$

From the transition probability matrix, we can see that the probability of moving from phase 1 to phase 3 is 15% and to approval is 6%; the probability from phase 2 to NDA approval is 0.12. However, no one will use the naive transition probabilities to make "Go/No-Go" decisions. Instead, the results from the clinical trials have to be considered in the decision-making process. In the Bayesian approach, the naive probability is the prior probability $P(A)$, which will be combined with observed data $P(B|A)$ to calculate the posterior probability $P(A|B)$ using Bayes' law:

$$P(A|B) = \frac{P(B|A)P(A)}{P(B)}.$$

Since each step moving forward in drug development is one step closer to success, it can be considered as a reward in the process. When a Markov process is attached with rewards, it becomes a *Markov decision process* (MDP).

11.2.3 Markov Decision Process

A *Markov decision process* is similar to a Markov chain, but there are also actions and utilities involved (rewards or gains, see Figure 11.3). MDPs provide a powerful mathematical framework for modeling the decision-making process in situations where outcomes are partly random and partly under the control of the decision-maker. They are useful in studying a wide range of optimization problems solved via dynamic programming and reinforcement learning. Since the 1950s when MDPs were first applied (Bellman, 1957; Howard, 1960), it is now widely used in robotics, automated control, economics, business management, nursing systems, and manufacturing (Gosavi, 2003; Bertsekas, 1995; Nunen, 1976).

There are two types of MDPs, formulated for finite and infinite horizon problems (Powell, 2007, p. 53–56). For a finite horizon problem, the transition probabilities (from state s at time t to state s' at time $t + 1$) vary over time, i.e.,

$$T\left(s, a, s'\right) = \Pr\left(s_{t+1} = s' | s_t = s, a_t = a\right). \tag{11.2}$$

In an infinite horizon MDP problem, the system reaches a steady state and the transition probability only depends on the two states and the action taken, a, and is independent of the time when the transition occurs. Thus the transition probability can be simplified as

$$T\left(s, a, s'\right) = \Pr\left(s' | s, a\right). \tag{11.3}$$

In the remainder of this chapter we will, for the sake of simplicity, discuss infinite MDPs only.

Let's denote a dynamic system (Figure 11.3) that moves over states $s_i \in S, i = 1, 2, ..., N$ and the motion is controlled by choosing a sequence of actions $a_i \in A, i = 1, 2, ...N$. There is a net numerical gain (immediate reward or cost) $g_i\left(\alpha_i\right)$ associated with each action a_i. The goal is to find a policy (strategy) which maximizes the total expected gain. This problem can be formalized as Markov decision process.

A infinite horizon Markov decision process is a four-tuple of four element $\{S, A, \Pr\left(S, A\right), g\left(S, A\right)\}$ with the following properties:
(1) Finite (N) set of states S.
(2) Set of actions A (can be conditioned on the state s).
(3) Policy/strategy/action rule — action mapping to state s.
(4) Discount rate for future rewards $0 < \gamma < 1$.
(5) Immediate reward (utility or gain function) $g : S \times A \to \Re$
$\quad g\left(s, a\right) =$ immediate fixed reward by reaching the state s and taking action a.
(6) Transition model (dynamics) $T : S \times A \times S \to [0, 1]$.

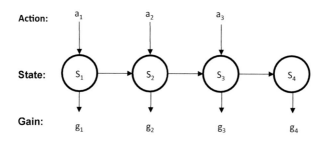

FIGURE 11.3
Sequential Decision-Making (Dynamic System)

$$T\left(s, a, s'\right) = \text{probability of going from } s \text{ to } s' \text{ under action } a:$$

$$T\left(s, a, s'\right) = \Pr\left(s'|s, a_i = a\right).$$

(7) The goal is to find a policy $\pi^* = \{a_i^*, i = 1, ..., N\}$ that maximizes the total expected reward over the course of all possible actions, which may be subject to some initial condition and other constraints. Note that a often associates with state s. We may write a_s for the action at state s.

Here are examples of Markov decision processes in clinical trials (Figure 11.3):

(1) $s_1 = $ Drug Discovery phase, $s_2 = $Preclinical phase, $s_3 = $ Clinical Development phase, and $s_4 = $ Regulatory approval and marketing. Here, a_i represents the general decision rule or action rule for moving from the ith phase to the next phase in drug development.

(2) $s_1 = $ Preclinical phase, $s_i = $ the ith phase clinical trial ($i = 1, 2, 3$), and a_i has the similar meaning as in (1).

(3) $s_1 = $ initiation of a clinical trial, $s_i = $ interim and the final analyses for the $(i-1)^{th}$ phase trial ($i = 2, 3, 4$); actions are the stopping and adaptive rules.

11.2.4 Dynamic Programming

Bellman's optimality principle in dealing with a deterministic sequential decision problem says that an optimal policy has the property that, whatever the initial state and initial decision are, the remaining decisions must constitute an optimal policy with regard to the state resulting from the first decision. This principle is applied to a stochastic decision problem here, but the optimal policy is regarding the expected gain (loss) as opposed to loss in a deterministic case.

Computationally, stochastic decision problems are usually solved using the so-called Bellman equations:

$$V(s) = \max_{a_s \in A} \left(g(s,a) + \gamma \sum_{s' \in S} \Pr(s'|s,a_s) \cdot V(s') \right). \tag{11.4}$$

The optimal policy $\pi^* = \{a_1^*, a_2^*, ..., a_N^*\}$ for the MDP is a vector, whose components are defined by

$$a_s^* = \arg\max_{a_s} \left\{ g(s,a_s) + \gamma \sum_{s' \in S} \Pr(s'|s,a_s) \cdot V(s') \right\}. \tag{11.5}$$

(11.4) can be written in the matrix form:

$$\boldsymbol{V}_\pi = \boldsymbol{g}_\pi + \gamma \boldsymbol{P}_\pi \boldsymbol{V}_\pi. \tag{11.6}$$

The solution to (11.5) can be written in a matrix form:

$$\boldsymbol{V}_\pi = (\boldsymbol{I} - \gamma \boldsymbol{P}_\pi)^{-1} \boldsymbol{g}_\pi. \tag{11.7}$$

The standard family of algorithms used to calculate the policy requires storage for two arrays indexed by state: value V, which contains real values, and policy π which contains actions. The Bellman equations can be used to develop the dynamic programming (a backward induction algorithm). The iteration is going from the last state to the first state (Algorithm 11.1).

Algorithm 11.1: Dynamic Programming for Markov Decision Problem (*Value Iteration*, Bellman, 1957; Powell, 2007)
Objective: return the optimal policy and expected total rewards (a^*, V^*).
Input precision ε.
$V_0(s) := 0 \ \forall s \in S$.
$i := 1$
$\varepsilon_i := 2\varepsilon$
While $\varepsilon_i > \varepsilon$:
    For Each $s \in S$ compute
        $V_i(s) := \max_{a_s \in A} \left(g(s,a_s) + \gamma \sum_{s' \in S} \Pr(s'|s,a_s) \cdot V_{i-1}(s') \right)$
    Endfor
    $\pi^* :=$ solution vector for action rule a to the above equation or (11.4)
    $\varepsilon_i := \max_s ||V_i(s) - V_{i-1}(s)||$
    $V^* = V_i(s)$
    $i := i + 1$
Endwhile
Return (a^*, V^*)

The solution to a Markov decision process can be expressed as a policy π, i.e., a function from states to actions such that, for any given policy, the action for each state is fixed and the MDP behaves just like a Markov chain. Therefore, in addition to the value iteration approach, there is also a policy iteration approach, in which finite numbers of policies (strategies) are identifiable and denoted by

$$\pi_i(s, a), i = 1, ..., m. \tag{11.8}$$

The Bellman equation for evaluating policy $\pi(s, a)$ is given by

$$V(s) = \max_a \pi(s, a) \left\{ g(s, a_s) + \gamma \sum_{s' \in S} \Pr(s'|s, a_s) \cdot V_{s'} \right\}. \tag{11.9}$$

Algorithm 11.2: Dynamic Programming for Markov Decision Problem (*Policy Iteration*, Howard, 1960; Powell, 2007)

    Objective: calculate optimal policy using policy iteration
    Input precision ε, and policy set $\pi = \{\pi_1, ..., \pi_n\}$
    Select an initial policy π_0
    For $i := 0$ **To** n
        Compute transition probability matrix $\boldsymbol{P}_{\pi_i}$ for policy π_i
        Compute gain vector $\boldsymbol{g}_{\pi_i}$ for policy π_i
        Let $v_{i+1} :=$ the solution to equation

$$(\boldsymbol{I} - \gamma \boldsymbol{P}_{\pi_i})v = \boldsymbol{g}_{\pi_i}.$$

    Find a policy $\boldsymbol{\pi}_{i+1}$, whose action elements are defined by

$$a^*_{s,i+1} = \arg\max_{a_s}(g(s, a_s), \gamma P_\pi v_{i+1})$$

    This requires computing an action for each state s.
    If $a^*_{s,i} = a^*_{s,i-1}$ for all states s **Then**
        $\pi^* = \pi_{i+1}$
        Exitfor
    Endif
    Endfor
    Return π^*

    The procedure-wise difference between value iteration and policy iteration is that in value iteration, we find the maximum expected reward for a given state and then loop over all states, whereas in policy iteration, for a given policy, we calculate the total expected reward from all states and then loop over all possible policies.

    In general, either value iteration or policy iteration can solve the problem at hand, but their efficiency will depend on particular situations. If there are many actions or if we already have a fair policy, we choose policy iteration. If there are few actions, acyclic networks, we use value iteration. It is also possible to use a mix of value iteration and policy iteration (van Nunen, 1976; Puterman and Shin, 1978). Another commonly used method for solving the problem is linear programming.

11.3 Pharmaceutial Decision Process

11.3.1 Model for Clinical Development Program

The success of a pharmaceutical company depends on integrating scientific, clinical, regulatory, and marketing approaches to the development and commercialization of therapies. Clinical development program design offers several important benefits (Pharsight.com): (1) It eliminates unnecessary or redundant clinical trials used for internal decision-making; (2) It identifies and addresses critical path issues that could delay development timeliness; (3) It ensures that clinical programs focus quickly and unambiguously on key attributes of the compound.

RL with Monte Carlo simulations can provide a rational basis for decision-making and help in optimizing a compound's regulatory strategy and determining its commercial position and value. Simulation of the clinical development program (CDP) can increase the confidence in decision-making and help to define and track critical success factors and their uncertainties.

Let's study how to model a clinical development program using a Markov decision process.

A typical drug development program includes phase 1 to phase 3 clinical trials conducted in sequence. At the end of each phase the decision is made regarding whether to stop the program (the "No-Go" decision) or carry on to the next phase (the "Go" decision). Each action associates with an estimated net gain (Figure 11.4). Formally it can be modeled by the following MDP:

1. A finite set of states S

   For the case in Figure 11.4, the state space is given by

   $S = \{\text{phase 1, phase 2, phase 3, NDA approval}\}.$

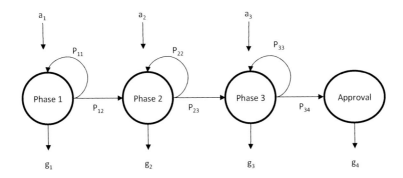

FIGURE 11.4
Markov Decision Process for Clinical Trials

2. A set of actions A_s (state-dependent action space)

   The choice of action set can be $A_s = \{$Go, No-Go$\}$ and/or sample size n_i for the ith phase.

3. Action rules:

   The action rule or decision rule is typically a set of values, $A_i = \{a_{ij}, j = 1, 2, ...m, i = 1, 2, 3\}$, where a_{ij} can be a set of sample size options $(j = 1, ..., m)$ for the ith phase trial, or $A_i = [a_{min}, a_{max}]$ as the range of the cutpoint for the "Go, No-Go" decision-making. The action space can also be a vector, e.g., corresponding to efficacy and safety requirements or to point estimate and confidence interval of a parameter (or p-value).

   The goal is to find the optimal policy $A_i^* = \{a_{1*}, a_{2*}, a_{3*}\}$ that maximizes the expected gain.

4. A discount factor for future rewards $0 < \gamma \leq 1$.

5. Immediate reward:

   In the simplest case, the immediate reward $g_i (a_i = a)$ or $g_i (a)$ for simplicity at the ith state (phase) could be a cost c_{i+1} and a reward from NDA approval or the value of out-licensing of the NME.

6. Transition dynamics $T : S \times A \times S \rightarrow [0, 1]$.

   The transition probability of going from the ith phase to the $(i+1)^{th}$ phase is a function of action rule A_i written as

$$T (s_i, a, s_{i+1}) = \Pr (s_{i+1}|s_i, A_i). \qquad (11.10)$$

We now use a numerical example to illustrate the process.

Example 11.2: Optimal Sample Size for Trials in Clinical Development Plan

Suppose the CDP team want to develop a clinical development program for a disease indication, which include phase 1 to 3 trials (Figure 11.4). Phase 1 is a single arm toxicity study; phase 2 and 3 trials are two-group efficacy studies with two parallel groups. One common and important question is how to determine the optimal sample size for each trial so that the success or the expected overall gain for the CDP is maximized. Let's build, step by step, a Markov decision process for this problem.

(1) State space

$$S = \{\text{phase 1, phase 2, phase 3, NDA approval}\}.$$

(2) Action space
  In practice, there are limits for sample size. For simplicity's sake, assume the limits (i.e., the action space) are independent of state (the phase):

$$A = \{10, ..., 1000\}.$$

(3) Discount rate

Based on the typical trial durations, assume the discount rate

$$\gamma = 0.95.$$

(4) Immediate reward and cost

In calculating immediate gains in phase 1 to 3 trials, the costs are approximately proportional to the corresponding sample size. After approval for marketing, there is a commercial/advertising cost. The net gains (cost) for phases 1 to 3 can be written as a function of sample size (or may simply be the sample size):

$$g_i(n_i) = n_i, i = 1, 2, 3, \tag{11.11}$$

and net gain for marketing the drug less commercial cost is assumed to be $g_4 = 10,000$.

(5) Decision rules and transition probabilities

Assume the decision rules for the phase 1 trial is that if the toxicity rate $r_1 \le \eta_1 = 0.2$, the study will continue to phase 2; otherwise stop. The decision rules for phase 2 are based on a hypothesis test for treatment difference in efficacy. Specifically, if the one-sided p-value p_2 for the test satisfies $p_2 \le \eta_2 = 0.1$, the study will continue and a phase 3 trial will be launched; otherwise, the study will stop after phase 2. Similar to phase 2, the phase 3 is a parallel two-group active-control trial. The decision rules are: if the one-sided p-value p_3 for the efficacy testing satisfies $p_3 < \eta_3 = 0.025$, the drug will be approved for marketing, otherwise the NDA will fail.

From these decision rules, we can determine the transition probabilities. The transition probability from phase 1 to phase 2 can be calculated using the binomial distribution.

$$p_{12}(n_1) = \sum_{i=0}^{\lfloor \eta_1 n_1 \rfloor} B(i; p_0, n_1) = \sum_{i=0}^{\lfloor \eta_1 n_1 \rfloor} \binom{n_1}{i} p_0^i (1 - p_0)^{n_1 - i}, \tag{11.12}$$

where p_0 is toxicity rate of the NME, $B(\cdot; \cdot, \cdot)$ is a binomial p.m.f., and the floor function $\lfloor x \rfloor$ gives the integer part of x.

The transition probability p_{23} from phase 2 to phase 3 is the power of the hypothesis test in the phase 2 trial at the one-sided alpha level of 0.1, i.e.,

$$p_{23}(n_2) = \Phi\left(\frac{\sqrt{n_2}}{2}\delta - z_{1-\eta_2}\right), \tag{11.13}$$

where δ is normalized treatment difference, Φ is the standard normal c.d.f.

Similarly, the transition probability from phase 3 to NDA approval is given by

$$p_{34}(n_3) = \Phi\left(\frac{\sqrt{n_3}}{2}\delta - z_{1-\eta_3}\right). \tag{11.14}$$

Note that there is no repeated trial, even though p_{ii} in Figure 11.4 seems to be representing a transition loop. Keep in mind that in this simple case that each state has only one immediately preceeding state. The rest of the transition probabilities can be easily calculated as follows:

$$p_{11} = 1 - p_{12}, \tag{11.15}$$
$$p_{22} = 1 - p_{23},$$
$$p_{33} = 1 - p_{34}.$$

(6) Dynamic programming

We are now ready to use dynamic programming to solve this problem. Start with the Bellman's equation:

$$\begin{cases} V_4 = g_4 \\ V_i = \max_{n_i \in A} \{g_i(n_i) + \gamma p_{i,i+1}(n_i) V_{i+1}\}, i = 3, 2, 1. \end{cases} \tag{11.16}$$

Note that compared to (11.4), there is no summation $\sum$ in (11.16). This is because for the MDP in Figure 11.4, there is only one state leading in and one coming out from each state.

(7) Training Monte Carlo algorithm

The value iteration algorithm, Algorithm 11.1, can be directly used.

Example 11.3: Optimal Cutpoints for Trials in Clinical Development Plan

In Example 11.2, we were only concerned with the optimization of sample size, but in practice, it is also important to optimize the cutpoints $\{\eta_1, \eta_2, \eta_3\}$ simultaneously. Let's build a MDP for this decision problem:

(1) State space

$$S = \{\text{phase 1, phase 2, phase 3, NDA approval}\}.$$

(2) Action Space

$$A_i = \boldsymbol{u}_i \times \boldsymbol{v}_i, \tag{11.17}$$

where the sample-size space and cutpoint space are given, respectively, by

$$\boldsymbol{u}_i = \{n_{i,\min}, ..., n_{i,\max}\} \text{ and } \boldsymbol{v}_i = [\eta_{i,\min}, \eta_{i,\max}], i = 1, 2, 3.$$

(3) Discount rate

Let $\gamma = 1$ to simplify matters.

(4) Immediate reward and cost

$$g_i(a_i) = c_{0i} + n_i, i = 1, 2, 3, \tag{11.18}$$

where c_{0i} is constant.

The net gain for marketing the drug less commercial cost is assumed to be a linear function of the normalized treatment effect δ:

$$g_4 = g_0 + b\delta,$$

where g_0 and b are constants.

(5) Decision rules and transition probabilities

Assume the decision rules are similar to those in Example 11.3, and specified as follows

Phase 1: if the observed toxicity rate r_1, satisfies $r_1 \leq \eta_1$, the study will continue to phase 2; otherwise stop.

Phase 2: if the one-sided p-value p_2 for the efficacy test satisfies $p_2 \leq \eta_2$, launch a phase 3 trial; otherwise, stop.

Phase 3: if the one-sided p-value p_3 for the efficacy test satisfies $p_3 < 0.025$ (0.025 is a regulatory requirement for approval, therefore the bound is not considered a variable), the drug will be approved for marketing; otherwise the NDA will fail.

The expressions for the transition probabilities are the same as in Example 11.3, but for clarity, we rewrite p_{ij} as a function of both sample size n_i and the cutpoint η_i.

$$p_{12}(n_1, \eta_1) = \sum_{i=0}^{\lfloor \eta_1 n_1 \rfloor} B(i; p_0, n_1), \tag{11.19}$$

where p_0 is the toxicity rate of the NME.

The transition probability p_{23} from phase 2 to phase 3 is the power of the hypothesis test in the phase 2 trial, i.e.,

$$p_{23}(n_2, \eta_2) = \Phi\left(\frac{\sqrt{n_2}}{2}\delta - z_{1-\eta_2}\right), \tag{11.20}$$

where δ is the normalized treatment difference, and Φ is the standard normal c.d.f.

Similarly, the transition probability from phase 3 to approval is given by

$$p_{34}(n_3) = \Phi\left(\frac{\sqrt{n_3}}{2}\delta - z_{1-0.025}\right). \tag{11.21}$$

The remaining transition probabilities can be calculated using (11.15).

(6) Dynamic programming

We are now ready to use dynamic programming to solve this problem. Start with the Bellman's equation:

$$\begin{cases} V_4 = g_4 \\ V_i = \max_{n_i \in u_i, \eta_i \in v_i} \{g_i(n_i) + p_{i,i+1}(n_i) V_{i+1}\}, i = 3, 2, 1. \end{cases} \tag{11.22}$$

(7) Monte Carlo algorithm

Algorithm 11.1 can be used with the consideration of the initial condition $s(t = 0) = s_0$, the financial, time, patient population, and other constraints.

Example 11.4: Optimal Trials in Clinical Development Plan Considering Efficacy and Safety

So far, we have not considered action rules (policy) that consider efficacy and safety jointly. In most realistic settings, they have to be considered simultaneously at some point, e.g., in phase 2 and phase 3 trials. The key to constructing the transition probabilities is in using the policy with the efficacy and safety components.

Let's first consider the situation in which efficacy and safety are independent. The action rules are specified in the following way (Figure 11.4).

For the phase 1 trial, if the observed toxicity rate satisfies $r_1 \leq \eta_{11}$, a phase 2 trial will be launched with sample size n_2; for the phase 2 trial, if the observed toxicity rate satisfies $r_2 \leq \eta_{21}$ and if the p-value for efficacy test satisfies $p_2 < \eta_{22}$, an phase 3 trial will be initiated with sample size n_3. The drug approval for marketing requires the observed toxicity rate $r_3 \leq \eta_{31}$ and one-sided p-value $p_3 < 0.025$. Based on these action rules, the transition probabilities can now be calculated using following formulations:

$$p_{12}(n_1) = \sum_{i=0}^{\lfloor \eta_1 n_1 \rfloor} B(i; p_0, n_1) = \sum_{i=0}^{\lfloor \eta_1 n_1 \rfloor} \binom{n_1}{i} p_0^i (1 - p_0)^{n_1 - i}, \qquad (11.23)$$

$$p_{23}(n_2) = \Phi\left(\frac{\sqrt{n_2}}{2} \delta - z_{1 - \eta_2} \right) \sum_{i=0}^{\lfloor \eta_2 n_2 \rfloor} B(i; p_0, n_2), \qquad (11.24)$$

$$p_{34}(n_3) = \Phi\left(\frac{\sqrt{n_3}}{2} \delta - z_{1 - \eta_3} \right) \sum_{i=0}^{\lfloor \eta_3 n_3 \rfloor} B(i; p_0, n_3). \qquad (11.25)$$

In these formulations, we have assumed that toxicity and efficacy are independent responses in any given patient with a given dose, which is often reasonable — we know that it is not necessary that a patient with high efficacy have high toxicity. We should not confuse this with the common concept that efficacy and toxicity are correlated in the sense that an increase in dosage will likely increase both efficacy and toxicity for the same patient.

Note that we have not considered the efficacy criterion in the action rule for the phase 1 trial. The reason is that in an early phase trial, there are usually no observable clinical efficacy endpoints. In such a case, we have to use a partially observable Markov decision process (POMDP), as discussed in the next section.

Example 11.5: Optimal Clinical Development Plan

We now consider another common situation, i.e., multiple optional paths in a clinical development plan (Figure 11.5). One path is to follow the traditional paradigm: single ascending dose (SAD) study $\rightarrow$ Dose finding study (DFS) $\rightarrow$ Pivotal phase-3 studies (PPS) $\rightarrow$ NDA Approval. The second path is an innovative approach with adaptive design: SAD $\rightarrow$ Adaptive seamless trial

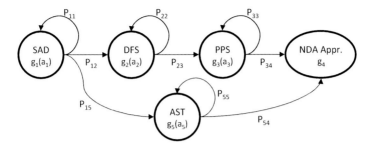

FIGURE 11.5

Clinical Development Options

(AST) → NDA Approval. The adaptive seamless trial combines DFS and PPS as a single (seamless) trial.

Let's build a MDP for this problem:

The power of a trial is dependent on the true but unknown treatment effect, and therefore the transition probabilities will be estimated from the data available at the decision point. Consequently, the same transition probability (e.g., from the phase 2 to the phase 3 trial) may have different values when it is calculated at the end of the phase 1 trial and the end of the phase 2 trial.

(1) State space

$$S = \{\text{SAD, DFS, PPS, NDA approval, AST}\}.$$

(2) Action Space

$$A_i = \boldsymbol{u}_i \times \boldsymbol{v}_i, \tag{11.26}$$

where the sample-size space and cutpoint space are given, respectively, by

$$\boldsymbol{u}_i = \{n_{i,\min}, ..., n_{i,\max}\} \text{ and } \boldsymbol{v}_i = [\eta_{i,\min}, \eta_{i,\max}], i = 1, 2, 3, 5,$$

where $\eta_{3,\min} = \eta_{3,\max}$.

(3) Discount rate

Let $\gamma = 1$ for simplicity.

(4) Immediate reward and cost

$$\begin{cases} g_i\,(a_i) = c_{0i} + n_i, i = 1, 2, 3, 5 \\ \\ g_4 = g_0 + b\delta. \end{cases} \tag{11.27}$$

The net gain can also be a random function.

(5) Decision rules and transition probabilities

Assume the decision rules are similar to those in Example 11.4, i.e.:

Phase 1: if the observed toxicity rate r_1 is bounded within $\eta_1 < r_1 \leq \eta_5$, the phase 2 trial will start; if $r_1 \leq \eta_1$, the adaptive trial will initiate; otherwise stop. (Questions to students: What will happen to the decision rules if $\eta_1 = \eta_5$? What will happen if $\eta_1 < 0$?)

Phase 2: if the one-sided p-value for the efficacy test is $p_2 \le \eta_2$, launch a phase 3 trial; otherwise, stop.

Phase 3: if the one-sided p-value for the efficacy test is $p_3 < 0.025$, the drug will be approved for marketing, otherwise the NDA will fail.

The expressions for the transition probabilities are

$$p_{12}(n_1, \eta_1, \eta_5) = \sum_{i=\lfloor \eta_1 n_1 \rfloor + 1}^{\lfloor \eta_5 n_1 \rfloor} B(i; p_0, n_1), \tag{11.28}$$

$$p_{15}(n_1, \eta_1) = \sum_{i=0}^{\lfloor \eta_1 n_1 \rfloor} B(i; p_0, n_1), \tag{11.29}$$

$$p_{23}(n_2, \eta_2) = \Phi\left(\frac{\sqrt{n_2}}{2}\delta - z_{1-\eta_2}\right), \tag{11.30}$$

and

$$p_{34}(n_3) = \Phi\left(\frac{\sqrt{n_3}}{2}\delta - z_{1-0.025}\right). \tag{11.31}$$

The transition probability p_{54} is the power of the adaptive trial (see Chapter 6), and simulation is required to calculate power for a given sample size. However, at the macro-simulation level, an approximation based on the corresponding classical design can be used. In this case, we can write

$$p_{54} \approx \zeta p_{34}(n_3),$$

where the coefficient ζ varies from one adaptive design to another, but for two-group trials is about $\zeta = 0.9$.

The rest of the transition probabilities are straightforward calculations.

Another possible decision rule that can be used at phase 1 involves choosing transition probabilities p_{12} and p_{15} based the expected Q-value (net gain) of the traditional approach (Q_T) or the innovation approach (Q_I). There are two common ways: (1) $Q_T < Q_I$, set $p_{12} = 0$, $p_{12} = 0$; otherwise set $p_{15} = 0$. (2) Choose the probability proportional to the corresponding Q-value, i.e., $\frac{p_{12}}{p_{15}} = \frac{Q_T}{Q_I}$. Therefore, $p_{12} = \frac{Q_T}{Q_T + Q_I}(1 - p_{11})$.

6) Dynamic programming

We are now ready to use dynamic programming to solve this problem using the Bellman?s equation:

$$\begin{cases} V_4 = g_4 \\ V_5 = \max_{n_5 \in \mathbf{u}_5, \eta_5 \in \mathbf{v}_5} \{g_5(n_5) + p_{5,4}(n_4)V_4\} \\ V_i = \max_{n_i \in \mathbf{u}_i, \eta_i \in \mathbf{v}_i} \{g_i(n_i) + p_{i,i+1}(n_i)V_{i+1}\}, i = 3, 2 \\ V_1 = \max_{n_1 \in \mathbf{u}_1, \eta_1 \in \mathbf{v}_1} \{g_1(n_1) + p_{1,2}(n_1)V_2 + p_{1,5}(n_1)V_5\}. \end{cases} \tag{11.32}$$

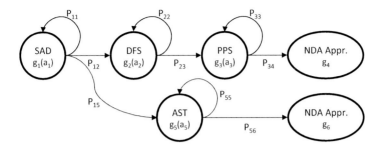

FIGURE 11.6
MDP Conversion to Stochastic Decision Tree

11.3.2 Markov Decision Tree and Out-Licensing

So far we have not considered the value of adaptive design that may shorten the development time or time-to-market, which is a significant factor driving the CDP. If an adaptive trial shortens the time-to-market, the value can be reflected in the gain g_4 from drug marketing in Example 11.5. Thus the gain g_4 in Figure 11.5 becomes path-dependent. To solve this problem, we can convert the stochastic decision problem (not a Markov decision process anymore) to a stochastic decision tree (SDT). In a stochastic decision tree, there can be several paths out from a decision node, but only one parent decision node from which the path comes in. Furthermore, each state in a stochastic decision tree can be visited at most once. Figure 11.6 is a SDT converted from Figure 11.5 in Example 11.6.

After the conversion, the transition probabilities can be calculated similarly as in MDP. The expected net gain for each path can easily calculated using the transition probability $p_{ij}(a_i)$ and the net gains g_i associated with that path. Note that stochastic decision tree is a simple MDP with a single path in for each decision node. However, when converting to SDT from a MDP, the total number of paths in a stochastic decision tree will grow very fast. If each node has $k = 2$ action-lines connected and the tree has $m = 4$ layers of nodes, there will be $n = 2^4 = 16$ possible paths to reach the end-nodes. If $k = 3, m = 6, n = 3^6 = 729$; if $k = 6, m = 14, n = 6^{14} = 78\,,364\,,164\,,096$. The computational time quickly becomes overwhelming or intractable.

Example 11.6: Optimal Clinical Development Program with Out-Licensing Option

In this example, we consider the option of out-licensing a NME at a certain stage (e.g., end of phase 2 trial) or joint venture (shared risk and profit) after a promising result from phase 2. This option is often considered when internal resource is limited. To include out licensing (OL) and joint

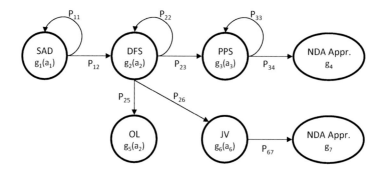

SAD = single ascending, DFS = Dose finding study, OL = out-licensing, JV = Joint venture, PPS = Pivotal phase 3 study

FIGURE 11.7
MDP with Out-Licensing and Joint Venture

venture (JV) options in the CDP in Example 11.3, the MDP is plotted in Figure 11.7.

From various examples in this section so far, we can see that action rules are critical to the MDP. For one thing, they affect the transition probabilities. We can simplify the steps for building our MDP as follows (assuming the discount rate $\gamma = 1$).

(1) Draw the MDP diagram (see e.g., Figure 11.7).

(2) Specify a family of action (decision) rules.

A family of decision rules or a policy is a collection of options for choosing different decision rules. In the previous several examples, "if $p_3 < 0.025$, drug approved" is an action rule, whereas "If $p_2 < \eta_2$, launch a phase 3 trial" is a family of action rules because each value of η_2 determines an action rule. For this reason, the goal of using a MDP is to determine an optimal action rule for each state.

Note that a family of action rules doesn't have to be a simple smooth function; it is often expressed using a combination of several equations, as in the case of in Example 11.5.

The expected gain from OL is dependent on the results at stage 2, i.e., $g_5(a_2)$. The net gain (cost) $g_6(a_6) > g_3(a_3)$ because of risk sharing, meanwhile $g_7 < g_4$ because of profit sharing in the JV.

(3) Determine the transition probabilities.

The transition probabilities are determined by the action rules, and numerical values are usually obtained through Monte Carlo simulation.

(4) Apply dynamic programming algorithms.

For an example of a MDP in R & D portfolio optimization see Chang (2010).

11.4 *Q*-Learning

We have seen in the previous examples some ways to extend MDPs: (1) the reward may be a function of the action as well as the state, $g(s, a)$; (2) the reward may be a function of the resulting state as well as the action and state, $g(s, a, s\prime)$; or (3) the action space may be different at each state, so that it is A_s rather than A. These extensions only complicate the notation, but make no real difference to the problem or the solution method. In this section we will discuss several important extensions that are demanded by practice.

The MDPs we have discussed so far have known transition probabilities, so that calculations can be carried out. What if the probability is unknown or too complicated, and the problem becomes intractable? In our previous examples for clinical development, the transition probabilities, and consequently the expected overall gain, are heavily dependent on the toxicity rate p_0 and normalized treatment effect δ, the two unknown parameters. If the assumptions about the two parameters change, the overall expected gain will change. In other words, the optimal action rules will depend on the assumptions around the parameters. To solve this dilemma there are several methods available, including Q-learning and the Bayesian stochastic decision process (BSDP). We are going to discuss these methods in this and the next sections.

Q-learning (Watkins, 1989, Gosavi, 2003, Chang, 2010) is a form of model-free reinforcement learning. It is a forward-induction and asynchronous form of dynamic programming. The power of reinforcement learning lies in its ability to solve the Markov decision process without computing the transition probabilities that are needed in value and policy iteration. The key algorithm in Q-learning is the recursive formulation for the Q-value:

$$Q_i = \begin{cases} (1 - \alpha_i)Q_{i-1}(s, a) + \alpha_i \left[g_i + \gamma V_{i-1}(s_i') \right] & \text{if } s = s_i, a = a_i \\ Q_{i-1}(s, a), \text{ otherwise,} \end{cases} \tag{11.33}$$

where the learning rate $0 \leq \alpha_i < 1$ is chosen a constant chosen and

$$V_{i-1}(s_i') = \max_b \left\{ Q_{i-1}(s', b) \right\}. \tag{11.34}$$

When the Bayesian learning paradigm is applied to uncertainty of the Q-value, it becomes Bayesian Q-learning (Dearden, Friedman and Russell, 1998).

11.5 Bayesian Stochastic Decision Process

So far we assume that the parameters of the NME to characterize the toxicity rate and treatment effect are known. In reality, the transition probability

$\Pr(s'|s, a; \delta)$ is not a fixed function; rather it is dependent on the cumulative information about δ and others. In Bayesian terms, there is a need to differentiate the prior and posterior distributions. We rewrite the transition probability $\Pr(s'|s, a; \delta)$ to include the parameter δ (normalized treatment effect) to stress the δ-dependency of the transition probability. In the frequentist paradigm, the parameter δ is an unknown but fixed number. However, people make decisions based on the knowledge (estimate) about δ, not δ itself because the true value of δ is unknown.

Because $\Pr(s'|s, a; \delta)$ and δ is updated based on cumulative data from previous states, the stochastic process is usually not stationary or Markovian anymore. This implies that the backwards induction algorithm cannot be applied. However, an approximate solutions is to use the posterior distribution in Bellman's equation. That is,

$$V(s) = \max_{a_s \in A} \left(g(s, a) + \gamma E \sum_{s' \in S} \Pr\left(s'|s, a_s; \tilde{\delta}_s\right) \cdot V(s') \right), \qquad (11.35)$$

where $\tilde{\delta}_s$ is the posterior estimate of δ at stage s, and E denotes the expectation taken with respect to the posterior distribution of $\tilde{\delta}_s$.

We can see that the Bayesian stochastic decision process (BSDP) is similar to a MPD except for the calculation of the transition probability. In the MDP, parameter δ is a fixed number, whereas in the BSDP, the posterior $\tilde{\delta}_s$ is used in the probability calculation.

For Example 11.3, the p_{12} for the BSDP will be same as for the MPD, where the prior toxicity rate p_0 is used. For the calculation of $p_{23}(n_2)$ is also the same for the two methods, where the prior effect size $\delta = \delta_0$ is used. However, the calculation for $p_{34}(n_3)$ is different for BSDP from that for MDP. For the BSDP, the transition probability is given by

$$p_{34}(n_3) = \Phi\left(\frac{\sqrt{n_3}}{2} \tilde{\delta}_3 - z_{1-\eta_3} \right), \qquad (11.36)$$

where the posterior is

$$\tilde{\delta}_3 \sim N\left(\frac{n_0 \delta_0 + n_2 \hat{\delta}_2}{n_0 + n_2}, \frac{1}{n_0 + n_2} \right). \qquad (11.37)$$

Note that δ_0 and n_0 are the prior (normalized mean treatment effect) and the sample size, and $\hat{\delta}_2$ is the observed treatment effect (normalized) in the phase 2 trial.

We can see that the calculation of the BSDP is much more intensive due to both the calculations of the posterior distribution and the expectation. Approximations may be used to reduce the computational burden, e.g.,

$$V(s) = \max_{a_s \in A} \left(g(s, a) + \gamma \sum_{s' \in S} \int \Pr(s'|s, a_s; \delta) \pi(\delta) \cdot V(s') \right), \qquad (11.38)$$

where $\pi(\delta)$ is the prior distribution.

> **Algorithm 11.3**: Bayesian Stochastic Decision Process
> Objective: Optimization with Bayesian Stochastic Decision Process
> **While** precision$>\varepsilon$:
>   Record the current path and best path with maximum $V(s)$.
>   Calculate the posterior distribution of the model parameters.
>   Calculate p_{ij} based on action and posterior parameters.
>   Use the Bayesian Bellman Eq. (11.58) for value iteration.
>   Stop if the predefined precision ε is reached.
> **Endwhile**

Instead of taking the step towards the maximum expected reward, the agent can randomly choose an action with the selection probability proportional to the potential (expected) gain from each action. The idea behind this approach is to modify the randomization probability gradually in such a way that when our initial judgement about expected gains is wrong, the action will be corrected over time. However, this approach is effective only when the same experiments can be repeated many times.

11.6 Partially Observable Markov Decision Processes

So far we have assumed that the state s is known at the time when action is to be taken; otherwise policy $\pi(s)$ cannot be calculated. However, in an early stage of drug development, the clinical effects of a NME are usually not directly observable; instead, observations are made on biomarkers, which supposedly have a correlation with the definitive clinical endpoint. This leads to the so-called partially observable Markov process (POMDP).

A POMDP is a generalization of a Markov decision process. In robotics, a POMDP models an agent decision process in which it is assumed that the system dynamics are determined by a MDP, but the agent cannot directly observe the underlying state. Instead, it must infer a distribution over the state based on a model of the world and some local observations. The framework originated in the operation research community and has been spread into artificial intelligence, automated planning communities, and the pharmaceutical industry (Lee, Chang, and Whitemore, 2008).

In some cases in early drug development, the efficacy of a NME is completely nonobservable. In such a case, the motion can be modeled by a hidden Markov chain (HMC). Table 11.1 summarizes the differences between MsC, MDPs, POMDPs, and HMs with respect to the state observability and motion controllability.

Recall that a Markov chain has three properties: (1) a set of states S, (2) transition probabilities: $T(s, s') = P(s_{t+1} = \acute{s}|s_t = s)$, and (3) the

TABLE 11.1
Comparison of Stochastic Processes

| Markov | | Transition controllable? | |
Models		**Yes**	**No**
State	**Yes**	MDP	MC
Observable?	**No**	POMDP	HMC

starting distribution $P_0(s) = P(s_0 = s)$. For HMC, two additions are required: (4) set of observations Z, and (5) observation probabilities: $O(s, z) = P(z_t = z | s_t = s)$, where s is not observable.

The computations of a POMDP are too complex to cover here without a significant increase of the size.

Case Study 11.1: Classical and Adaptive CDPs for CABG patients

In coronary heart disease (CHD), the coronary arteries become clogged with calcium and fatty deposits. Plaques narrow the arteries that carry blood to the heart muscle and can cause ischemic heart disease (too little blood and oxygen reaching the heart muscle). Coronary artery bypass graft (CABG) surgery is a treatment option for ischemic heart disease. CABG surgery is used to create new routes for blood to flow around the narrowed and blocked arteries so that the heart muscle will receive needed oxygen and nutrients. According to the American Heart Association, 427,000 coronary artery bypass graft (CABG) surgeries were performed in the United States in 2004.

The problem is that during a heart bypass procedure, a substance called complement is activated by the body. This complement activation causes an inflammation that can lead to side-effects such as chest pain, heart attacks, stroke, heart failure, impairment of memory, language and motor skills, or death.

A test drug XYZ was brought to the development phase by company A, which was expected to block the complement activation and thus reduce side-effects. Meanwhile another company, company B, was at the stage of designing their phase 2 trial for their test compound ABC for the same indication. The concern for company A was that at the time of designing phase 3 trial for XYZ, ABC could be approved for marketing, so that company A would not only lose the time and majority of the market, but also would be required to run an active-control trial (using ABC as the control), instead of a placebo-control trial. As a result, the required phase 3 trial for XYZ would be much larger and more costly. During brainstorming, the CDP team in company A proposed another option: using seamless adaptive design that combines phase 2 and phase 3 trials as a single trial so that when the competitor started its phase 3 trial, company A would start its seamless phase 2–3 trial. With that strategy, company A would be able to catch up the time and conduct the placebo-controlled trial since no drug would have been approved for marketing yet at the time. The team summarized their comparisons between the classical and adaptive approaches as follows.

The classical design with separated phase 2 and 3 trials is economically infeasible: (1) a potential third in class position, (2) the pivotal trial would be for non-inferiority to *ABC* with a larger sample size, and (3) disadvantages in the US market would not be fully offset by the market gains in EU which could potentially be realized by the classical CDP paradigm.

Seamless Adaptive design is preferable, having these advantages: (1) large combined trial, protracted recruitment period, (2) a short treatment period, good timing for registration in the US, and (3) placebo controlled trial with reduced cost.

11.7 Summary

Reinforcement learning involves sequential decision processes. The rewards are received throughout the processes. Therefore, MDP is a common model for RL. Many problems in everyday life can be viewed as sequential decision processes in search of an optimal solution, either minimization of the loss or maximization of the gain. The optimum can be obtained using a backward induction algorithm derived from Bellman's optimality principle. This principle states that an optimal policy has the property that, whatever the initial state and initial decision are, the remaining decisions must constitute an optimal policy with regard to the state resulting from the first decision. There are two commonly used backward induction algorithms: Value-iteration and Policy-iteration. Another approach consists of combining these two algorithms.

The deterministic sequential design process is not very practical in drug development without the inclusion of probabilities in the model. A Markov decision process (MDP) is an extension of a Markov chain, in which actions and rewards are attached to the MC states. A MDP can have a finite or an infinite horizon. In finite horizon MPD, the transition probabilities are time-dependent, whereas in infinite horizon MPD, the system has reached its steady state and the transition probabilities are independent of time.

MDPs can be used in Clinical Development Planning, as illustrated in examples 11.3-11.7. A major task in application of MDP to CDP is the calculation of transition probabilities, which is often based on power calculations in clinical trials with respect to various decision (stopping and adaptive) rules. MDPs can also be used in pharmaceutical R & D portfolio optimization.

One of the challenges in application of a MDP in drug development is that the transition probability is dependent on the model parameters (e.g., treatment effect) that are unknown. Different estimation procedures of the parameters can be used, but they are not very convincing. Alternatively, Q-learning, Bayesian Q-learning, and the one-step forward approach can be used. All these methods are forward induction methods as opposed to backward induction methods that are based on the Bellman optimality principle.

The second challenge is that in the early stage of drug development, a definitive clinical endpoint such as survival usually can't be measured, but assessments on certain biomarkers can be made and used to serve as a marker for the clinical endpoint. In such a case, a partially observable Markov decision process can be used.

11.8 Problems

11.1: Develop computer pseudocode for the MDP of Example 11.3.

11.2: In Example 11.3, assume the sample size at each phase is determined, but you are asked to determine the optimal cutpoints $\{\eta_1, \eta_2, \eta_3\}$ to maximize the expected overall gain for the CDP. Develop a MDP for this problem.

11.3: In Example 11.5, consider a model with a bi-logistic model (safety & efficacy), where the parameters for safety and efficacy have a joint distribution. Develop computer pseudocode for the CDP problem.

11.4: Formulation (11.24) provides a backward induction algorithm for the multi-path CDP problem. Develop computer pseudocode based on (11.24).

11.5: Figure 11.5 can represent different models depending on how one defines p_{15} and p_{25}. Can you elaborate these models?

11.6: In Example 11.9, suppose there are $n = 10$ NMEs in the Preclinical pipeline; formulate a MDP for the problem using a pick-winner strategy. Please make additional assumptions as necessary.

12

Swarm and Evolutionary Intelligence

12.1 Swarm Intelligence—Artificial Ants

12.1.1 Artificial Swarm Intelligence

A *social animal* or *social insect* is an organism that is highly interactive with other members of its species to the point of having a recognizable and distinct society, in which interactions and cooperation go beyond familiar members.

Swarm intelligence can be found in social insects. Colonies of ants organize shortest paths to and from their foraging sites by leaving pheromone trails, forming chains of their own bodies to create bridges and to pull and hold leaves together with silk, dividing this work between major and minor ants. Similarly, bee colonies regulate hive temperature and work efficiently through specialization (division of labor).

Social insects organize themselves without a leader. The modeling of such insects by means of self-organization can help design artificial swarm-intelligent systems. For instance, ant colony optimization and clustering models are inspired by observed food foraging and cleaning activities in ant colonies; a particle swarm optimization model imitates the social behavior of fish schools and bird flocks; and the bird flocking model for documentation clustering was inspired by bird flock flying behaviors. In a natural swarm intelligence system, direct interaction and communication occurs through food exchange, visual contact, and chemical contact (pheromones). Such interactions can be easily imitated in artificial swarm intelligence. In addition to direct interaction there can be indirect interactions among members where individual behavior modifies the environment, which in turn modifies the behavior of other individuals (stigmergy).

A swarm is a large number of homogeneous, uncomplicated agents collaborating locally among themselves and their circumstances, with no central control to allow a global interested behavior to emerge. Swarm-based algorithms are nature-inspired, population-based algorithms that are clever in producing optimal solutions for several complex and combinatorial problems (Rajeshkumar and Kousalya, 2017).

Ant Routing Algorithm for Optimization

An ant routing algorithm, introduced by Marco Dorigo's (1992), was inspired by the food foraging behavior of ants hunting the shortest or fastest route. Its key algorithm can be described in this way:

1. Ants lay pheromones on the trail when they move food back to nest.

2. Pheromones accumulate with multiple ants using the same path, evaporating when no ants pass by.

3. Each ant always tries to choose trails having higher pheromone concentrations.

4. In a fixed time period, ant agents are launched into a network, each agent going from a source to a destination node.

5. The ant agent maintains a list of visited nodes and the time elapsed in getting there. When an ant agent arrives at its destination, it will return to the source following the same path by which it arrived, updating the digital pheromone value on the links that it passes by. The slower the link, the lower the pheromone value will be.

6. At each node, the ant colony (data package) will use the digital pheromone value as the transitional probability for deciding the ant (data) transit route.

Ant Clustering Model for Data Clustering (Figure 12.1)

The size of the internet has been doubling in size every year. Organizing and categorizing the massive number of documents is critical for search engines. The operation of grouping similar documents into classes can also be used to obtain an analysis of their content; data clustering is the task that seeks to identify groups of similar objects based on the value of their attributes. The ant clustering algorithm categorizes web documents into different domains of interest. Ant routing algorithms can be used to better route traffic on telecommunications systems and the internet, roads, and railways to reduce congestion that efficient routing algorithms may provide. Southwest Airlines is actually putting the ant colony research to work, with an impressive payback.

In nature, ants collect and pile dead corpses to form "cemeteries." The corresponding agent (ant) action rules are: agents move randomly, only recognizing objects immediately in front of them, picking up or dropping an item based on the pickup probability, $P_p = \left(\frac{c_p}{c_p + f}\right)$ and drop-off probability, $P_d = \left(\frac{c_d}{c_d + f}\right)$, where f is the fraction of the items in the neighborhood. c_p and c_d are constants.

In PSO (particle swarm optimization), individuals are the "particles" and the population is the "swarm." PSO is based on simple social interactions, but can solve high-dimensional, multimodal problems. In PSO systems, every

FIGURE 12.1
Ant Clustering Model in Action

particle is both teacher and learner. The communication structure determines how solutions propagate through the population. After a solution has been learned by the population, resilience allows the particles to start exploring again, should they receive substantially new information. Resilience versus efficiency is often a trade-off.

12.1.2 Applications

Swarm intelligence-based applications include complex interactive virtual environment generation in the movie industry, cargo arrangement in airline companies, route scheduling for delivery companies, packet routing in telecommunication networks, power grid optimization controls, data clustering and data routing in sensor networks, unmanned vehicle control in the U.S. military, and planetary mapping and micro-satellite control as used by NASA (Xiaohui Cui, ppt).

A central part of the rational drug development process is the prediction of the complex structure of a small ligand with a protein, the so-called protein-ligand docking problem, used in virtual screening of large databases and lead optimization. Korb et al. (2006) developed a new docking algorithm called PLANTS (Protein-Ligand ANTSystem), which is based on ant colony optimization, to structure-based drug design. An artificial ant colony is employed to find a minimum energy conformation of the ligand in the protein's binding site. The algorithm showed better efficacy than a genetic algorithm.

Molecular docking is critically important for a ligant binding to the intended site, which is essential for a small molecular drug to take effect. Fu et al. (2015) studied a new approach for flexible molecular docking based on swarm intelligence. They compute the interactions of 23 protein-ligand complexes. The experimental results show that their approach leads to substantially lower docking energy and higher docking precision in comparison to the Lamarckian genetic algorithm and the QPSO algorithm alone. This suggests that the novel algorithm may be used to dock ligands with many rotatable bonds with high accuracy. Rajeshkumar and Kousalya (2017) present a view on applications of swarm-based intelligence algorithms in the pharmaceutical industry, including drug design, pharmacovigilance, and alignment of sequence.

Soulami et. Al. (2017) used a particle swarm optimization (PSO)–based algorithm for detection and classification of abnormalities in mammographic images using texture features and support vector machine (SVM) classifiers.

Protein essentiality is fundamental to comprehending the function and evolution of genes. The prediction of protein essentiality is pivotal in identifying disease genes and potential drug targets. Fang et. al. (2018) presented a novel feature selection called the elite search mechanism-based flower pollination algorithm (ESFPA) used to determine protein essentiality. ESFPA uses an improved SI algorithm for feature selection and selects optimal features for protein essentiality prediction. The first step is to collect numerous features with the highly predictive characteristics of essentiality. The second step is to develop a feature selection strategy based on a swarm intelligence algorithm to obtain an optimal feature subset. Furthermore, an elite search mechanism is adopted to further improve the quality of the feature subset. The experimental results show that the SI method is competitive with some well-known feature selection methods.

The *metaheuristicOpt* package in *R* includes many optimization algorithms:

1. ALO: Optimization using Ant Lion Optimizer
2. DA: Optimization using Dragonfly Algorithm
3. FFA: Optimization using Firefly Algorithm
4. GA: Optimization using Genetic Algorithm
5. GOA: Optimization using Grasshopper Optimisation Algorithm
6. GWO: Optimization using Grey Wolf Optimizer
7. HS: Optimization using Harmony Search Algorithm
8. metaOpt: metaOpt The main function to execute algorithms for getting optimal solutions
9. MFO: Optimization using Moth Flame Optimizer
10. PSO: Optimization using Particle Swarm Optimization

11. SCA: Optimization using Sine Cosine Algorithm

12. WOA: Optimization using Whale Optimization Algorithm

The particle swarm optimization (PSO) algorithm was proposed by Kennedy and Eberhart in 1995, inspired by the behavior of social animals/particles, such as a flock of birds in a swarm. The inertial weight proposed by Shi and Eberhart is used to increase the performance of PSO (Kennedy and Eberhart, 1995; Shi and Eberhart, 1998). Here is simple example of PSO from community.rstudio.com.

Example 11.1: Particale Swarm for Optimizing Sphere Function

```
# Artificial ants for optimizing sphere function
library(metaheuristicOpt)
# define sphere function as objective function
sphere <- function(X){ return(sum(X^2)) }
## Define parameter
Vmax = 2; ci = 1.5; cg <- 1.5; w <- 0.7; numVar <- 5
rangeVar <- matrix(c(-10,10), nrow=2)
## calculate the optimum solution using Particle Swarm Optimization
Algorithm
resultPSO <- PSO(sphere, optimType="MIN", numVar, numPopulation=20,
 maxIter=100, rangeVar, Vmax, ci, cg, w)
## obtain the optimum (minimum) value of the sphere function
optimum.value <- sphere(resultPSO)
optimum.value
```

See the package documentation for details at https://community.rstudio. com/t/hello-are-there-any-packages-in-r/14086.

12.2 Evolutionary Intelligence

12.2.1 Genetic Algorithm

A *genetic algorithm* (GA), a name inspired by Darwin's theory of evolution, is an AI algorithm designed to solve an optimization problem. John Holland introduced genetic algorithms in 1960 and his student David E. Goldberg extended the GA idea in 1989. A typical genetic algorithm requires: (1) a genetic representation of the solution domain, a fitness function to evaluate the solution domain, and crossover and mutation operations.

Initialization: Generate an initial population, usually by random selection.

Natural Selection: During each successive generation, a portion of the existing population is selected to breed a new generation. Individual solutions are selected through a fitness-based random selection process. The fitness function is defined over the genetic representation and measures the quality of the represented solution. The fitness function is problem-dependent.

Genetic Operators: Crossover and mutation are known as the main types of genetic operators, but it is possible to use other operators such as colonization-extinction and migration in genetic algorithms. The mutation probability is usually smaller than the crossover probability. A very small mutation rate may lead to genetic drift, which is non-ergodic in nature. A recombination rate that is too high may lead to premature convergence of the genetic algorithm. A mutation rate that is too high may lead to loss of good solutions, unless elitist selection is employed.

Termination: Termination can occur if one of the following conditions is met:

1. A good-enough solution is found.

2. The maximum number N_{max} of generations is reached. N_{max} can be determined by the budget or time available.

3. The highest-ranking solution's fitness is reaching or has reached a plateau such that successive iterations cannot make significant improvement.

12.2.2   Genetic Algorithm for Infertility

The National Institutes of Health (NIH) defines infertility as an inability to achieve pregnancy after 1 year of having regular, unprotected sex (6 months if the woman is more than 35 years old). According to The Centers for Disease Control and Prevention (CDC), about 10 percent (6.1 million) women in the United States aged 15 to 44 have difficulty becoming pregnant owing to equal proportions of infertility (1/3) for female, male, and both sexes. Dealing with infertility involves an awareness of medical/procedure options and success rates, costs, and the time required to address the issue.

Infertility treatment options include the following.

Ovulation Induction (OI): A process of stimulation of oocyte developments to release mature egg(s), followed by a fertilization step, which can be OI/IUI Oral, OI/IUI Gnd, or IVF.

OI/IUI Oral: *Ovarian stimulation* by oral medications (letrozole or clomiphene) with timed intercourse or intrauterine insemination (IUI), a type of *artificial insemination* for placement of sperm inside a woman's uterus to increase the chance of fertilization.

OI/IUI Gnd: The stimulation of ovulation by gonadotropin injections with or without oral medications (letrozole or clomiphene) with timed intercourse or IUI.

In Vitro Fertilization (IVF): A process of controlled hyperstimulation of the ovaries by injectable medications to develop many eggs and retrieve them, followed by fertilization of the eggs with sperm outside the body.

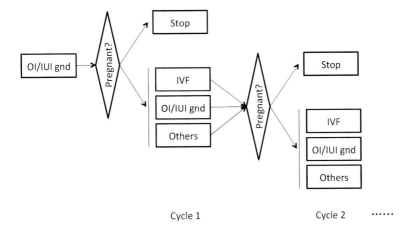

FIGURE 12.2
Infertility Treatment Strategies

Frozen Embryo Transfer (FET) involves an additional process of warming a previously frozen embryo and transferring it into the woman's uterus.

Chaharsough (2019) studied the causal effect of dynamic fertility treatment strategies on the probability of pregnancy using marginal structural models = based on a patient's electronic medical record (EMR), a digital version of charts which includes the information collected when a patient visits a clinic or a hospital. This may include diagnoses, medication the patient already uses, as well as medications prescribed at each visit, lab records, and medical conditions.

Genetic Algorithm for Optimizing Infertility Treatment Strategy

In treating women infertility there are several options. The sequence of intervention is believed to be important, in addition to the demographic and baseline characteristics. There are many possible treatment courses (Figure 12.2), but randomly or exhaustively trying out the treatment sequences is inefficient or impossible. It is believed that if a long sequence of treatments is effective, then a partial sequence will likely retain partial effectiveness. If this assumption is true, then a GA will be a better way to find the best or nearly the best treatment sequence. The basic idea is to view the a treatment sequence as a (short) DNA sequence and apply a GA.

1. **Initialization**: Select sequences (strategies) of treatments based on the likelihood of their effectiveness (prior knowledge) or at random, for example, $T1 \rightarrow T3 \rightarrow T5 \rightarrow T4$ and $T1 \rightarrow T2 \rightarrow T3 \rightarrow T4$ (Figure 12.3).

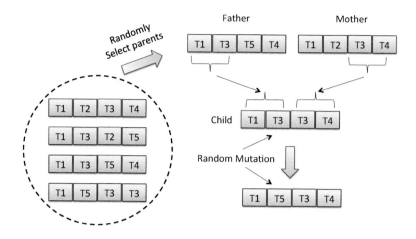

FIGURE 12.3
Genetic Algorithm for Treatment Strategy Optimization

2. Obtain the *fitness* defined as treatment outcome (e.g., pregnant or not).

3. Fitness-based random selection of two sequences and mate or *crossover*.

4. Randomly select a treatment for mutation (*point mutation*).

For infertility treatment we can use response-adaptive randomization, which is equivalent to the randomization probability that is proportional to the fitness function. The fitness function is simply the probability of success.

Function Optimization with GA Package in R

https://cran.r-project.org/web/packages/GA/vignettes/GA.html

```
library(GA) # genetic algorithm
# One-dimentional function optimization
f <- function(x) (x^2+x)*sin(x)
lbound <- -10; ubound <- 10
curve(f, from = lbound, to = ubound, n = 1000)
GA <- ga(type = "real-valued", fitness = f, lower = c(th = lbound),
upper = ubound)
summary(GA)
curve(f, from = lbound, to = ubound, n = 1000)
points(GA@solution, GA@fitnessValue, col = 2, pch = 20)
```

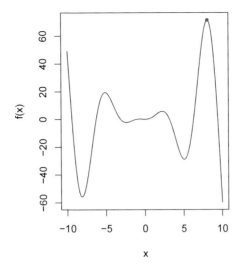

FIGURE 12.4
Function Optimization with Genetic Algorithm

The output of the function optimization with the genetic algorithm is shown in Figure 12.4.

12.2.3   Genetic Programming

Genetic programming (GP), inspired by our understanding of biological evolution, is an evolutionary computation (EC) technique that automatically solves problems without requiring the user to know or specify the form or structure of the solution in advance. At the most abstract level, GP is a systematic, domain-independent method for getting computers to solve problems automatically starting from a high-level statement of what needs to be done (Poli, Langdon, and McPhee, 2008). The idea of genetic programming is to evolve a population of computer programs. The aim and hope is that, generation by generation, GP techniques will stochastically transform populations of programs into new populations of programs that will effectively solve problems under consideration. Like evolution in nature, GP has in fact been very successful at developing novel and unexpected ways of solving problems.

(1) Representation: *Syntax Tree*

To study GP, it is convenient to express programs using syntax trees rather than as lines of code. For example, the programs $2.2 - (x/11)$, $7 * \sin(y)$, and $(2.2 - (x/11)) + 7 * \sin(y)$ can be represented by the three syntax trees in the diagram (Figure 12.5). The variables and constants in the program (x, y, 2.2, 7, and 11) are leaves of the tree, called *terminals*, while the arithmetic operations ($+$, $\times$, and sin) are internal nodes, or *functions*. The sets of allowed functions and terminals together form the primitive set of a GP system.

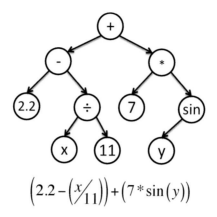

$$\left(2.2 - \left(\frac{x}{11}\right)\right) + \left(7 * \sin(y)\right)$$

FIGURE 12.5
Genetic Programming

(2) Reproductive Mechanism

For programs to propagate, GP must have well-defined mechanisms of re-production. Here, there are two common ways to generate offspring: crossovers and mutations. *Crossover* is the primary way to reduce the chance of chaos because crossovers lead to much similarity between parent and child. A crossover is a selection of a subtree for crossover, whereas a mutation randomly generates a subtree to replace a randomly selected subtree from a randomly selected individual. Crossover and mutation are illustrated in the diagram, where the trees on the left are actually copies of the parents; their genetic material can freely be used without altering the original individuals.

(3) Survival Fitness

The second mechanism required for program evolution is *survival fitness*. There are many possible ways to define a measure of fitness. As an example, for the problem of finding a function $g(x)$ to approximate the target function $f(x)$, the mean square error between the two functions can be used as the fitness measure.

The usual termination criterion is that an individual's fitness should exceed a target value, but could instead be a problem-specific success predicate, or some other criterion. Typically, a single best-so-far individual is then harvested and designated as the result of the run.

Koza (Banzhaf et al., 1998) studied the symbolic regression:

$$y = f(x) = \frac{x^2}{2}, \text{ where } x \text{ ranges 0 to 1.}$$

Using as the terminal set x ranging from -5 to 5 on the syntax tree, a function set consisting of $+, -, \times$, and protected division %, and 500 individuals in each generation, (Koza, et al 2003) was able to obtain the best individual (function) f_i in generation i, where $f_0 = \frac{x}{3}$, $f_1 = \frac{x}{6-3x}$,...., and $f_3 = \frac{x^2}{2}$. Therefore, at generation 3, the correct solution is found. However, as the generation number increases, the best fit function starts to expand again. Javascript code for GP is provided by Chang (2010).

Genetic programming and genetic algorithms are very similar in that they both use evolution by comparing the fitness of each candidate in a population of potential candidates over many generations. In each generation, new candidates are found by randomly changing (mutating) or swapping parts (crossover) of other candidates. The least-fit candidates are removed from the population.

The main difference between them is the representation of the algorithm/program. A genetic algorithm is represented as a list of actions and values, often a string. For example: $x * *2/2$. A parser has to be written for this encoding, to understand how to turn this into a function. A genetic program is represented as a tree structure of actions and values, usually a nested data structure. Genetic algorithms inherently have a fixed length, meaning the resulting function has bounded complexity. Genetic programs inherently have a variable length.

12.2.4 Application

Genetic programming does not have to be for mathematical function fitting. In fact it has proliferated with applications in many fields, including: RNA structure prediction (van Batenburg, Gultyaev, and Pleij, 1995), molecular structure optimization (Wong et al. 2011; Willett, 1995; Wong, Leung, and Wong, 2010), code-breaking, finding hardware bugs (Iba et al., 1992), robotics (Li et al. 1996; Michail, 2009), mobile communications infrastructure optimization, mechanical engineering, work scheduling, the design of water distribution systems, natural language processing (NLP), the construction in forensic science of facial composites of suspects by eyewitnesses (Li et al., 2004), airlines revenue engagement, trading systems in the financial sector, and in sound synthesis in the audio industry.

Ghosh and Jain (2005) collected articles across a broad range of topics on the applications of evolutionary AI in drug discovery, GAs for data mining and knowledge discovery, strategies for scaling up evolutionary instance reduction algorithms for data mining, the construction and selection of features with evolutionary computing, multi-agent data mining using evolutionary computing, a rule extraction system with class-dependent features, knowledge discovery via GA, diversity and neuro-ensemble, unsupervised niche clustering, evolutionary computation in intelligent network management, GP in data mining for drug discovery, microarray data mining with evolutionary computation, and an evolutionary modularized data mining mechanism for financial distress forecasts.

Panagiotis Barmpalex et al. (2011) has used symbolic regression via genetic programming in the optimization of a controlled release pharmaceutical formulation and compared its predictive performance to artificial neural network (ANN) models. Two types of GP algorithms were employed: 1) standard GP, where a single population is used with a restricted or an extended function set, and 2) multi-population (island model) GP. Optimal models were selected by minimization of the Euclidian distance between predicted and optimum release parameters. Their results showed that the prediction ability of GP on an external validation set was higher compared to that of the ANNs, with the multi-population and standard GP combined with an extended function set, showing slightly better predictive performance. Ghaheri et al. (2015) introduced the genetic algorithm and its applications in medicine. They reviewed the applications of the genetic algorithm in disease screening, diagnosis, treatment planning, pharmacovigilance, prognosis, and health care management.

Ghaheri et al. (2019) provided a comprehasive review of the applications of genetic algorithms in medicine, in 15 different areas: radiology(Karnan and Thangavel, 2007; Sahiner and Chan et al., 1996), oncology (Ooi, Tan, 2003; Tan et al., 2011), cardiology (Vinterbo, Ohno-Machado, 1999; Khalil et al., 2006), endocrinology (Nguyen et al., 2013), obstetrics and gynecology (Hoh et al, 2012; Marinakis et al., 2009; Boisvert, 2012), pediatrics (Ocak, 2013; Latkowski T., Osowski, 2015; Lin et al., 2013), surgery (Jefferson et al., 1997), infectious diseases (Elveren and Yumuşak, 2011; Castiglione et al., 2004), pulmonology (Güler et al., 2005; Engoren et al., 2005), radiotherapy (Nazareth et al., 2009; Ezzell and Gaspar, 2000; Wu and Zhu 2000; Yu et al., 1997), rehabilitation medicine (Pei et al., 2011; Aminiazar, 2013), orthopedics (Ishida et al., 2011; Hsu et al., 2006; Amaritsakul et al., 2013), neurology (Khotanlou and Afrasiabi, 2012; Koer and Canal, 2011), pharmacotherapy (Koh et al., 2008; Zandkarimi et al., 2014), and health care management (Zhang et al., 2013; Du et al., 2013; Boyd and Savor, 2001).

12.3   Cellular Automata

A *cellular automaton* (CA) is one of a class of simple computer simulation methods used to model both temporal and spatiotemporal processes. CAs normally consists of large numbers of identical cells that form a lattice (like a chessboard) with defined interaction rules.

Cellular automata were invented in the late 1940s by von Neumann and Ulam (von Neumann, 1966) and have been used to model a wide range of processes seen in artificial intelligence, image processing, virtual music creation, and physics (Wolfram, 2002). Cellular automata also have a long history in biological modeling. Indeed, one of the first and most interesting CA simulations in biology is Conway's Game of Life (Berlekamp, Conway, and Guy, 1982). CA is simple but can do virtually everything that any computer can do. For example, there is a finite initial state such that any paragraph of English prose, when properly coded as a sequence of gliders, will result in a "spell-checked" paragraph of English prose (again coded as a sequence of gliders).

The rule of CA can be defined in many ways. Here is a simple example (Chang, 2011):

An occupant of a cell with fewer than two neighbors will, sadly, die of loneliness; with two or three neighbors, it will continue into the next generation; with four or more neighbors, it will die of over-excitement.

The objects (cells or proteins) in a CA simulation usually do not move: they only appear, change properties, or disappear. Thus, objects' properties, attributes, and information are the only things that "move." Variations on the CA model, known as dynamic cellular automata (DCA), actually enable objects to exhibit motion (Wishart et al. 2005). We can apply random walks or other stochastic processes to DCA. Depending on the implementation of the DCA algorithm, molecules can move one or more cells in a single time step. DCA models permit considerably more flexibility in simulating biological processes (Materi and Wishart, 2007).

There are CA applications in the pharmaceutical industry. A few examples are: drug release in bio-erodible microspheres (Zygourakis and Markenscoff, 1996), lipophilic drug diffusion and release (Kier et al., 1997; Wishart et al., 2005; Laaksonen et al., 2009; Fathi et al., 2013), drug-carrying micelle formation (Kier et al., 1996), the progression of HIV/AIDS, HIV treatment strategies (Peer et al., 2004, Sloot et al., 2002), and the simulation of different drug therapies or combination therapies. Some CA models have the capacity to model extreme time scales (days to decades) efficiently and to simulate the spatial heterogeneity of viral infections.

Example 12.2: Cellular Automata with R

Aschinchon (2014) gives the following example implemented in R.

Imagine a linear grid that extends to the left and right. The grid consists of cells that may be in only one of the two states: On or Off. At each time step, the next state of a cell is computed as a function of its left and right neighbors and the current state of the cell itself, according to the following rules.

(1) Given a cell, if the three cells (both immediate neighbors and the cell itself) are all On or all Off, the next state of the cell is Off. Otherwise, the next state is On.

(2) Time flows in a downward direction; thus, the cell immediately below another cell represents the next state, and cells on one edge are neighbors of the cells on the opposite edge.

The ultimate result is very similar to the well-known Sierpinski triangle. The following R code is modified slightly from Aschinchon (2014).

```
# Cellular Automaton
library(sp)
width <- 2^5
depth <- width
gt = GridTopology(cellcentre=c(1,1),cellsize=c(1,1),cells=c(width, depth))
gt = SpatialGrid(gt)
z <- data.frame(status=sample(0:0, width, replace=T))
z[width/2, 1] <- 1
z[width/2+1, 1] <- 1
for (i in (width+1):(width*depth))
{
    ilf <- i-width-1
    iup <- i-width
    irg <- i-width+1
    if (i%%width==0) irg <- i-2*width+1
    if (i%%width==1) ilf <- i-1
    if((z[ilf,1]+z[iup,1]+z[irg,1]>0)&(z[ilf,1]+z[iup,1]+z[irg,1]<3))
        {st <- 1} else {st <- 0}
    nr<-as.data.frame(st)
    colnames(nr)<-c("status")
    z<-rbind(z,nr)
}
sgdf = SpatialGridDataFrame(gt, z)
image(sgdf, col=c("white", "black"))
```

The output is shown in Figure 12.6.

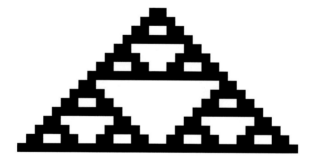

FIGURE 12.6
Cellular Automata in Action

12.4   Summary

Systems in which organized behavior arises without a centralized controller or leader are often called self-organized systems. The intelligence possessed by a complex system is called swarm intelligence (SI) or collective intelligence. SI is characterized by micro motives and macro behavior.

Evolutionary algorithms (EAs) include genetic algorithms (GAs) and genetic programming (GP). A typical EA requires: a genetic representation of the solution domain, a fitness function to evaluate the solution domain, and crossover and mutation operations.

GP techniques will, generation by generation, stochastically transform populations of computer programs into new populations of programs that will effectively solve problems under consideration.

Genetic programming and genetic algorithms are very similar in that they both use evolution by comparing the fitness of each candidate in a population of potential candidates over many generations. The main difference between them is the representation of the algorithm or program. A genetic algorithm is represented as a list of actions and values, often a string, while a genetic program is represented as a tree structure of actions and values, usually a nested data structure.

Swarm intelligence and evolutionary AI have found applications in drug discovery and development and in artificial general intelligence.

12.5   Problems

12.1: What are the differences between swarm intelligence and evolutionary intelligence?

12.2: What are the differences between reinforcement learning and evolutionary learning?

13

Applications of AI in Medical Science and Drug Development

We have discussed many AI applications for different AI methods In this chapter we will provide a review of AI development in terms of application areas: quantitative structure-activity relationships (QSARs) in drug discovery, cancer prediction using microarray data, deep learning for medical image analysis, healthcare, clinical trials, and drug safety monitoring.

13.1 AI for QSARs in Drug Discovery

The applications of AI, especially deep learning, in drug discovery include (1) new drug molecule identification and protein engineering using QSARs, (2) gene expression data analysis, and (3) pharmacokinetics and pharmacodynamics modeling (Hughes et al., 2015; Lu Zhang et al., 2017). Examples of AI application are precision medicine development (LIAng et al., 2015), sequence specification prediction (Alipanahi et al., 2015), genomics modeling for drug repurposing (Aliper et al., 2016), and drug-protein interactions (Wen et al., 2017). Examples of AI in computational biology include predicting whether a protein is secreted or not from its amino acid sequence, predicting whether a tissue is healthy from a gene-profiling experiment, or predicting whether a chemical compound can bind a given target or not from its structure. In this section we will focus on AI applications in QSARs.

13.1.1 Deep Learning Networks

Drug discovery protocols in the pharmaceutical industry have for many years mainly relied on high-throughput screening (HTS) methods for rapidly ascertaining the biological or biochemical activity of a large number of drug-like compounds. Various problems, including the efficacy, activity, toxicity, and bioavailability of the designed compounds, are frequently encountered during the discovery process (Ghasemi et al., 2018b). With high-throughput virtual screening (HTVS), those involving quantitative structure-activity relationships (QSARs) have proved their applicability in modern drug discovery protocols. Various AI methods have been used in discovering QSARs from biological data related to designed molecules.

Hiller et al. (1973) indicated that ANNs could be helpful for the classification of molecules into two categories: active and inactive. Later, Rose et al. (1991) applied an unsupervised neural network method known as Kohenen topology-preserving mapping to QSAR analysis based on the 2-D representation of compound similarities. Hinton et al. (2006) proposed a fast learning algorithm used for deep belief nets. Lowe et al. (Lowe et al., 2014; Xu et al., 2017) proposed multi-task neural networks for QSAR predictions. Eri et al. (2012) used an ANN with adjustment of the relative importance of descriptors to identify drug-like ligands.

The Kaggle competition promoted by Merck in 2012 led to the advent of deep learning (DL) in drug discovery. DL researchers won the competition mainly through using DL in QSARs to capture complex statistical patterns among thousands of descriptors extracted from numerous compounds.

Gupta et al. (2015) used RNNs and LSTMs for De Novo Drug Design. The learned pattern probabilities can be used for de novo SMILES generation without the need for virtual compound library enumeration. Ghasemi et al. (2018b) provides a review of neural network and deep-learning algorithms used in QSAR studies. They pointed out that challenges remain: arcane descriptors can affect the results of biological activity prediction or classification, although the models cannot be interpreted, whereas simpler interpretable descriptors cannot make good models for diverse data sets. By contrast, the major disadvantages of ANNs in QSAR studies are the issues of overfitting and local optima, which makes prediction or classification unreliable. Proposed remedies include regularization (introducing a penalty for overfitting), convolutional neural networks (CNNs) with dropout weights, deep belief networks (DBNs), autoencoders, rectified linear units (ReLUs) instead of the sigmoid function, and the conditional RBM (Mnih et al., 2012).

Pereira et al. (2016) proposed a novel deep learning-based virtual screening (VS) method, DeepVS. They performed docking with 40 receptors and 2950 ligands, and compared the results with 95,316 decoys. The docking outputs were used to train a deep CNN that could rank the list of ligands for each receptor. DeepVS shows encouraging results. DL has also been used to generate focused molecule libraries (Segler et al., 2017) or new molecular fingerprints (Kadurin et al., 2016).

13.1.2   Network Similarity-Based Machine Learning

Many similarity measures have been implemented using biological, chemical, or topological properties of the targets, drugs, and known interactions. Hu and Agarwal (2009) assembled a disease-disease, drug-drug, and disease-drug networks by matching molecular profiles of disease and drug expression profiles. Performance and prediction power varied according to the similarity measures used and availability of the data.

Biological relevance is established when at least one of the three following criteria is met: (1) the new targets contribute to the primary activity of the drug; (2) they mediate drug adverse effects; or (3) they are unrelated by sequence, structure, and function to the canonical targets. Guney et al. (2016) developed a network-based method based on a new proximity measure that combines six different topological measures and uses topological structures called "disease modules." A disease module is formed by genes associated with a given disease (Menche et al., 2015). The authors hypothesized that a drug is effective against a disease if it targets proteins in the close vicinity of the related disease module. The proximity measure performs better than six of the most common similarities. Furthermore, this method is capable of taking into account the elevated number of interactions of targets, and as a result, it is not biased regarding either the number of targets a drug has or their degrees; however, this improvement requires access to disease genes, drug targets, and drug-disease annotations (Vanhaelen, Mamoshina et al., 2017)

Zhou et al. (2010) and Chen et al. (2011) proposed a method purely based on topology measures to predict drug-disease associations or recommend diseases for a drug by mining data on the properties of a drug-disease bipartite network of experimentally verified drug-disease associations. There are two different ways the network similarity measure can be used: (1) if a drug interacts with a target, then other drugs similar to the drug will be recommended to the target, and (2) if a drug interacts with a target, then the drug will be recommended to other targets with similar sequences to the target. To improve the prediction, Alaimo et al. (2013) developed the DT-hybrid algorithm using a similarity matrix to directly plug the domain-dependent biological knowledge into the model. The similarity matrix is obtained as a linear combination of a structure similarity matrix and a target similarity matrix. This method performs better for the prediction of biologically significant interactions and outperforms the methods presented in (Vanhaelen, Mamoshina et al., 2017).

13.1.3 Kernel Method and SVMs

Unlike small molecules, knowledge of protein folding has a profound impact on understanding the heterogeneity and molecular function of proteins, further facilitating drug design. Predicting the 3-D structure (fold) of a protein is a key problem in molecular biology (Wei and Zou, 2016). Determination of the fold of a protein mainly relies on molecular experimental methods, such as X-ray crystallography, and nuclear magnetic resonance spectroscopy. With the development of next-generation sequencing techniques, the discovery of new protein sequences has been rapidly increasing, exceeding capability of traditional experimental techniques to determine protein folding. Computational methods for protein-fold recognition can be template-based methods or template-free methods. In template-based methods, proteins of known structures retrieved from public protein structure databases (e.g., Protein Data Bank (PDB), UniProt, DSSP, SCOP, SCOP2, and CATH) are used as template proteins for a query protein sequence. To make template-based prediction fast and reliable, a simplified database is usually employed, in which the sequence similarity is less than 50% to 70%. Multi-alignment algorithms are then adopted to exploit evolutionary information by encoding amino acid sequences into profiles. Finally, to determine the optimal alignments, the scoring functions (Z-score and E-value) are used as measures to evaluate the similarity between the profiles derived from a query protein and those of template proteins with known structures. The 3-D structure models based on optimal query-template alignments can be further optimized using energy minimization and loop modeling.

Recent research efforts have focused on the development of template-free methods, which try to predict protein structures solely based on amino acid sequences rather than on known structural proteins as templates. The template-free methods include hidden Markov Models, genetic algorithms, artificial neural networks, support vector machines (SVMs), and ensemble classifiers. Machine learning aims to build a prediction model by learning the differences between different protein-fold categories and using the learned model to automatically assign a query protein to a specific protein-fold class. This approach is thus more efficient for large-scale predictions and can examine a large number of promising candidates for further experimental validation.

SVMs as a special case of kernel methods have been used in protein remote homology detection, protein structural classification, and DNA-binding protein prediction. SVM-based protein-fold recognition methods (Shamim et al., 2007, 2013; Damoulas and Girolami, 2008; Dong et al., 2009; Yang and Chen, 2011). The main difference among these SVM-based methods is their feature representation algorithms. Wei and Zou (2016) provided a review of the recent progress in machine learning-based methods for protein-fold recognition prior to 2011. Poorinmohammad et al. (2014) combined the SVM approach with pseudo amino acid composition descriptors to classify anti-HIV peptides, with a prediction accuracy of 96.76%.

13.1.4   Decision-Tree Method

Decision trees (DTs) are a transparent and interpretable machine-learning strategy. Generally, there are two essential steps for the construction of decision trees: selecting attributes and pruning. DTs have the problem that upstream errors can quickly propagate to the downstream nodes, causing failures. There are multiple solutions to the problem. For example, using a random forest (RF) is an ensemble modeling approach that operates by constructing multiple DTs as base learners. By introducing a random selection of features and the "bagging" idea (Breiman, 2001), each base learner further increases the "test" nodes and is trained by random sampled subsets instead of by the original data set. The final outcome is a consensus score from all individual DT outputs. Compared with DTs, an RF is less likely to overfit the data. RFs have been widely used for bioactivity classification (Singh et al.,

2015), toxicity modeling (Mistry et al., 2016), protein-ligand binding affinity prediction (Wang et al., 2015), and drug target identification (Kumari et al., 2015), among other purposes. For example, Mistry et al. (2016) used RF and DTs to model the drug-vehicle toxicity relationship for the first time. Their data set included 227,093 potential drug candidates and 39 potential vehicles. The resulting model predicted the toxicity relief of drugs by specific vehicles. Wang et al. (2015) used the RF method to model the protein-ligand binding affinity between 170 complexes of HIV-1 proteases, 110 complexes of trypsin, and 126 complexes of carbonic anhydrase. Furthermore, Kumari et al. have constructed an improved RF by integrating bootstrap and rotation feature matrix components that successfully discriminated human drug targets from nondrug targets (Kumari et al., 2015).

13.1.5 Other AI Methods

Traditional QSAR approaches suffer from overfitting and from the active cliffs problem, leading to the failure of predicting new compounds. Linear discriminant analysis (LDA) has been used to predict drug-drug interactions, identify new compounds (Marrero et al., 2015), and detect adverse drug events (Vilar et al., 2015). To identify antiviral drugs, Weidlich et al. (2013) applied a kNN integrated with a simulated annealing method and RF to 679 drug-like molecules. The kNN approach has also been applied to predict other bioactivities of drug-like molecules (Jaradat et al., 2015). Li et al. (2013) recently proposed an RF-based protein-fold recognition method called PFP-RFSM. The framework of PFP-RFSM involves a comprehensive feature representation algorithm that can capture distinctive sequential and structural information representing features from seven perspectives: amino acid composition, secondary structure contents, predicted relative solvent accessibility, predicted dihedral angles, PSSM matrix, nearest-neighbor sequences, and sequence motifs (Lu Zhang et al., 2017). Lampros et al. (2014) proposed a protein-fold classification method based on a Markov chain trained with the primary structure of proteins and on a reduced state-space hidden Markov Chain. This method has proven to be effective in protein-fold categorization.

13.1.6 Comparisons with Different Methods

Wei and Zou (2016) review the recent progress in ML methods for protein-fold recognition and provided comparisons with different methods on a benchmark dataset. They examine the effectiveness of existing machine learning-based methods in the literature for protein-fold recognition a public and stringent dataset. This dataset, referred to as to DD, has been widely used in several studies. The DD dataset is comprised of a training dataset and a testing dataset, both of which cover 27 protein-fold classes in the SCOP database. The training dataset contains 311 protein sequences with $\leq 40\%$ residue identity, while the testing dataset contains 383 protein sequences with $\leq 35\%$ residue identity. Wei and Zou evaluated and compared the 20 representative methods published in the 10 years from 2006 to 2016 on the DD dataset. The compared 20 methods are first modeled by the training dataset of the DD dataset, and then they are tested on the testing dataset of the DD dataset. The prediction accuracies range from 61.1% in 2006 to 76.2% in 2016, a great improvement over time.

Interestingly enough, Simon Smith (2019) reviewed 127 startups using artificial intelligence in drug discovery in March 2019 and classified them into 14 different categories:

1. Aggregate and Synthesize Information

2. Understand Mechanisms of Disease

3. Establish Biomarkers

4. Generate Data and Models

5. Repurpose Existing Drugs

6. Generate Novel Drug Candidates

7. Validate and Optimize Drug Candidates

8. Design Drugs

9. Design Preclinical Experiments

10. Run Preclinical Experiments

11. Design Clinical Trials

12. Recruit for Clinical Trials

13. Optimize Clinical Trials

14. Publish Data

Readers can visit https://www.scienceboard.net/index.aspx?sec=sup&sub =drug&pag=dis&ItemID=11 for more information on the startups.

13.2 AI in Cancer Prediction Using Microarray Data

13.2.1 Cancer Detection from Gene Expression Data

Cancer is a serious worldwide health problem usually associated with genetic abnormalities. These abnormalities can be detected using microarray techniques which measure the expression and the activity of thousands of genes. Microarray technology is a commonly used tool for analyzing genetic diseases. A standardized microarray dataset consists of thousands of gene expressions and a few hundred samples. Each expression measures the level of activity of genes within a given tissue, so comparing the genes expressed in abnormal cancerous tissues with those in normal tissues gives a good insight into the disease pathology and allows for better diagnosis and predictions for future samples (Daouda and Mayo, 2019). However, the high dimensionality of the gene expression profiles makes it challenging to use such data for cancer detection. AI methods have been either used for (1) dimension reduction, i.e., selecting or creating the most relevant discriminating features and eliminating the non-relevant dependent features in a prior step to prediction; (2) predicting the existence of cancer, cancer type, or the survivability risk; and (3) clustering for unlabeled samples.

13.2.2 Feature Selection

The commonly used public data repositories for AI applications in cancer prediction and clustering include the TCGA, UCI, NCBI Gene Expression Omnibus (GEO), and Kentridge biomedical databases. Removing the genes that have low expression value across all samples is one of the simplest and most straightforward preprocessing techniques (Kumardeep et al., 2017; Padideh et al., 2016; Yuan, 2016). Principle component analysis (PCA) has also been used for feature selection (Zhang et al., 2018; Rasool et al., 2013). The PCA method transforms the dataset features into a lower-dimensional space. Neural network filtering methods are used for extracting representations that best describe the gene expressions without any direct consideration to the prediction goal as unsupervised learning (Tan, J. et al., 2015; Rasool et al., 2013). Convolutional neural networks have been used for feature extraction by scanning a set of weight matrices across the input (Steve et al., 1997). Generative adversarial networks, as discussed earlier, consisting of a generative network and a discriminator network, have also been used for feature selection.

13.2.3 Cancer Prediction

Cancer prediction using AI includes predicting the existence of cancer, cancer type, and survivability risk. Different types of autoencoders have been recently used for filtering microarray gene expressions, such as stacked denoising autoencoders (Jie et al., 2015), contractive autoencoders (Macías-García et al., 2017), sparse autoencoders (Rasool, 2013), regularized autoencoders (Kumardeep et al., 2017), and variational autoencoders (Way and Greene, 2017). Danaee et al. (2016) used a deep architecture of four layers with 15,000, 10,000, 2000, and 500 neurons, respectively, and a stochastic gradient descent optimization algorithm. MLPs are used to predict the survivability of lung cancer patients: very low, low, normal, high, and very high (Chen et al., 2014, 2015). The models outperform in classification accuracy compared to Bayesian networks, SVMs, and KNNs. A different MLP network was used by Mandal and Banerjee (2015) and tested using two different datasets for breast and lung cancer with 15 genes. The best accuracy level (94.0%) was achieved when the number of hidden layers was three. A deep generative model called

DeepCancer for binary (cancerous/non-cancerous) classification has been proposed by Rajendra et al. (2017). DeepGene, a convolutional feedforward model was proposed in [84] for somatic point mutation-based cancer-type classification. The ANN model with four hidden layers, ReLU activation function, softmax output layer and logarithmic loss function is tested on classifying 12 types of cancers and compared with SVM, KNN and naïve Bayes methods. Liu et al. (2017) used a CNN with 7 layers (input, 2 convolutional layers, 2 max pooling layers, one fully connected layer and an output layer) using the 1-D gene expression dataset. Kumar (2018) employ an MLP model having one hidden layer with 20 neurons and compared the performance with SVM, logistic regression, a naïve Bayes method, classification trees, and KNNs, in classifying a leukemia dataset into two groups. No preprocessing, filtering, or feature selection was applied on the 6817 genes in the comparisons.

In addition to a CNN, an AdaBoost classifier was used by Zhang et al. (2018) to classify breast cancer patients into good prognosis and poor prognosis groups according to whether distant metastasis had occurred within 5 years or not. A model combining genetic algorithms with an ANN was proposed by Hou et al. (2018). The model consisted of 3 layers, 1000 nodes as the input layer, and 1 node in the output layer. The role of the genetic algorithm was to select the best number of input variables that maximizes the classification accuracy in clinical phenotypes. The population size for the genetic algorithm was set to 100, the maximum evolutionary generations was 50, and the algorithm selected 15 candidate input variables. An ensemble approach that incorporates 5 different machine-learning models were proposed by Xiao et al. (Xiao, et al., 2018). Informative gene selection was applied on differentially expressed genes. Then, a deep learning method was employed to assemble the outputs of the 5 classifiers KNNs, SVMs, decision trees, random forests, and gradient boosting decision trees. The model of 5 hidden layers was used for binary (tumor/normal) classification, hence 1 output unit was used in the output layer. ReLU activation functions and a SGD optimizer was used to minimize the MSE similarity function; regularization, to constrain the magnitudes of the weights, was added to the function to overcome the overfitting problem (Daouda and Mayo, 2019).

13.2.4 Clustering

ANNs can be used in gene-related clustering (samples-to-cluster or gene-to-cluster assignment). To overcome the high-dimensionality problem, two approaches can be used: (1) clustering ensembles: repeatedly running a single clustering algorithm such as SOM or neural gas (NG) with different initializations or numbers of parameters, and (2) projective clustering: only considering a subset of the high-dimensional features (Daouda and Mayo 2019). Yu et al. (2017) suggested projective clustering ensemble (PCE), which combines the advantages of both projective clustering and ensemble clustering.

13.3 Deep Learning for Medical Image Analysis

13.3.1 Deep Learning for Medical Image Processing

Medical imaging includes those processes that provide visual information of the human body. The purpose of medical imaging is to aid radiologists and clinicians to make the diagnostic and treatment process more efficient. Medical imaging is a predominant part of the diagnosis and treatment of diseases (Qayyuma et al., 2018). The Institute of Medicine at the National Academies of Science, Engineering and Medicine reports that "diagnostic errors contribute to approximately 10% of patient deaths," and account for 6% to 17% percent of hospital complications (Sennaar, 2018). The cause of diagnostics errors is attributed to a variety of factors, including information technology.

Innovations in deep learning and the increasing availability of large annotated medical image datasets are leading to dramatic advances in automated disease detection and organ and lesion segmentation. Computer-aided diagnosis (CAD) in medical imaging has flourished over the past several decades. For medical imaging, CAD has focused predominantly on radiology, cardiology, and pathology. Examples in radiology include the automated detection of microcalcifications and masses on mammography, lung nodules on chest X-rays and CT scans, and colonic polyps on CT colonography (Giger et al., 2008). In cardiology, examples include CAD for echocardiography and angiography (Willems et al., 1991). In digital pathology, examples include detection of cellular components, such as nuclei and cells, and diseases such as breast, cervical, and prostate cancers (Xing and Yang, 2016; Ranald and Summers, 2017, in Lu et al., p. 3, 2017).

For radiologic images, the precise location and extent of the abnormality traditionally are determined by trained experts. Such hand annotation by an expert is time-consuming and expensive. As an example, the NLST study involved over 53,000 patients and cost over $250 million. In the medical image processing community, there is a great need for false positive (FP) reduction for automated detection tasks. Early work indicates that substantial FP reduction or improved image retrieval is possible with off-the-shelf deep learning networks (Roth, 2016). Yet even these reductions do not reach the specificity obtained by practicing clinicians. Clearly, further work is required, not just on improving machine learning, but also on the more traditional feature engineering (Ranald and Summers, 2017, in Lu et al., p. 3, 2017). Among the main applications of deep neural networks for medical image analysis are detection, segmentation, and classification in heterogeneous imaging modalities.

Segmentation is a process of dividing an image into multiple non-overlapping regions based on specific criteria i.e., sets of pixels or intrinsic features such as color, contrast and texture for meaningful information extraction. Such information extraction involves shape, volume, relative position of organs, and abnormalities detection. For instance, a brain tumor segmentation algorithm can be used in cascaded deep CNNs and a different method for segmentation of glioma tumors is used for deep CNNs.

Anwar et al. (2018) present a state-of-the-art review in medical image analysis using CNNs. The application area of CNNs covers the whole spectrum of medical image analysis including detection, segmentation, classification, and computer aided diagnosis.

CNNs are used for the classification of lung diseases (Anthimopoulos, 2014) using two databases of interstitial lung diseases (ILDs) and CT scans, each having dimension of 512×512. After the feature extraction procedure, image patches are represented by feature vectors which are fed to a random forest classifier for classification. The network is trained by using 14,696 image patches derived from the original scan with a classification accuracy of 85.5%. A deep learning method based on convolutional classification restricted Boltzmann machine for lung texture classification and airway detection in CT images was proposed by Tulder and Bruijne (2016) using two different CT lungs datasets. The network is trained on image patches of size 32×32 voxels along a gird with a 16-voxel overlap; a patch is kept if it has 75% of voxels belonging to the same class. In both applications, a combination of learning objectives outperformed purely discriminative or generative learning, increasing, for instance, the lung tissue classification accuracy by 1 to 8 percentage points.

A dual pathway, 11-layer deep, 3-D CNN for the challenging task of brain lesion segmentation was proposed by Kamnitsas et al. (2016). The devised architecture is the result of an in-depth analysis of the limitations of current networks proposed for similar applications. An efficient dense training scheme is used to overcome the computational burden of processing 3-D medical scans. For post-processing of the network's soft segmentation, the authors used a 3-D fully connected conditional random field which effectively removes false positives. The deep learning approach (pipeline) was extensively evaluated for three challenging tasks of lesion segmentation in multi-channel MRI patient data associated with traumatic brain injuries, brain tumors, and ischemic stroke, demonstrating its computational efficiency.

13.3.2   Deep Learning Methods in Mammography

Breast cancer is one of the most common types of cancer affecting the lives of women worldwide. Recent statistical data published by the World Health Organisation (WHO) estimates that 23% of cancer-related cases and 14% of cancer-related deaths among women are due to breast cancer. Recent studies show that this manual analysis has a sensitivity of 84% and a specificity of 91%. A second reading of the same mammogram either from radiologists or from computer-aided diagnosis (CAD) systems can improve this performance (Carnerio et al., 2017, in Lu et al., p. 11, 2017).

An AI system that can analyze breast lesions using mammograms usually comprises three steps: (1) lesion detection, (2) lesion segmentation, and (3) lesion classification. Lesion detection involves identifying a large number of candidate regions, usually based on the use of traditional filters, such as morphological operators. These candidates are then processed by a second stage that aims at removing false positives using machine learning approaches (e.g., region classifiers) (Kozegar et al., 2013). Lesion segmentation is helpful because lesion shape is an important feature in lesion classification in the final stage analysis. The challenges involved in these steps are mainly due to the low signal-to-noise ratio present in the imaging of the lesion, and the lack of a consistent location, shape, and appearance of lesions (Tang et al., 2009; Oliver et al., 2010). Current advances in deep learning bring the hope that these challenges can be overcome. However, the main challenge faced by deep learning methods is overfitting: too many data with too few data points. The need for large annotated training sets given the scale of the parameter space, usually on the order of 100 parameters, whereas annotated training sets rarely have more than a few thousand samples. A special CNN, called a Region Convolutional Neural Network (R-CNN) was proposed by Dhungel et al. (2015) for a breast mass detection methodology that consists of a cascade of classifiers. A deep learning–based mass detection method consisting of a cascade of deep learning models was proposed by Ertosun and Rubin (2015).

13.3.3 Deep Learning for Cardiological Image Analysis

Cardiovascular disease is the number-one cause of death in developed countries, and it claims more lives each year than the next seven leading causes of death combined. Various imaging modalities, such as computed tomography (CT), magnetic resonance imaging (MRI), ultrasound, and nuclear imaging, are widely available technologies used in clinical practice to generate images of the heart, and different imaging modalities meet different clinical requirements. Ultrasound is most widely used for analysis of cardiac function due to its low cost and lack of radiation dose; nuclear imaging and MRIs are used for myocardial perfusion imaging to measure viability of the myocardium; a CT reveals the most detailed cardiac anatomical structures and is routinely used for coronary artery imaging; fluoroscopy/angiography is the workhorse imaging modality for cardiac interventions (Lloyd-Jones et al., 2009).

Cardiovascular structures are composed of the heart (cardiac chambers and valves) and vessels arteries and veins. A cardiac image analysis involves detection, segmentation, motion tracking, quantification, and disease diagnosis (Gulsun et al., 2016).

Deep learning networks such as CNNs (Avendi et al., 2016) and DBNs (Carneiro et al., 2012; Ngo et al., 2016) have recently been used in cardiac imaging, including left/right ventricle segmentation (Zhen et al., 2016), retinal vessel segmentation (Wang et al., 2015; Chandrakumar and Kathirvel, 2016).

A new convolutional deep belief network is proposed by Zhen et al. (2016) for direct estimation of a ventricular volume from images without performing segmentation at all. Other work in deep learning for CV can be found: a system that automatically learns the most effective application-specific vesselness measurement from an expert-annotated dataset (Zheng et al., 2011, 2013), a CNN for pixel-wise classification of retinal vessels (Wang et al., 2015; Wu et al., 2016), deep learning for detecting retinal vessel microaneurysms (Haloi, 2015), and the classification of diabetic retinopathy (Chandrakumar and Kathirvel, 2016) via fundus imaging.

The recent book edited by Lu et al. (2017) covers broad applications of deep learning in image studies with titles such as

1. Efficient False Positive Reduction in Computer-Aided Detection Using Convolutional Neural Networks and Random View Aggregation

2. Robust Landmark Detection in Volumetric Data with Efficient 3D Deep Learning

3. A Novel Cell Detection Method Using Deep Convolutional Neural Network and Maximum-Weight Independent Set

4. Deep Learning for Histopathological Image Analysis: Towards Computerized Diagnosis on Cancers

5. Interstitial Lung Diseases via Deep Convolutional Neural Networks

6. Three Aspects on Using Convolutional Neural Networks for Computer-Aided Detection in Medical Imaging

7. Cell Detection with Deep Learning Accelerated by Sparse Kernel

8. Fully Convolutional Networks in Medical Imaging: Applications to Image Enhancement and Recognition

9. On the Necessity of Fine-Tuned Convolutional Neural Networks for Medical Imaging

10. Fully Automated Segmentation Using Distance Regularized Level Set and Deep-Structured Learning and Inference

11. Combining Deep Learning and Structured Prediction for Segmenting Masses in Mammograms

12. Deep Learning Based Automatic Segmentation of Pathological Kidney in CT: Local Versus Global Image Context

13. Robust Cell Detection and Segmentation in Histopathological Images Using Sparse Reconstruction and Stacked Denoising Autoencoders

14. Automatic Pancreas Segmentation Using Coarse-to-Fine Superpixel Labeling

13.4 AI in Healthcare

13.4.1 Paradigm Shift

Artificial intelligence is bringing a paradigm shift to healthcare. The increasing availability of healthcare data and rapid development of big data (structured and unstructured) analytic methods has made possible the recent successful applications of AI in healthcare. In addition to the popular deep learning neural network approaches for structured data, natural language processing for unstructured data such as physicians' notes also has a great value in healthcare.

The goal of AI research in healthcare is not to replace human physicians in the foreseeable future but to assist physicians to make better clinical decisions, or even to replace human judgment in certain functional areas of healthcare (e.g., radiology). A few examples are: an AI system that can assist physicians by providing up-to-date medical information from journals, textbooks, and clinical practices to inform proper patient care, reducing diagnostic and therapeutic errors, and extracting useful information from a large patient population to assist in making real-time inferences for health risk alerts and health outcome prediction.

According to Sennaar (2018), a healthcare work system which, by design, does not adequately support the diagnostic process. To provide additional context, a review of 25 years of malpractice claims payouts in the U.S. by Johns Hopkins researchers showed that diagnostic error claims had a higher occurrence in outpatient settings (68.8%) vs. inpatient settings (31.2%). However, those which occurred in an inpatient setting were roughly 11.5% more likely to be lethal. Total payouts over the 25-year period amounted to a substantial sum of \$38.8 billion. To address these challenges many researchers and companies are leveraging artificial intelligence to improve medical diagnostics. With the in vitro diagnostics (IVD) market size projected to reach \$76 billion by 2023, the aging population, prevalence of chronic diseases, and emergence of personalized medicine are contributing factors to this growth. National health expenditures are estimated to have reached \$3.4 trillion in 2016, and the health share of the GDP is projected to reach nearly 20% by 2025. Sennaar (2018) summarizes current applications of AI in dedical diagnostics into four categories:

Chatbots: Companies are using AI-chatbots with speech-recognition capability to identify patterns in patient symptoms to form a potential diagnosis, prevent disease, and/or recommend appropriate courses of action.

Oncology: Researchers are using deep learning to train algorithms to recognize cancerous tissue at a level comparable to trained physicians.

Pathology: Pathology is the medical specialty that is concerned with the diagnosis of disease based on the laboratory analysis of bodily fluids and tissues. Machine vision and other machine learning technologies can enhance the efforts traditionally left only to pathologists with microscopes.

Rare Diseases: Facial recognition software is being combined with machine learning to help clinicians diagnose rare diseases. Patient photos are analyzed using facial analysis and deep learning to detect phenotypes that correlate with rare genetic diseases.

In the U.S., there are approximately 5.4 million new skin cancer diagnoses each year, and early detection is important for a greater rate of survival. The Face2Gene app, an ambitious application of facial recognition software, is aims to help clinicians diagnose rare diseases (in this case, from facial dysmorphic features). Patient photos can be analyzed using facial analysis and deep learning to detect phenotypes that correlate with rare genetic diseases. Readers can try the app for evaluation.

The IBM Watson system is a pioneer in the field of medical imaging. The system includes both ML and NLP modules, and has made promising progress in oncology. In a study (Lohr, 2016), 99% of the treatment recommendations for cancer treatment from Watson are coherent with physician decisions. Watson uses genetic data to successfully identify the rare secondary leukemia caused by myelodysplastic syndromes in Japan (Otake, 2017). Watson collaborated with Quest Diagnostics to offer AI Genetic Diagnostic Analysis (Jiang et al., 2019).

The cloud-based CC-Cruiser (Long et al., 2017) is an AI platform with a front-end data input and back-end clinical actions. CC-Cruiser is an AI agent involving three functional networks: (1) identification networks for screening for CC in populations, (2) evaluation networks for risk stratification among CC patients, and (3) strategist networks to assist in treatment decisions by ophthalmologists. The AI agent showed promising accuracy and efficiency *in silico.*

13.4.2   Disease Diagnosis and Prognosis

Among the many AI applications in radiology are medical image segmentation, registration, digital health records, computer-aided detection and diagnosis, brain function or activity analysis, neurological disease diagnosis from FMR images, content-based image retrieval systems for CT or MRI images, and text analysis of radiology reports using NLP, epidemic outbreak prediction, and surgical robotics. Specific AI methodologies and applications in healthcare can be found in *Machine Learning in Healthcare Informatics*, a recent book edited by Dua et al. (2014). These include: wavelet-based machine learning techniques for ECG signal analysis, applications of fuzzy logic control for regulation of glucose level of diabetic patients, the application of genetic algorithms for unsupervised classification of ECG, pixel-based machine learning in computer-aided diagnosis of lung and colon cancer, understanding foot function during the stance phase of walking by Bayesian network-based causal inference, rule learning in healthcare and health services research, machine learning techniques for AD/MCI diagnosis and prognosis, using machine learning to plan rehabilitation for home care clients, clinical utility of machine learning and longitudinal EHR data, rule-based computer aided decision-making for traumatic brain injuries, supervised learning methods for fraud detection in healthcare insurance, feature extraction by quick reduction algorithms, assessing the neurovascular pattern of migraine sufferers from NIRS signals, and a selection and reduction approach for the optimization of ultrasound carotid artery images segmentation.

The majority of deep learning is used in imaging analysis, followed by electrodiagnosis, genetic diagnosis, and clinical laboratory testing. In medical applications, the commonly used deep learning algorithms include CNNs, RNNs, and DBNs. CNNs have been recently implemented in the medical area to assist disease diagnosis. For instance, Gulshan et al. (2016) applied a CNN to detect referable diabetic retinopathy through retinal fundus photographs. The sensitivity and specificity of the algorithm are both over 90%, which demonstrates the effectiveness of the technique in the diagnosis of diabetes. Long et al. (2017) used CNN architecture to diagnose congenital cataract disease through learning ocular images with 90% accuracy in diagnosis and treatment recommendations. Esteva et al. (2017) identify skin cancer from clinical images using CNN deep learning. Their method provides over 90% sensitivity and specificity in the diagnosis of malignant lesions and benign lesions.

Stroke is a common and frequently occurring disease that affects more than 500 million people worldwide. The major AI applications in stroke care include: early disease diagnosis, prognosis, treatment recommendation. Stroke, 85% of the time, is caused by a thrombus in a blood vessel in the brain leading to a cerebral infarction. However, for lack of judgment of early stroke symptoms, only a few patients can receive timely treatment (Jiang et al., 2017). To this end, a movement-detecting device for early stroke prediction has been developed using PCA and a genetic fuzzy finite state machine (Villar et al., 2015). A SVM was used by Rehme et al. (2015) in interpreting resting-state functional MRI data, by which endophenotypes of motor disability after stroke were identified and classified. The SVM correctly classifies stroke patients with 87.6% accuracy. An alert of stroke is activated once the recognized movement of the patient is significantly different from the normal pattern. The AI methodology is proposed by Maninini et al. (2016) based on the extracted features from hidden Markov models (HMMs) and SVMs. The device is tested on gait data recorded on two pathological populations (Huntington's disease and post-stroke subjects) and healthy elderly controls using data from inertial measurement units placed at shank and waist. 90.5% of subjects were assigned to the correct group after leave-one-subject-out cross-validation and majority voting.

Machine Intelligence for Healthcare, a recent book by Campion et al. (2017) presents 10 topics in AI in healthcare, focusing on case studies without methodological details. In machine learning and machine intelligence, two case studies involve the application of PCA of microarray and TDA micro-biome analysis; in clinical variation and hospital clinical pathways the two case studies presented are total knee replacement and colectomy surgery; in machine intelligence for precision medicine, the two case studies discussed are diabetes subtypes and patient stratification-revealed molecular validation for asthma subtypes; in population health management, the following cases are presented: the Syria polio vaccine campaign, smartphone gait data elucidating parkinsonism, and population health transformation, and in machine intelligence for revenue cycle and payment integrity the three cases discussed are: denials management, detecting Medicare overpayments, and fraud, waste and abuse modeling. For AI methodology studies, readers can refer to the early chapters in this book and books by Clifton and Dua et al. (2014).

13.4.3   Natural Language Processing in Medical Records

In additional to structured data such as image, EP and genetic data, large proportions of clinical information are in the form of narrative text, such as physical examinations, clinical laboratory reports, operative notes and discharge summaries, most of which are unstructured and incomprehensible for computer programs. In such situations, NLP targets extracting useful information from the narrative text to assist clinical decision-making.

An NLP pipeline comprises two main components: (1) text processing and (2) classification. Through text processing, the NLP identifies a series of disease-relevant keywords in the clinical notes based on historical databases. Then a subset of the keywords are selected through examining their effects on the classification of the normal and abnormal cases. The validated keywords then enter and enrich the structured data to support clinical decision-making (Afzal et al., 2017; Jiang et al., 2019).

NLP pipelines have been developed to assist clinical decision-making for alerting treatment arrangements, monitoring adverse effects, and other aspects of care (Jiang, F. et al., 2019). For instance, Fiszman et al. (2000) show that introducing NLP for reading chest X-ray reports would assist the antibiotic assistant system to alert physicians to the possible need for anti-infective therapy. Miller et al. (2017) use NLP to automatically monitor laboratory-based adverse effects. Castro et al. (2017) employed NLP to identify 14 cerebral aneurysm disease-associated variables in the clinical notes. The resulting variables are successfully used for classifying normal patients (with 95% accuracy) and the patients with cerebral abnormalities (86% accuracy). Afzal et al. (2017) implemented NLP to extract peripheral arterial disease-related keywords from narrative clinical notes, achieving over 90% accuracy.

13.5   AI for Clinical Trial and Drug Safety Monitoring

13.5.1   Necessary Paradigm Shift in Clinical Trials

In the chapter on reinforcement learning and in Chang's books (Chang, 2010; Chang et al., 2019), we have discussed the reinforcement learning in clinical trials, especially, in clinical development program via stochastic decision process, the game theory, and Q-learning. We discussed the backward induction (Bellman's principle) and dynamic programming for optimization problem, including the policy-iteration and value-iteration algorithms. In another Chang book (Chang, 2011), *Modern Issues and Methods in Biostatistics*, we have discussed data mining in pharmacovigilance. In Chapter 4, Similarity-Based Learning (see also, Chang, 2018; Hwang and Chang, 201), we discussed SBML and its application in clinical trials, e.g., rare disease trial. However, the applications of AI in clinical trials are limited today, mainly due to current statistical strategies implemented in the regulatory guidance, which are predominately based on frequentist's hypothesis testing that is focused on type-I error control. Adopting this approach, we are facing even bigger challenges than we ever faced before. On one hand, as the available drugs become ever more effective, the diminished efficacy margin for improvement requires an impractically larger sample size for clinical trials so as to control the type-I error and maintain a sufficient power for the hypothesis test. On the other hand, as awareness of disease heterogeneities and needs for individualized medicine increase, the sample size available for each homogeneous group becomes even smaller. Realizing the ever-increasing challenges, regulatory authorities have issued guidance on drugs for rare diseases, breakthrough designations, adaptive trials, master protocol design, multiregional trials, and the use of real-world data (FDA, 2018, 2019). To solve this problem effectively requires a paradigm shift: (1) a focus on prediction of drug effects (efficacy and safety) instead of type-I error control, (2) smaller clinical trials with possible adaptive designs, (3) synergizing trial data with real-world evidence, and (4) stage-wise or adaptive marketing authorization and enhanced pharmacovigilance.

Prediction of drug effect is the problem to be solved: we want to know the effect of the drug and how much based on data. The p-value is the quantity measuring the probability of an outcome *assuming* the null hypothesis is true. This quantity does not consider the other sources of evidence and does not directly answer the underlying scientific questions about the effectiveness of the drug. With the hypothesis approach and p-value (p), if you ask someone if A is better than B, the answer would be something like this: if A is the same as B, you will have p probability of this or better data. Does this event answer your question at all?

Smaller clinical trials will not only make the trials feasible, but also enable us to deliver drugs earlier with a reduced cost. Adaptive designs allow us to learn the potential benefits and risks sooner and either provide early evidence of drug effectiveness or mitigate the risk of potentially unsafe drugs. Adaptive designs with emphasis on prediction allow learning multiple things with data in a single clinical trial without too much worrying about multiplicity. Learning is a progressive process; we should not expect in one instance to obtain an absolute conclusion and never change it.

Clinical trial data are important evidence regarding drug effects, but they are only partial data and are often obtained under ideal/unrealistic settings. Therefore, real-world evidence is necessary. The combination of clinical trials and real-world data will provide more realistic assessment of treatment effect and at the same time reduce the size of clinical trials.

When there is reasonable evidence in supporting the efficacy of a safe drug candidate, the drug can be approved for marketing with limited amounts allowable for sales and strong pharmacovigilance. After a period of use in real clinical settings, if the data continues to show the drug's effectiveness and safety, it can be further approved for full marketing. Otherwise, the marketing authorization can be at a reduced scale or be removed. This stage-wise or adaptive marketing authorization can reduce drug cost and deliver effective drugs to patients earlier. It will allow us to use real-world experience for better prescribing drugs and developing precision medicine by fine-tuning the set of target patients based on post-marketing data.

Although predictive models are used in clinical trials, they are used mainly for the purpose of obtaining p-values, not for prediction. The predictions (point estimate, confidence interval, posterior distribution, creditable interval) from a model are only served as secondary measurements. In Chapter 4, we proposed a general similarity-based machine learning (SBML) method and applied it to a rare disease trial (Chang, 2018; Hwang et al., 2019; Hwang and Chang, 2019).

Developing medicine or evaluating the effects of a clinical intervention essentially is a form of learning from scientific evidence. Therefore, it is concerned with the fundamental question: what constitute scientific evidence and learning paradigms? The surrounding controversies have been discussed in earlier chapters, including Simpson's paradox, subgroup analyses, multiregional trials, all of which explicitly or implicitly involve similarity grouping. Yet different beliefs lead to different learning paradigms, as we shall see in the next section.

13.5.2   Learning Paradigms

Learning from facts is not to simply state exactly what has been observed, rather it is to make inferences regarding causal relationships and make predictions about the future. This involves differentiating scientific truths (evidence) from false discoveries. We are discussing four different learning paradigms: frequentism, likelihoodism, Bayesianism, and similaritism.

Let's begin with a familiar example. Imagine two people, John and Lisa, who want to decide if a coin is a fair coin with equal probability of heads and tails when tossed. Their student, Bill, is the experimenter who tosses the coin.

John's intention is to stop the experiment at 12 trials and draw his conclusion based on the outcomes. Lisa's intention is to stop the experiment as soon as there are 3 heads. Interestingly, at the 12th trial, the outcome is 3 heads and 9 tails. John and Lisa both call to stop the trials. Now it is up to John and Lisa to draw their conclusions.

John, a frequentist, draws his conclusion using a binomial distribution for the response; since the number of trials, 12, is fixed, the p-value is

$$\Pr\left(X \geq 9 | H_o\right) = \sum_{x=9}^{12} \binom{12}{x} 0.5^x 0.5^{12-x} = 0.073.$$

According to Bill, the null hypothesis cannot be rejected at a one-sided level of $\alpha = 0.05$.

For Lisa, the number of responses, $n = 3$, is predetermined and the experiment continues until 3 responses are observed; then X follows the negative binomial $NB(3; 1 - p)$ and the frequentist p-value of the test is given by

$$\Pr\left(X \geq 9 | H_o\right) = \sum_{x=9}^{\infty} \binom{3 + x - 1}{2} 0.5^x 0.5^3 = 0.0327.$$

Therefore, for Lisa the null hypothesis is rejected at a one-sided level of $\alpha = 0.05$.

However, Bill thinks the probability of heads should only depend on the coin itself, not the intention of the researcher. He further argues that the likelihood ratio is a constant: the likelihood in John's case is $l_1(x|p) = \binom{12}{9} p^9 (1 - p)^3 = 220 p^9 (1 - p)^3$, and the likelihood in Lisa's case is $l_2(x|p) = \binom{3+9-1}{2} p^9 (1 - p)^3 = 55 p^9 (1 - p)^3$, therefore, based on the *likelihood principle*, we should not draw different conclusions! Bayesians and likelihoodists believe the *likelihood principle*, which can be stated as: The information contained by an observation x about θ is entirely contained in the likelihood function. Moreover, if x_1 and x_2 are two observations depending on the same parameter and the ratio of their corresponding likelihood functions is a constant, then x_1 and x_2 must lead to identical inferences. However, Bayesians also believe individual prior should also play a role in drawing a conclusion, while a likelihoodist will draw conclusions purely based on likelihood principle (e.g., likelihood ratio) without consideration of prior at all.

Frequentists (John and Lisa) draw conclusions based on the observed data and imaginary data from future repeated experiments (but not the prior knowledge); a different intention will lead to different future data and different conclusions. In contrast, a Bayesian (Bill) draws a conclusion based on individual prior knowledge and the observed data (but not any unperformed repeated experiment); different priors can lead him to a different conclusion.

I, a similaritist, believe that prior knowledge will not only determine what experiment is most appropriate to perform, but will also guide us in deciding what model (analysis) is best to use for the data and how to interpret the outcomes and draw a conclusion. We have to decide what the relevant data is based on the similarity grouping, and the similarity grouping is based on an individual's prior knowledge. Therefore, not only the prior knowledge but also the relevant data are subjectively defined. As we have seen, in Simpson's paradox and a multi-regional global trial, different similarity grouping methods can lead to different conclusions.

Bayesians and likelihoodists all have their prior beliefs that intention has no effect on the experimental conclusion. The experimenter's intention may impact future data collection and thus lead to a different conclusion in the future. Also, an experimenter's intention can change over time, so there is no reason to use the current intention for future unperformed experiments and imaginary data. A similaritist who believes "intention's effect" will define the similarity groups differently from a similaritist who does not believe "intention's effect," therefore, they will likely have different conclusions even when the experimental data are the same. Similaritists may combine data from experimenters with different intentions and use experimenters' intention as one of the attributes in calculating the similarity and let data determine the effect of intention via SBML.

One of the differences between AI and classical statistics is found in multiplicity: in AI we learn many things from a single data source without multiplicity penalties, but we verify the findings later, over time. As a simple example, we learn the phrases "artificial intelligence" and "drug development" as well as the sentence structure from a single English sentence: "Artificial intelligence is the future of drug development," without the alpha-penalty.

13.5.3   AI in Pharmacovigilance

Pharmacovigilance (PV), also known as drug safety, is the pharmacological science relating to the collection, detection, assessment, monitoring, and prevention of adverse effects with pharmaceutical products. As noted earlier, PV can be extended to collect and further evaluate drug efficacy after marketing, so that the size required for clinical trials can be reduced in the proposed paradigm of adaptive (stagewise) drug marketing approval. As stated by Sparkes (2018), in pharmacovigilance, many pharmaceutical companies and major regulatory agencies are currently evaluating AI to reduce the burden of case processing, enabling them to improve compliance while reducing case processing costs. Examples of AI being developed within PV include auto-narrative generation; narrative analysis (including case extraction and creation); QC assessment; causality assessment; and "touchless" case processing, where non-serious cases are received, verified, coded, processed, and submitted without any human intervention. In all of these examples, AI is being used to significantly reduce case processing costs and/or improve compliance.

With the emergence of electronic health records, researchers are exploring AI techniques to develop disease models, probabilistic clinical risk stratification models, and practice-based clinical pathways (Huang et al., 2015; Zhang et al., 2015). A considerable number of studies have focused on information extraction using NLP and text mining to gather insights from largely unstructured data sources, such as drug labels, scientific publications, and postings on social media (Schmider et al., 2019; Segura-Bedmar and Martinez, 2015).

Text mining techniques have also been combined with rule-based and certain machine-learning classifiers, such as the U.S. Vaccine Adverse Event Reporting System (Botsis et al., 2011). Many researchers focus on extracted information from multiple sources for different purposes, including detection of patient safety events (Fong et al., 2015) from patient records. Others have focused on developing and/or improving approaches using natural language processing to recognize and extract information from various medical text sources (Abacha and Zweigenbaum, 2011; Fong et al., 2015) from patient records and from drug-drug interactions reported in the biomedical literature (Abacha et al., 2015; Kim, et al., 2015), and from disease information in emergency department free-text reports (Pineda et al., 2015). Natural language processing techniques have also been applied to the extraction of information on adverse drug reactions from the growing amounts of unstructured data available via the discussion and exchange of health-related information between health consumers on social media (Yang et al., 2015; Liu and Chen, 2015).

AI technologies have been implemented for pharmacovigilance in big pharmaceutical companies such as Pfizer and Bayer, though they are still in their earlier stages. Pfizer's AI pilot was undertaken to test the feasibility of using artificial intelligence and robotic process automation to automate the processing of adverse event reports (Schmider and Kumar, 2019). The pilot paradigm was used to simultaneously test proposed solutions of three commercial vendors. Several different machine-learning algorithms were used to extract data from the digitized documentation: (1) table pattern recognition for predicting if a specific table cell contained a certain type of information of interest (e.g., patient name or case narrative), (2) sentence classification for predicting if a sentence within a case narrative was related to AEs, and (3) named entity recognition used to predict AEs at a token level. A conditional random field (sequence labeling) model was used to detect adverse drug reactions from case narratives (Nikfarjam et al., 2015), and a fourth machine-learning algorithm was applied, rule-based pattern matching using various predefined rules to extract information of interest, including patient name (initials), sex, age, and date of birth. Validity was established by the presence of four elements (i.e., an AE (suspected adverse drug reaction), putative causal drug, patient, and reporter), which had to be extracted and specifically coded into the respective fields. Three different criteria were used for the performance evaluation: (1) overall accuracy of information extraction, (2) case-level accuracy, and (3) case-level validity. The result confirmed the feasibility of using artificial intelligence-based technology to support extraction from adverse event source documents and evaluation of case validity.

13.6 Summary

We have provided a comprehensive review of AI applications in QSARs in drug discovery, in cancer prediction using microarray data, and in deep learning for medical image analysis, healthcare, clinical trials, and drug safety monitoring.

The main applications of AI in drug discovery include (1) new drug molecule identification/design and protein engineering using QSARs and (2) gene expression data analysis. The similarity principle is the backbone of QSAR analysis. Varieties of AI methods have been successfully used with QSARs in the discovery of newdrugs. AI methods for gene expression data analysis can be used for disease diagnosis and prognosis, and have provided opportunities for early effective treatment and optimal treatment for individual patients based on their disease stage and other characteristics. Cancer is a worldwide genetic-related disease, which imposes significant mortality and cost. Cancer has been characterized as a heterogeneous disease consisting of many different subtypes. The early diagnosis and prognosis of a cancer type have become a necessity in cancer research, as it can facilitate the subsequent clinical management of patients (Kourou et al., 2014). The importance of classifying cancer patients into high- or low-risk groups has led much research in AI methods to model the progression and treatment of cancerous conditions.

AI research in healthcare can help physicians make better clinical decisions or even replace human judgment in certain functional areas of healthcare. The increasing availability of healthcare data and the rapid development of big data analytic methods have made possible the recent successful applications of AI in healthcare. In addition to the popular deep learning neural network strategies for structured data, natural language processing for unstructured data, such as that in physicians' notes, also has a great value in healthcare. AI systems can provide up-to-date medical information from publications and large patient populations to assist physicians making real-time inferences, for health risk alerts and health outcome prediction, and in reducing diagnostic and therapeutic errors.

The applications of AI in clinical trials include stochastic decision process for clinical development programs, Q-learning and similarity-based machine learning, but are limited today mainly due to current statistical strategies implemented in the regulatory guidance, which are predominately based on frequentist's hypothesis testing that focuses on type-I error control. To solve this problem effectively requires a paradigm shift: (1) a focus on the prediction of drug effects (efficacy and safety) instead of type-I error control, (2) smaller clinical trials with possible adaptive designs but synergizing with real-world evidence, (3) stage-wise or adaptive marketing authorization and enhanced pharmacovigilance. Similarity-based machine learning (SBML) and its recursive versions provide a powerful AI approach for small and big data in clinical trials and in scientific discoveries generally.

Figure 13.1 is the summary of AI in the medical field.

Last but not least, data collection, fusion, and sharing all play an integral and critical part in successful applications of AI technologies.

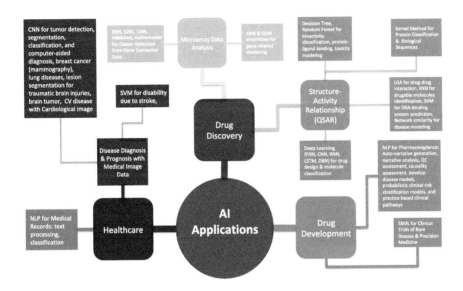

FIGURE 13.1
List of AI Applications

14

Future Perspectives—Artificial General Intelligence

Artificial general intelligence (AGI), strong AI, or full AI, is the intelligence of a machine that could successfully perform any intellectual task that a human being can. AGI refers to machines capable of experiencing consciousness, discovery, creativity, self-awareness, and evolution, collaborative intelligence, and creation of other machines like themselves. Simply put, AGI represents another race of human beings. In contrast, narrow or weak AI refers to the use of software to study or accomplish specific problem-solving or reasoning tasks. Weak AI does not attempt to attain the full range of human cognitive abilities. For this reason, weak AI is efficient so far as the specific task is concerned, and its performance is often more efficient than a human's performance.

The term artificial general intelligence was introduced by Mark Gubrud in 1997 in a discussion of the implications of fully automated military production and operations. As yet, most AI researchers have devoted little attention to AGI, with some claiming that human intelligence is too complex to be completely replicated in the near term; only a small number of computer scientists are active in AGI research. The research is extremely diverse and often pioneering in nature. Some of it may not be real AGI as proponents have claimed. Organizations explicitly pursuing AGI include the Swiss AI lab IDSIA, the OpenCog Foundation, Adaptive AI, LIDA, and Numenta and the associated Redwood Neuroscience Institute. In addition, organizations such as the Machine Intelligence Research Institute and OpenAI have been founded to influence the development path of AGI. Finally, projects such as the Human Brain Project have the goal of building a functioning simulation of the human brain. A 2017 survey of AGI categorized forty-five known active AGI projects including DeepMind, the Human Brain Project, and OpenAI.

Since modern AI research began in the mid 1950s, the first generation of AI researchers were convinced that artificial general intelligence was possible and that it would exist in just a few decades. Later, in the 1960s, AI pioneers predicted that within a generation the problem of creating artificial intelligence would substantially be solved. Particularly, in 1965 Herbert Simon predicted that "machines will be capable, within twenty years, of doing any work a man can do." However, not long thereafter, such overoptimism about AI had a first major setback when researchers discovered that the single-layer perceptron could not even model simple XOR behaviors. Funding agencies became

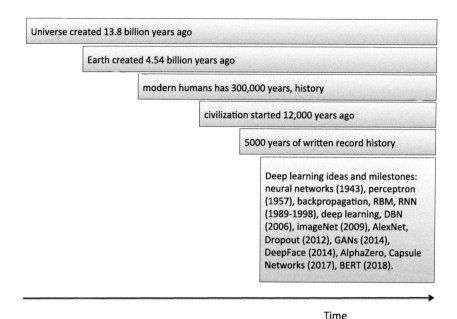

FIGURE 14.1
Historical Perspective of Development

skeptical of AGI and put researchers under increasing pressure to produce narrow AI. As the 1980s began, Japan's Fifth Generation Computer Project revived interest in AGI, setting out a ten-year timeline. Researchers were attracted to the first knowledge discovery in databases (KDD). Both industry and government pumped money back into the field. But confidence in AI spectacularly collapsed in the late 1980s when, for the second time in 20 years, AI researchers who had predicted the imminent achievement of AGI had been shown to be fundamentally mistaken. In 1993 Vernor Vinge published "The Coming Technological Singularity," in which he predicted that "within thirty years, we will have the technological means to create superhuman intelligence. Shortly after, the human era will be ended." While some experts still think that human-level AI is centuries away, most AI researches at the 2015 Puerto Rico Conference guessed that it would happen before 2060. If we look from a historical perspective (Figure 14.1) and challenges we are facing, as I think we should, AGI will take at least 200 years to acquire something close to humans' general intelligence, to become like us in appearance and behavior and, perhaps, to be as likable as a human – something that is fully accepted by human beings as another race (machine-race humans).

In the 1990s and the early 21st century, mainstream AI had achieved far greater commercial success and academic respectability by focusing on specific application problems where it could produce verifiable results and commercial applications, such as convolution neural networks in computer vision and voice-recognition. These applied AI systems are now used extensively throughout the technology industry, and research in this regard is very heavily funded in both academia and industry. Research in self-driving cars (not even true AGI) has achieved a great success using big data. However, this approach is not practical in the foreseeable future since it will be very expensive. Historically, cars became popular and a necessary transportation tool in daily life because we built systems of traffic lights and created commonly agreed-upon traffic rules. Similarly, to put self-driving cars into use as a common transportation tool, we need to have new equipment that plays similar roles to traffic lights and a new set of traffic rules. Such equipment can be, for example, a memory chip that everyone carries for easy communication with our self-driving cars. This will make these vehicles much safer and cost-effective and useable as common transportation in the near future.

Will AI surpass humankind? This is the same type of question as "Which came first, the chicken or the egg?" During the long future course of AGI's development, we will become respectful of the AGI agent as we would a human being; we will respect the agent's intelligence. We will develop emotions towards AGI agents when they live with us on daily basis. We will not discriminate against "anyone" because of race, color, gender, or sexual orientation; all that matters are time and interactions. The concept (connotation and extension) of a human being, like all other concepts, is subject to dynamics and change over time. We cannot even well-define what a human being is, though everyone probably thinks they have a clear concept of a human being that is similar to everyone else's. Here are two paradoxes to elaborate my point.

Can an AGI agent make scientific discoveries? To answer this question, we have to clarify the difference between a discovery and an invention. Since determination of a discovery or invention is dependent on whether it initially exists outside of a human mind or not, it is critical to clarify the connotation and denotation of a human or human identity. Puzzles about identity and persistence can be stated as: under what conditions does an object persist through time as one and the same object? If the world contains things which endure and retain their identity in spite of undergoing alteration, then somehow those things must persist through changes (Chang, 2012, 2014). We replace malfunctioning organs with healthy ones. We do physical exercises to improve our health. We like to learn new knowledge, and the quicker the better. We try hard to forget sad memories as soon as we can, so maybe we'll be able to use medical equipment to erase undesirable memories in the future. As these processes continue, we are making a human-machine mixed race, so then when does the person lose their identity? If we cannot clearly define what is a person, how can we make clear differentiation between discovery and invention at a fundamental level? You may ask the question: could an AI agent discover or rediscover the theory of evolution? But what does the term "discover" mean here? If an AI agent (say, a random text generator) did generate the exact same text as Darwin's evolution theory but before Darwin did, would it mean that the AI had "discovered" the theory? And since a "discovery" is presumably made by its "discoverer", would this be the AI or the human reader? Should we explain the text differently from Darwin's text just because it was generated by a machine? If we do explain the machine-generated text the same way as we did for Darwin's text, should we call it a discovery? If we do, we in fact have created an AI that can carry out scientific discovery, because such an AI is just a random text generator.

The second paradox discussed here is about identity. Suppose Professor Lee is getting old. He expresses his wish to have a younger and healthy body, whereas a healthy young student, John, truly admires Professor Lee's knowledge. After they learn each other's wishes, they decide to switch their bodies or knowledge, however you wish to say it. In the operation room, the professor's knowledge (information) is removed from his brain and transferred to John's brain. At the same time whatever original information that was in John's brain is removed and transferred to Doctor Lee's brain. The operation is carried out using the "incredible machine." Now the question is: Who is who after the operation? Would they both be happy with the operation? Please think, and then think again before you decide upon your answer (Chang, 2012).

In my view (Chang, 2012, 2014), we humans have not yet understood some fundamental concepts in the making of AGI agents, such as the meaning of understanding, emotion, creativity and knowledge, discovery versus invention, and causality. I believe interpretations of these concepts in mechanical terms are essential before we can implement AGI. As an example, Chang and Chang (2017; US patent: US 9,734,141 B2) take a quite different approach to analyzing the connotations of understanding using recursive concept-mapping (*iWordnet*), and conclude that understanding is nothing but word-mapping or concept-mapping. "One explains a concept using other concepts that are further explained by other concepts, and so on. Since concepts are limited for any individual at any time, we will eventually come to circular definitions." If we connect the words/concepts used in the definitions in sequence, they will form a wordnet, called *iWordnet* by the authors (Figure 14.2). Here "*i*" in the term emphasizes the ties of the network to the individual person. In the paper, an individual's knowledge or IQ in relationship to the global topological properties of his *iWordnet* were analyzed, and the association was shown to be statistically significant and predictive in a small pilot study of 20 subjects. The paper and patent further proposed the concept of *path of understanding*, that is, a trace of sentences on the person's *iWordnet*. The topological properties such as degrees and centrality on the path of understanding characterize numerically the person's understanding; thus understanding of the sentences can be analyzed numerically via the path of understanding.

You see, we as human being are trying to make an AI agent that has the ability to understand of concepts before we ourselves have a common understanding of the term "understanding" and to make an agent that is capable

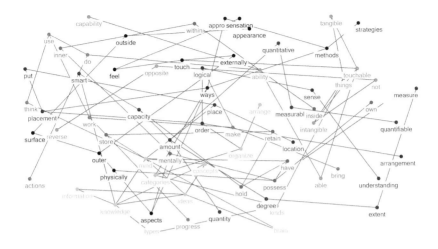

FIGURE 14.2
An Example of iWordnet in a Pilot Study

of knowledge, discovery and invention without a common understanding of these terms.

In my view, our major efforts in AGI and robotics should focus on how our technologies can provide emotionally appropriate assistance and companion ship for seniors, since the problem of population aging is coming much sooner than we thought and the impact is bigger than we can imagine. Another area of research should be the general learning mechanism of AI which should be simple, sequential, hierarchical, and recursive. After positing the importance of the similarity principle as the most general, most fundamental principle for scientific discovery, and which characterizes causal relationships(Chang, 2012, 2014, 2016, 2017), I believe AGI should start with a zero-data based, language-independent strategy as opposed to a big-data based, expert-knowledge-dependent, and language-specific approach. An effective way of studying AGI has to start with zero data methodology before adopting a big data approach. In the year 2010, I experimentally made a prototype of zero-data-based and language-independent Agents, named ZDA (male) and LIA (female), that can understand simple concepts without any built-in data and language. The agents can learn the language through interactions with you no matter what language you are using. In the next several years, my main efforts will focus on developing recursive algorithms that enable ZDA and LIA to learn more complicated things, progressively. Here are the main differences and similarities between my approaches for ZDA and LIA and mainstream AI research:

1. Goal: Achieving human-like behaviors but not necessarily scientifically correct responses vs. Maximizing robot capabilities to serve human beings.

2. Data: Zero-data based approach vs. Big-data based approach.

3. Language: Language-independent approach vs. Built-in languages required.

4. Conceptualization: Understanding a concept is a tuning process over time for an individual vs. A concept has a fixed correct meaning, or meanings, for everyone.

5. Principle of Learning: The similarity principle vs. Lack of a general principle (at the moment).

6. Learning Architecture: No built-in terms for concepts in any language; learning is a recursive process of concept-to-concept or concept-to-action mapping, moving from simplicity to complexity vs. Concepts are built-in terms in some languages, and learning is just command-to-action mapping.

7. Learning-Engine: Mainly through curiosity-driven and purpose-driven active learning through asking smart questions and a feedback mechanism vs. Passive learning through "human" feed-in training data and reinforcement learning.

8. Learning Pathway: Learning through extensive interactions – just as we take 20 some years to teach a person until a college graduate – through teacher-student, peer to peer interactions observation of others, mimicry and analogy (via genetic operations), and reinforcement learning through interaction with the environment vs. Big data feed-in and teacher-student, peer-to-peer interactions, and reinforcement learning through interaction with environment.

9. Response-Engine: Adaptive and proactive response rules based on a time-dependent maximum expected utility rule and a basic mechanism of game theory vs. Task-driven responses.

10. Knowledge Discovery, Invention, Creativity: All through genetic operations, analogy, and feedback mechanics vs. Data mining, genetic algorithms.

11. Conscientiousness: ZDA's conscientiousness will be dependent on how he will be treated vs. Built-in fixed social norms (some).

12. Inheritance: ZDA and LIA have some 14 built-in intrinsic abilities and the mechanism of the similarity principle, everything else being obtained through learning or interaction with the environment over time vs. None.

13. Logical Reasoning: Learning of deduction and other reasoning approaches through induction and the similarity principle vs. Built-in laws of logical reasoning.

14. Advanced skills: Math and AI can be learned by ZDA vs. Built-in math operations and AI (genetic) algorithms.

15. Sensor Organs Coordination: For learning enhancement vs. For command-to-action response.

16. Swarm Intelligence: SI exists and can promote social collaboration vs. SI exists and can promote social collaboration.

17. Evolution: Darwin's laws of evolution vs. Darwin's laws of evolution.

18. Cloning: ZDA and LIA can be cloned at any age in their lifetimes vs. Agents can be cloned at any time.

15

Appendix

15.1 Data for Learning Artificial Intelligence

Kaggle

Inside Kaggle you'll find all the code and data you need to do your data science work. Use over 19,000 public datasets and 200,000 public notebooks are ready to conquer any analysis in no time.

www.kaggle.com

UCI Center for Machine Learning and Intelligent Systems

http://archive.ics.uci.edu/ml/datasets.php

15.2 AI Software Packages

R is a natural choice for a statistician. Advanced users can write C, C++, and Python code to manipulate R objects directly. Python has been leading in terms of implementing deep learning easily using its famous deep learning libraries, such as *Keras*, *Tensorflow*, etc. However, due to the recent launch of the *keras* library in R with Tensorflow (CPU and GPU compatibility) at the backend, for statisticians or data scientists, R would be the first choice.

About TensorFlow (Alex, MIT YouTube)

1. What is it: Deep Learning Library (facts: open source, Python, Google)

2. Community: 117,000+ GitHub starts; TensorFlow.org: Blogs, Documentation, DevSummit, YouTube talks

3. Ecosystem:

    (a)   Keras: high-level API

    (b)   TensorFlow.js: in the browser

    (c)   TensorFlow Lite: on the phone

    (d)   Colaboratory: in the cloud

    (e)   TPU: optimized hardware

    (f)   TensorBoard: visualization

    (g)   TensorFlow Hub: graph modules

4. Extras:

    (a)   Swift for TensorFlow

    (b)   TensorFlow Serving

    (c)   TensorFlow Extended (TFX)

    (d)   TensorFlow Probability

    (e)   Tensor2Tensor

5. Alternatives: PyTorch, MXNet, CNTK, Deeplearning4j

Installation of Keras in R
Install or update Python3 from Mac terminal prompt $.

```
$ python3 -m pip install
```

To install keras (https://www.youtube.com/watch?v=3rEeBk6cyzs), you need to install tensorflow that is used as backend in keras. Enter the following command in the terminal prompt:

```
$ pip3 install tensorflow
$ pip3 install –upgrade tensorflow
$ sudo python3 -m pip install keras
$ python3
>>> import keras
```

15.3   Derivatives of Similarity Functions

Cosine similarity for continuous variables and mixed variables: In the definition of cosine similarity, the value of attribute V_j cannot be directly used since the importance of each variable in defining similarity depends on the problem under consideration. Therefore, we use a weight or attribute-scaling factor R_k to scale each variable, $U_k = R_k V_k$ and $u_{jk} = R_k v_{jk}$.

$$S_{mj} = \cos(\theta) = \frac{\sum_{k=1}^{K} u_{mk} u_{jk}}{||u_m|| \cdot ||u_j||} = \frac{\sum_{k=1}^{K} R_k v_{mk} R_k v_{jk}}{\sqrt{\sum_{k=1}^{K} R_k^2 v_{mk}^2} \sqrt{\sum_{k=1}^{K} R_k^2 v_{jk}^2}} \qquad (A.1)$$

where R_k will be determined through training.

$$\frac{\partial S_{ij}}{\partial R_m} = \frac{2R_m x_{im} x_{jm}}{||\mathbf{S}_i|| \cdot ||\mathbf{S}_j||} - \frac{\left(R_m x_{jm}^2 ||\mathbf{S}_i||^2 + R_m x_{im}^2 ||\mathbf{S}_j||^2\right) \sum_{k=1}^{K} R_k^2 x_{ik} x_{jk}}{\left(||\mathbf{S}_i|| \cdot ||\mathbf{S}_j||\right)^3}$$

$$= \frac{2v_{im} v_{jm}}{R_m ||\mathbf{S}_i|| \cdot ||\mathbf{S}_j||} - \frac{\left(v_{jm}^2 ||\mathbf{S}_i||^2 + v_{im}^2 ||\mathbf{S}_j||^2\right) \sum_{k=1}^{K} v_{ik} v_{jk}}{R_m \left(||\mathbf{S}_i|| \cdot ||\mathbf{S}_j||\right)^3}$$

$$= \frac{2v_{im} v_{jm}}{R_m ||\mathbf{S}_i|| \cdot ||\mathbf{S}_j||} - \frac{\left(v_{jm}^2 ||\mathbf{S}_i||^2 + v_{im}^2 ||\mathbf{S}_j||^2\right) \omega_{ij}}{R_m \left(||\mathbf{S}_i|| \cdot ||\mathbf{S}_j||\right)^3} \tag{A.2}$$

$$\omega_{ij} = \sum_{k=1}^{K} v_{ik} v_{jk}, ||\mathbf{S}_i| = \sqrt{\sum_{k=1}^{K} v_{ik}^2} \tag{A.3}$$

15.4 Derivation of Backpropagation Algorithms for ANN

The loss function $L = E + \phi$ involves a penalty ϕ and the mean squared error,

$$E = \frac{1}{N_M} \sum_{i=1}^{N_M} (y_{Mi} - t_i)^2, \tag{A.4}$$

where y_i and t_i are the model and observed outputs of the ith node at the Mth layer, respectively.

The minimization of the loss function using the gradient method to determine the optimal weights w_{mij} involves the following derivatives.

At the rth iteration:

$$w_{mij}^{(r+1)} = w_{mij}^{(r)} - \alpha \frac{\partial E}{\partial w_{mij}} - \alpha \frac{\partial \phi}{\partial w_{mij}}, \tag{A.5}$$

where $\alpha > 0$ is the learning rate.

$$\frac{\partial E}{\partial w_{mij}} = \sum_{k=1}^{N_M} \frac{\partial E}{\partial y_{Mk}} \frac{\partial y_{Mk}}{\partial w_{mij}} = \frac{2}{N_M} \sum_{k=1}^{N_M} (y_{Mk} - t_k) \frac{\partial y_{Mk}}{\partial w_{mij}}. \tag{A.6}$$

We are going to derive the recursive formulations of the derivatives of E. Note that $H_M (\cdot)$ is identical function and $H_M' (\cdot) = 1$. $\tilde{\delta}_{ij} = 1$ if $i = j$ and 0 otherwise.

$$\frac{\partial y_{Mi}}{\partial w_{M-1,kl}} = H_M' (z_{Ml}) y_{M-1,k} = y_{M-1,k} \tag{A.7}$$

$$\frac{\partial E}{\partial w_{M-1,kl}} = \frac{2}{N_M} \sum_{i=1}^{N_M} (y_{Mj} - t_i) \frac{\partial y_{Mi}}{\partial w_{M-1,kl}}$$

$$= \frac{2}{N_M} (y_{Ml} - t_l) H'_M (z_{Ml}) y_{M-1,k}$$

$$= \left[\frac{2}{N_M} (y_{Ml} - t_l) \right] y_{M-1,k} = \delta_{Ml} \cdot y_{M-1,k}$$

$$\frac{\partial y_{Mi}}{\partial w_{M-2,kl}} = \frac{\partial}{\partial w_{M-2,kl}} H_M \left(\sum_{j=0}^{N_{M-1}} w_{M-1,ji} y_{M-1,j} \right)$$

$$= H'_M (z_{Mi}) \sum_{j=0}^{N_{M-1}} w_{M-1,ji} \frac{\partial}{\partial w_{M-2,kl}} y_{M-1,j}$$

$$= H'_M (z_{Mi}) \sum_{j=0}^{N_{M-1}} w_{M-1,ji} H'_{M-1} (z_{M-1,j})$$

$$\cdot \frac{\partial}{\partial w_{M-2,kl}} \sum_{t=0}^{N_{M-2}} w_{M-2,tj} y_{M-2,t}$$

$$\frac{\partial y_{Mi}}{\partial w_{M-2,kl}} = H'_M (z_{Mi}) \sum_{j=0}^{N_{M-1}} w_{M-1,ji} H'_{M-1} (z_{M-1,j}) \tilde{\delta}_{jl} y_{M-2,k}$$

$$= H'_M (z_{Mi}) H'_{M-1} (z_{M-1,l}) w_{M-1,li} y_{M-2,k}$$

$$= \left[H'_{M-1} (z_{M-1,l}) w_{M-1,li} \right] y_{M-2,k} = \delta_{M-2,li} \cdot y_{M-2,k}$$

$$\frac{\partial y_{Mi}}{\partial w_{M-3,kl}}$$

$$= H'_M (z_{Mi}) \sum_{j=0}^{N_{M-1}} w_{M-1,ji} H'_{M-1} (z_{M-1,j}) \frac{\partial}{\partial w_{M-3,kl}} \sum_{t=0}^{N_{M-2}} w_{M-2,tj} y_{M-2,t}$$

$$= H'_M (z_{Mi}) \sum_{j=0}^{N_{M-1}} w_{M-1,ji} H'_{M-1} (z_{M-1,j})$$

$$\cdot \sum_{t=0}^{N_{M-2}} w_{M-2,tj} \frac{\partial}{\partial w_{M-3,kl}} H_{M-2} \left(\sum_{v=0}^{N_{M-3}} w_{M-3,vt} y_{M-3,v} \right)$$

$$\frac{\partial y_{Mi}}{\partial w_{M-3,kl}} = H'_M(z_{Mi}) \sum_{j=0}^{N_{M-1}} w_{M-1,ji} H'_{M-1}(z_{M-1,j})$$

$$\cdot \sum_{t=0}^{N_{M-2}} w_{M-2,tj} H'_{M-2}(z_{M-2,t}) \tilde{\delta}_{tl} y_{M-3,k}$$

$$= H'_M(z_{Mi}) \sum_{j=0}^{N_{M-1}} w_{M-1,ji} H'_{M-1}(z_{M-1,j})$$

$$\cdot w_{M-2,lj} H'_{M-2}(z_{M-2,l}) y_{M-3,k}$$

$$= \left[H'_{M-2}(z_{M-2,l}) \sum_{j=0}^{N_{M-1}} \delta_{M-2,ji} w_{M-2,lj} \right] y_{M-3,k}$$

$$= \delta_{M-3,li} \cdot y_{M-3,k}$$

$$\frac{\partial y_{Mi}}{\partial w_{M-4,kl}} = H'_M(z_{Mi}) \sum_{j=0}^{N_{M-1}} w_{M-1,ji} H'_{M-1}(z_{M-1,j})$$

$$\cdot \sum_{t=0}^{N_{M-2}} w_{M-2,tj} \frac{\partial}{\partial w_{M-4,kl}} H_{M-2} \left(\sum_{v=0}^{N_{M-3}} w_{M-3,vt} y_{M-3,v} \right)$$

$$= \sum_{j=0}^{N_{M-1}} w_{M-1,ji} H'_{M-1}(z_{M-1,j}) \sum_{t=0}^{N_{M-2}} w_{M-2,tj} H'_{M-2}(z_{M-2,t})$$

$$\cdot \sum_{v=0}^{N_{M-3}} w_{M-3,vt} \frac{\partial}{\partial w_{M-4,kl}} y_{M-3,v}$$

$$\frac{\partial y_{Mi}}{\partial w_{M-4,kl}} = \sum_{j=0}^{N_{M-1}} w_{M-1,ji} H'_{M-1}(z_{M-1,j}) \sum_{t=0}^{N_{M-2}} w_{M-2,tj} H'_{M-2}(z_{M-2,t})$$

$$\cdot \sum_{v=0}^{N_{M-3}} w_{M-3,vt} \frac{\partial}{\partial w_{M-4,kl}} H_{M-3} \left(\sum_{q=0}^{N_{M-4}} w_{M-4,qv} y_{M-4,q} \right)$$

$$\frac{\partial y_{Mi}}{\partial w_{M-4,kl}} = \sum_{j=0}^{N_{M-1}} w_{M-1,ji} H'_{M-1}(z_{M-1,j}) \sum_{t=0}^{N_{M-2}} w_{M-2,tj} H'_{M-2}(z_{M-2,t})$$

$$\cdot \sum_{v=0}^{N_{M-3}} w_{M-3,vt} H'_{M-3}(z_{M-3,v}) \frac{\partial}{\partial w_{M-4,kl}} \sum_{q=0}^{N_{M-4}} w_{M-4,qv} y_{M-4,q}$$

$$\frac{\partial y_{Mi}}{\partial w_{M-4,kl}} = \sum_{j=0}^{N_{M-1}} w_{M-1,ji} H'_{M-1}\left(z_{M-1,j}\right) \sum_{t=0}^{N_{M-2}} w_{M-2,tj} H'_{M-2}\left(z_{M-2,t}\right)$$

$$\cdot \sum_{v=0}^{N_{M-3}} w_{M-3,vt} H'_{M-3}\left(z_{M-3,v}\right) y_{M-4,k}\tilde{\delta}_{lv}$$

$$\frac{\partial y_{Mi}}{\partial w_{M-4,kl}} = \sum_{j=0}^{N_{M-1}} w_{M-1,ji} H'_{M-1}\left(z_{M-1,j}\right)$$

$$\cdot \sum_{t=0}^{N_{M-2}} w_{M-2,tj} H'_{M-2}\left(z_{M-2,t}\right) w_{M-3,lt} H'_{M-3}\left(z_{M-3,l}\right) y_{M-4,k}$$

$$= \left[H'_{M-3}\left(z_{M-3,l}\right) \sum_{t=0}^{N_{M-2}} \delta_{M-3,ti} w_{M-3,lt} \right] y_{M-4,k}$$

$$= \delta_{M-4,li} y_{M-4,k}$$

$$\frac{\partial y_{Mi}}{\partial w_{M-4,kl}} = \sum_{j=0}^{N_{M-1}} w_{M-1,ji} H'_{M-1}\left(z_{M-1,j}\right) \sum_{t=0}^{N_{M-2}} w_{M-2,tj} H'_{M-2}\left(z_{M-2,t}\right)$$

$$\cdot \sum_{v=0}^{N_{M-3}} w_{M-3,vt} \frac{\partial}{\partial w_{M-4,kl}} H_{M-3}\left(\sum_{q=0}^{N_{M-4}} w_{M-4,qv} y_{M-4,q} \right)$$

$$\frac{\partial y_{Mi}}{\partial w_{M-4,kl}} = \sum_{j=0}^{N_{M-1}} w_{M-1,ji} H'_{M-1}\left(z_{M-1,j}\right) \sum_{t=0}^{N_{M-2}} w_{M-2,tj} H'_{M-2}\left(z_{M-2,t}\right)$$

$$\cdot \sum_{v=0}^{N_{M-3}} w_{M-3,vt} H'_{M-3}\left(z_{M-3,v}\right) \frac{\partial}{\partial w_{M-4,kl}} \sum_{q=0}^{N_{M-4}} w_{M-4,qv} y_{M-4,q}$$

$$\frac{\partial y_{Mi}}{\partial w_{M-4,kl}} = \sum_{j=0}^{N_{M-1}} w_{M-1,ji} H'_{M-1}\left(z_{M-1,j}\right) \sum_{t=0}^{N_{M-2}} w_{M-2,tj} H'_{M-2}\left(z_{M-2,t}\right)$$

$$\cdot \sum_{v=0}^{N_{M-3}} w_{M-3,vt} H'_{M-3}\left(z_{M-3,v}\right) y_{M-4,k}\tilde{\delta}_{lv}$$

$$\frac{\partial y_{Mi}}{\partial w_{M-4,kl}} \quad = \quad \sum_{j=0}^{N_{M-1}} w_{M-1,ji} H'_{M-1}\left(z_{M-1,j}\right)$$

$$\cdot \sum_{t=0}^{N_{M-2}} w_{M-2,tj} H'_{M-2}\left(z_{M-2,t}\right) w_{M-3,lt} H'_{M-3}\left(z_{M-3,l}\right) y_{M-4,k}$$

$$\frac{\partial y_{Mi}}{\partial w_{M-4,kl}} \quad = \quad \left[H'_{M-3}\left(z_{M-3,l}\right) \sum_{t=0}^{N_{M-2}} \delta_{M-3,ti} w_{M-3,lt} \right] y_{M-4,k}$$

$$= \quad \delta_{M-4,li} y_{M-4,k}$$

15.5   Similarity-Based Machine Learning in R

```
library(mvtnorm)
library(class)
getwd()
# Normalize data (by subtracting Mean and Dividing by SD of the RefData
normalizeData = function(DataToNormalize, RefData){
 normalizedData = DataToNormalize
 for (n in 1:ncol(DataToNormalize)){
 normalizedData[,n] = (DataToNormalize[,n] - mean(RefData[,n])) /
sd(RefData[,n])
 }
 return(normalizedData)
}
# Calculate initial scaling factors R0s # S0s = p-values
InitialRs = function (X, eta, S0s) {
 K = length(S0s); R0s = rep(0, K)
 for (k in 1:K) {
 # force <12 to avoid s=exp(d) overflow
 R0s[k] = min((-log(S0s[k])) ^(1/eta) / max(IQR(X[ ,k]), 0.0000001), 12)
 }
 return (R0s);
}
Distance = function (Rs, x1, x2) {
 # 3 vectors: Rs = scaling factors, x1 = point 1, x2 = point 2
 K = length(Rs)
```

```r
 d = 0; for (k in 1:K) { d = d + (Rs[k]* (x2[k]-x1[k]))^2 }
 return (d^0.5)
}
# Calculate Similarity Score.
SimilarityTrain = function(eta, Rs, Xtrain) {
 N1 = nrow(Xtrain); K = length(Rs);
 S = matrix(0, nrow=N1, ncol=N1)
 for (i in 1:N1) { for (j in i:N1) {
 d2 = 0; for (k in 1:K) {
 d2 = d2 + (Rs[k]* (Xtrain[i,k]-Xtrain[j,k]))^2}
 S[i,j] = exp(-d2^0.5)
 S[j,i]=S[i,j] # Use symmetry to reduce 50% CPU time.
 } } # End j and i loops
 return (S)
}
# Calculate Similarity Scores between training and test subjects.
Similarity = function(eta, Rs, Xtrain, Xtest) {
 N1 = nrow(Xtrain); N2=nrow(Xtest); K = length(Rs)
 S = matrix(0, nrow=N2, ncol=N1)
 for (i in 1:N2) { for (j in 1:N1) {
 d2 = 0; for (k in 1:K) {
 d2 = d2 + (Rs[k]* (Xtrain[i,k]-Xtrain[j,k]))^2}
 S[i,j] = exp(-d2^0.5)
 } } # End j and i loops
 return (S)
}
Weight = function(S) {
 N1= nrow(S); N2=ncol(S);
 W = matrix(0, nrow=N1, ncol=N2)
 for (i in 1:N1) {
 sum_S_row = sum(S[i, ])
 for (j in 1:N2) { W[i,j] = S[i,j] / max(sum_S_row,0.00000000001) }
 }
 return (W)
} # End of Weight
PredictedY = function (W, Ytrain, Ytest) {
 # Calculate predicted outcome and Error
 OutObj = list()
 OutObj$pred_Y = W %*% Ytrain # For binary outcome, pred_y = prob of
being 1.
 # Mean squared error
 OutObj$MSE = mean((OutObj$pred_Y - Ytest)^2)
 return (OutObj)
} # End of PredictionY
```

```
DerivativeE = function(eta, pred_Y, Rs, X, S, O) {
# Derivatives if the loss function
N = nrow(X); K =length(Rs)
der_S = matrix(0, nrow=K*N, ncol=N); der_W = matrix(0, nrow=K*N,
ncol=N)
dist = matrix(0, nrow=N, ncol=N); der_E = rep(0, K)
for (i in 1:N) { for (j in 1:N) {d2 = 0; for (k in 1:K) {
d2 = d2 + (Rs[k]*(X[i,k]-X[j,k]))^2
}
dist[i,j] = max (d2^0.5, 0.0000001)}}
for (m in 1:K) { for (i in 1:N) { for (j in 1:N) {
der_d = (Rs[m]/dist[i,j]) * (X[i,m]-X[j,m]) ^2
der_S[(m-1)*N+i, j] = -1 * S[i,j] * eta * (dist[i,j])^(eta-1) * der_d
} } } # End of j, i, and m loops
# Weight Derivative
for (m in 1:K) { for (i in 1:N) {
sum_der_S = sum(der_S[(m-1)*N+i, ]); sumSi = sum(S[i, ])
for (j in 1:N) {
der_W[(m-1)*N+i, j] = der_S[(m-1)*N+i, j] / sumSi - S[i,j]* sum_der_S /
sumSi^2
} } } # End of j, i, and m loops
# Derivatives of E
for (m in 1:K) { for (i in 1:N) {
err = (pred_Y[i] - O[i])
for (j in 1:N) { der_E[m] = der_E[m] + 2/N * err * O[j] * der_W[(m-1)*N+i,
j] }
} } # End of I and m loops
return (der_E)
} # End of DerivativeE
Learning = function (LearningRate, Lamda, Rs, der_E) {
# Update scaling factors, Rs.
K=length(Rs)
der_lossFun = der_E+2*Lamda*Rs # derivatives of loss function
Rs = Rs - LearningRate * der_lossFun
# force Rs<25 to avoid S=exp(d) overflow
for (m in 1:length(Rs)) { Rs[m] = min(max(0,Rs[m]),25) }
return (Rs)
} # End of Learning
# Training AI to obtain scaling factors, Rs ###
SBMLtrain = function(Epoch, S0, Lamda, LearningRate, eta, Xtrain, Ytrain) {
TrainObj=list(); OutObj0= list(); OutObj= list()
R0 = InitialRs(Xtrain, eta, S0);
S = SimilarityTrain(eta, R0, Xtrain)
OutObj0 = PredictedY (Weight(S), Ytrain, Ytrain)
```

```
Rs=R0; OutObj =OutObj0 # in case Epoch =0 for no learning
TrainMSE0 = OutObj0$MSE
preLoss = OutObj0$MSE+Lamda*sum(Rs^2)
# Learning
iter=0;
while (iter<Epoch) {
preRs=Rs
Rs = Learning(LearningRate, Lamda, Rs, DerivativeE(eta, OutObj0$pred_Y,
Rs, Xtrain, S, Ytrain))
OutObj = PredictedY (Weight(SimilarityTrain (eta, Rs, Xtrain)), Ytrain,
Ytrain)
iter=iter+1
Loss = OutObj$MSE+Lamda*sum(Rs^2)
if (Loss>preLoss) {Rs=preRs; iter=Epoch+1}
preLoss=Loss
}
TrainObj$Rs = Rs; TrainObj$Y = OutObj$pred_Y; TrainObj$MSE =
OutObj$MSE
return( TrainObj)
}
# Linear regression
# outcome = "gaussian" or "binomial", ...
GLMPv = function(Y, X, outcome) {
LMfull=glm(Y ~., data=X, family=outcome)
sumLM=coef(summary(LMfull))
pVals=sumLM[,4][1:ncol(X)+1]
# pVals <- summary(LMfull)$coef[, "Pr(>|t|)"]
return (pVals)
}
```

Bibliography

[1] Abacha, B.A. and Zweigenbaum, P. (2011). A hybrid approach for the extraction of semantic relations from MEDLINE abstracts. 12th International Conference on Computational Linguistics and Intelligent Text Processing, CICLing, Tokyo, Japan, 2011.

[2] Abacha, B., Chowdhury, A.M., et al. (2015). Text mining for pharmacovigilance: using machine learning for drug name recognition and drug-drug interaction extraction and classification. J. Biomed. Inform. 58, 122–132.

[3] Abdo, A. and Salim, N. (2008). Molecular Similarity Searching Using Inference Network. University of Technology Malaysia. www.UTM.my.

[4] Afzal, N., Sohn, S. (2017). Abram S., et al. Mining peripheral arterial disease cases from narrative clinical notes using natural language processing. J Vasc Surg. 65:1753–61.

[5] Ahn, L.V., Blum, M. et al. (2003). CAPTCHA: Using Hard AI Problems for Security. In Proceedings of Eurocrypt, Vol. 2656, 294–311.

[6] Aindow, T. (2019). https://www.kaggle.com/taindow/simple-lstm-with-r.

[7] Ajmani, S., Jadhav, K., Kulkarni, S.A. (2006). Three-dimensional QSAR using the k-nearest neighbor method and its interpretation, J. Chem. Inf. Model. 46, 24–31.

[8] Alexander, J. et al. (2018). A New Dimension of Breast Cancer Epigenetics. BIOINFORMATICS 2018 - 9th International Conference on Bioinformatics Models, Methods and Algorithms.

[9] Aliper, A. et al. (2016) Deep learning applications for predicting pharmacological properties of drugs and drug repurposing using transcriptomic data. Mol. Pharmaceut. 13, 2524–2530.

[10] Altman, N.S. (1992) An introduction to kernel and nearest–neighbor nonparametric regression. Am. Stat. 46, 175–185.

[11] Aminiazar, W., Najafi, F., Nekoui, M.A. (2013). Optimized intelligent control of a 2-degree of freedom robot for rehabilitation of lower

limbs using neural network and genetic algorithm. J Neuroeng Rehabil 2013;10:96.

[12] Amaritsakul, Y., Chao, C.K., Lin, J. (2013). Multiobjective optimization design of spinal pedicle screws using neural networks and genetic algorithm: mathematical models and mechanical validation. Comput Math Methods Med 2013:462875.

[13] Anastasiadis, A. et. al. (2005) New globally convergent training scheme based on the resilient propagation algorithm. Neurocomputing 64, 253–270.

[14] Anthimopoulos, M., Christodoulidis, S. et al. (2014). Classification of interstitial lung disease patterns using local dct features and random forest. In: 2014 36th Annual International Conference of the IEEE Engineering in Medicine and Biology Society (EMBC). IEEE, pp. 6040–6043, 2014.

[15] Anthimopoulos, M. et al. (2016) Lung pattern classification for interstitial lung diseases using a deep convolutional neural network. IEEE transactions on medical imaging, vol. 35, no. 5, pp. 1207–1216.

[16] Anwar, S.M., et al. (2018). Medical image analysis using convolutional neural networks: a review. Journal of Medical Systems, November, 42:226.

[17] Asadi, H., Dowling, R. et al. (2014). Machine learning for outcome prediction of acute ischemic stroke post intra-arterial therapy. PLoS One 2014;9:e88225.

[18] Aschinchon, M. (2014). https://www.r-bloggers.com/cellular-automata-the-beauty-of-simplicity/.

[19] Asgari, E., Mofrad, M.R.K. (2015). Continuous Distributed Representation of Biological Sequences for Deep Proteomics and Genomics. PLoS One. 10 (11): e0141287.

[20] Avendi, M.R., Kheirkhah, A., Jafarkhani, H. (2016). A combined deep-learning and deformablemodel approach to fully automatic segmentation of the left ventricle in cardiac MRI. Med Image Anal 30:108–119.

[21] Barmpalex, P. et al. (2011). Symbolic regression via genetic programming in the optimization of a controlled release pharmaceutical formulation. Chemometrics and Intelligent Laboratory Systems 107(1).

[22] Barrat, S., Barthelemy, M. et al. (2004). The architecture of complex weighted networks. Proceedings of the National Academy of Sciences. 101 (11): 3747–3752. arXiv:cond-mat/0311416. Bibcode:2004PNAS..101.3747B.

[23] Baum, S. (2017). A Survey of Artificial General Intelligence Projects for Ethics, Risk, and Policy. Global Catastrophic Risk Institute Working Paper 17-1 99 Pages Posted: 16 Nov 2017.

[24] Baudry, B. et al. (2005). Automatic test case optimization: a bacteriologic algorithm. IEEE Software. 22 (2): 76–82.

[25] BELLAVITE, P. et al. (1997). A scientific reappraisal of the principle of similarity. Medical Hypotheses, 49, 203–212.

[26] Bentley, P., Ganesalingam, J. et al. (2014). Prediction of stroke thrombolysis outcome using CT brain machine learning. Neuroimage Clin; 4:635–40.

[27] Birkner, M.D., Kalantri, S. et al. (2007). Creating diagnostic scores using data-adaptive regression: an application to prediction of 30-day mortality among stroke victims in a rural hospital in India. Ther Clin Risk Manag; 3:475–84.

[28] Blondel, V.D. et al. (9 October 2008). Fast unfolding of communities in large networks. Journal of Statistical Mechanics: Theory and Experiment. (10).

[29] Botsis, T., Nguyen, M.D., et al. (2011). Text mining for the vaccine adverse event reporting system: medical text classification using informative feature selection. J. Am. Med. Inform. Assoc. 18, 631–638.

[30] Boyd, J.C., Savory, J. (2001). Genetic algorithm for scheduling of laboratory personnel. Clin Chem. Jan;47(1):118–123.

[31] Boisvert, M.R., Koski, K.G. et al. (2012). Early prediction of macrosomia based on an analysis of second trimester amniotic fluid by capillary electrophoresis. Biomark Med. Oct;6(5):655–662.

[32] Böcker, A. et al. (2005). A Hierarchical Clustering Approach for Large Compound Libraries. J. Chem. Inf. Model., 2005, 45 (4): 807–815.

[33] Brandes, U. (2008). On variants of shortest-path betweenness centrality and their generic computation. Social Networks. 30 (2): 136–145.

[34] Breslow, N. E. and Day, N. E. (1980). Statistical Methods in Cancer Research. Volume 1: The Analysis of Case-Control Studies. IARC Lyon / Oxford University Press.

[35] Campion, F., Carlsson, G., Francis, X.C. (2017). Machine Intelligence for Healthcare. CreateSpace Independent Publishing Platform; First Edition.

[36] Carneiro, G., Nascimento, J.C., Freitas, A. (2012). The segmentation of the left ventricle of the heart from ultrasound data using deep learning architectures and derivative-based search methods. IEEE Trans Image Process 21(3):968–982.

[37] Casamitjana, A., Puch, S., Aduriz, A., Sayrol, E., and Vilaplana, V. (2016). 3d convolutional networks for brain tumor segmentation. Proceedings of the MICCAI Challenge on Multimodal Brain Tumor Image Segmentation (BRATS): 65–68.

[38] Castiglione, F., Poccia, F., D'Offi, G., Bernaschi, M. (2004). Mutation, fitness, viral diversity, and predictive markers of disease progression in a computational model of HIV type 1 infection. AIDS Res Hum Retroviruses. Dec;20(12):1314–1323.

[39] Castro, V.M., Dligach, D. et al. (2017). Large-scale identification of patients with cerebral aneurysms using natural language processing. Neurology; 88:164–168.

[40] Choi, E. et al. (2016). Doctor AI: Predicting Clinical Events via Recurrent Neural Networks. JMLR Workshop Conf Proc. August; 56: 301–318.

[41] Chaharsough, S.A. (2019). Estimating the causal effect of dynamic treatment strategies on pregnancy using electronic medical records or adaptive clinical trials. PhD. Dissertation, Boson University, Boston MA.

[42] Chandrakumar, T., Kathirvel, R. (2016) Classifying diabetic retinopathy using deep learning architecture. Int J Eng Res Technol 5(6):19–24.

[43] Chang, M. (2010). Monte Carlo Simulations for the Pharmaceutical Industry. CRC. Boca Raton, FL.

[44] Chang, M. (2007). Adaptive Design Theory and Implementation Using SAS and R. CRC Press/Taylor Francis Group. Boca Raton, FL.

[45] Chang, M. (2018). Innovative trial design; Similarity Principle; Bayesian; precision medicine; multiple testing; rare disease. JSM, 2018, Vancouver, Canada.

[46] Chang, M. (2011). Modern Issues and Methods in Biostatistics, Springer. N.Y.

[47] Chang, M. (2012). Paradoxes in Scientific Inferences. CRC Press/Taylor Francis Group. Boca Raton, FL.

[48] Chang, M. (2014). Principles of Scientific methods. CRC Press/Taylor Francis Group. Boca Raton, FL.

[49] Chang, M. and Chang, M. (2017). iWordNet: A New Approach to Cognitive Science and Artificial Intelligence. Advances in Artificial Intelligence. Volume 2017 (2017).

[50] Chaudhary, K., et al. (2018). Deep learning based multi-omics integration robustly predicts survival in liver cancer. bioRxiv 2017(2017). 114892.

[51] Chen, H. et al. (2015) Network-based inference methods for drug repositioning. Comput. Math. Methods Med: 130620.

[52] Cheng, F. et al. (2011) Prediction of drug–target interactions and drug repositioning via network-based inference. PLoS Comput. Biol. 8, e1002503.

[53] Chen, Y. et al. (2014). Risk classification of cancer survival using ANN with gene expression data from multiple laboratories. Comput Biol Me; 48:1–7.

[54] Chen, Y. et al. (2015). Cancer adjuvant chemotherapy strategic classification by artificial neural network with gene expression data: An example for non-small cell lung cancer. J Biomed Inform;56:1–7.

[55] Chen, Y., Dhar, R., et al. (2016). Automated quantification of cerebral edema following hemispheric infarction: application of a machine-learning algorithm to evaluate CSF shifts on serial head CTs. Neuroimage Clin;12:673–80.

[56] Clauset, A., Newman, M. E. J., Moore, C. (2004). Finding community structure in very large networks. Physical Review E. 70 (6): 066111.

[57] Cole, J.H. et al. (2016). Predicting brain age with deep learning from raw imaging data results in a reliable and heritable biomarker. arXiv preprint arXiv:1612.02572.

[58] Contrino, B. and Lazic, S.E. (2017). https://cran.r-project.org/web/packages/BayesCombo/vignettes/BayesCombo_vignette.html.

[59] Czibula, G. et al. (2015). A Reinforcement Learning Model for Solving the Folding Problem. Int.J.Comp.Tech.Appl, 171–182.

[60] Dahl, G.E. et al. (2014) Multi-task neural networks for QSAR predictions. arXiv 2014 14061231.

[61] Damoulas, T. and Girolami, M.A. (2008). Probabilistic multi-class multi-kernel learning: On protein-fold recognition and remote homology detection. Bioinformatics, 24, 1264–1270.

[62] Daouda, M. and Mayo, M. (2019). A survey of neural network-based cancer prediction models from microarray data. Artificial Intelligence In Medicine 97 (2019) 204–214.

[63] Dhungel, N., Carneiro, G., Bradley, A. (2015). Automated mass detection in mammograms using cascaded deep learning and random forests. In 2015 International Conference on Digital Image Computing: Techniques and Applications (DICTA), 1–8.

[64] Dong, J. and Horvath, S. (2007) Understanding network concepts in modules. BMC Systems Biology, June 1:24.

[65] Dong, J. and Horvath, S. (2008). Miyano, Satoru, ed. Geometric interpretation of gene coexpression network analysis. PLoS Computational Biology. 4 (8):

[66] Dong, Q., Zhou, S., Guan, J. (2009). A new taxonomy-based protein-fold recognition approach based on autocross-covariance transformation. Bioinformatic, 25, 2655–2662.

[67] Doppler, C. et al. (2017). Unsupervised Anomaly Detection with Generative Adversarial Networks to Guide Marker Discovery. Cornell University. https://arxiv.org/abs/1703.05921.

[68] Du, G., Jiang, Z., et al. (2013). Clinical pathways scheduling using hybrid genetic algorithm. J Med Syst 2013 Jun;37(3):9945.

[69] Dua, S. Acharya, U. R., Dua, P. (2014). Machine Learning in Healthcare Informatics, Springer, Berlin, Germany.

[70] Eichler, H.G. and Sweeney, F. (2008). The evolution of clinical trials: Can we address the challenges of the future? Clinical Trials 2018, Vol. 15(S1) 27–32.

[71] Ezzell, G.A. and Gaspar, L. (2000). Application of a genetic algorithm to optimizing radiation therapy treatment plans for pancreatic carcinoma. Med Dosim;25(2):93–97.

[72] Engoren, M., Plewa, M. et al. (2005). O'Hara D, Kline JA. Evaluation of capnography using a genetic algorithm to predict PaCO2. Chest, Feb;127(2):579–584.

[73] Elveren, E. and Yumuşak, N. (2011). Tuberculosis disease diagnosis using artificial neural network trained with genetic algorithm. J Med Syst. Jun;35(3):329–332.

[74] Erić, S. et al. (2012) Prediction of aqueous solubility of drug-like molecules using a novel algorithm for automatic adjustment of relative importance of descriptors implemented in counter-propagation artificial neural networks. Int. J. Pharm. 437, 232–241.

[75] Ertosun, M.G. and Rubin, D.L. (2015). Probabilistic visual search for masses within mammography images using deep learning. In: 2015 IEEE international conference on bioinformatics and biomedicine (BIBM). IEEE, 1310–1315.

[76] Esteva, A., Kuprel, B. et al. (2017). Dermatologist-level classification of skin cancer with deep neural networks. Nature, vol. 542, no. 7639:115–118.

[77] Fang, M. et al. (2018). Feature selection via swarm intelligence for determining protein essentiality. Molecules. Jul; 23(7): 1569.

[78] Farooq, S. M. A. (2017). A Deep CNN based Multi-class Classification of Alzheimer's Disease using MRI, presented at the IST, Beijing, China, 2017.

[79] Fathi, M. et (2013). Cellular automata modeling of hesperetin release phenomenon from lipid nanocarriers. Food and bioprocess technology. November 2013, Volume 6, Issue 11, pp 3134–3142.

[80] FDA (2018a). Adaptive designs for clinical trials of drugs and biologics. Guidance for industry. Center for drug evaluation and research (CDER) center for biologics evaluation and research (CBER), September 2018. www.fda.gov.

[81] FDA (2014). Guidance for industry expedited programs for serious conditions – drugs and biologics. Center for drug evaluation and research (CDER) center for biologics evaluation and research (CBER) May 2014.

[82] FDA (2016). Adaptive designs for medical device clinical studies. Guidance for industry and food and drug administration staff document. Center for devices and radiological health, center for biologics evaluation and research, issued on July 27, 2016.

[83] FDA (2018b). Master protocols: efficient clinical trial design strategies to expedite development of oncology drugs and biologics guidance for industry. Draft guidance. center for drug evaluation and research (CDER), center for biologics evaluation and research (CBER), oncology center of excellence (OCE), September 2018.

[84] Fiszman, M., Chapman, W.W. et al. (2000). Automatic detection of acute bacterial pneumonia from chest X-ray reports. J Am Med Inform Assoc; 7:593–604.

[85] Fong, A., Hettinger, A.Z. & Ratwani, R.M. (2015). Exploring methods for identifying related patient safety events using structured and unstructured data. J. Biomed. Inform. 58, 89–95 (2015).

[86] Forsyth, R. (1981). Beagle-a darwinian approach to pattern recognition. K. Kybernetes, 10(1981), 159–166.

[87] Frey, P. W. and Slate, D. J. (1991). Letter Recognition using Holland-style Adaptive Classifiers. Machine Learning Vol. 6/2 March.

[88] Fu, Y. et al. (2015). A New Approach for Flexible Molecular Docking Based on Swarm Intelligence. Mathematical Problems in Engineering. Volume 2015, Article ID 540186, 10 pages.

[89] Fukushima, K., Neocognitron, M. (1980). A self-organizing neural network model for a mechanism of pattern recognition unaffected by shift in position. Biological cybernetics 36. 4: 193–202.

[90] Gawehn, E., Hiss, J.A., and Schneider, G. (2015), Deep learning in drug discovery, Mol. Inf.2016,35, 3–14.

[91] Ghaheri, A. et al. (2015). The Applications of Genetic Algorithms in Medicine. Oman Med J. Nov; 30(6): 406–416.

[92] Ghasemi, F. et al. (2018a). Deep neural network in QSAR studies using deep belief network. Applied Soft Computing 62: 251–258.

[93] Ghasemi, F. et al. (2018b). Neural network and deep-learning algorithms used in QSAR studies: merits and drawbacks. Drug Discov Today. Oct;23(10):1784–1790.

[94] Ghesu, F.C., Krubasik. E., et al. (2016). Marginal space deep learning: efficient architecture for volumetric image parsing. IEEE Trans Med Imaging 35(5):1217–1228.

[95] Ghosh, A. and Lakhmi C.J. (2005, editors) . Evolutionary Computation in Data Mining. Springer-Verlag, Berlin Heidelberg.

[96] Gil Press (1983). https://www.forbes.com/sites/gilpress/2013/05/28/a-very-short-history-of-data-science/#337720355cfc.

[97] Giger, M.L., Chan, H.P., Boone, J. (2008). Anniversary paper: history and status of CAD and quantitative image analysis: the role of Medical Physics and AAPM. Med Phys 35(12):5799–5820.

[98] Giger, M.L., Pritzker, A. (2014). Medical imaging and computers in the diagnosis of breast cancer. In SPIE optical engineering + applications. International Society for Optics and Photonics: m 918–908.

[99] Gil Press (2016). https://www.forbes.com/sites/gilpress/2016/12/30/a-very-short-history-of-artificial-intelligence-ai/#649f5c616fba.

[100] Giuliani A. (2017). The application of principal component analysis to drug discovery and biomedical data. Drug Discov Today. 2017 Jul; 22(7):1069–1076.

[101] Goldberg, D.E. (1983), Computer-aided gas pipeline operation using genetic algorithms and rule learning. Dissertation presented to the University of Michigan at Ann Arbor, Michigan, in partial fulfillment of the requirements for Ph.D.

[102] Gottlieb, G. (April, 2019). Statement from FDA Commissioner Scott Gottlieb, M.D. on steps toward a new, tailored review framework for artificial intelligence-based medical devices. https://www.fda.gov/NewsEvents/Newsroom/PressAnnouncements/ucm635083.htm.

[103] Granovetter, M. (1973). The strength of weak ties. American Journal of Sociology. 78 (6): 1360–1380.

[104] Graves, A. et al. (2009). A Novel Connectionist System for Improved Unconstrained Handwriting Recognition (PDF). IEEE Transactions on Pattern Analysis and Machine Intelligence. 31 (5): 855–868.

[105] Griffis, J.C., Allendorfer, J.B., et al. (2016). Voxel-based gaussian naïve Bayes classification of ischemic stroke lesions in individual T1-weighted MRI scans. J Neurosci Methods 2016; 257:97–108.

[106] Gubrud, M. (November 1997), Nanotechnology and International Security, Fifth Foresight Conference on Molecular Nanotechnology, retrieved 7 May 2011.

[107] Güler, I., Polat, H., Ergün, U. (2005). Combining neural network and genetic algorithm for prediction of lung sounds. J Med Syst 2005 Jun;29(3):217–231.

[108] Gulsun, M.A., Funka-Lea, G., (2016) Coronary centerline extraction via optimal flowpaths and CNNpath pruning. In: Proceedings of international conference on medical image computing and computer assisted intervention. Computer Science Published in MICCAI. DOI:10.1007/978-3-319-46726-9_37.

[109] Gulshan, V., Peng, L., et al. (2016). Development and Validation of a Deep Learning Algorithm for detection of Diabetic Retinopathy in retinal fundus photographs. JAMA 2016; 316:2402–10.

[110] Guney, E. et al. (2016) Network-based in silico drug efficacy screening. Nat. Commun. 7, 10331.

[111] Gupta, A. et al. (2018). Generative Recurrent Networks for De Novo Drug Design. Mol. Inf. 2018, 37, 1700111.

[112] Hahnloser, R., Sarpeshkar, R., et al. (2000). Digital selection and analogue amplification coexist in a cortex-inspired silicon circuit. Nature. 405: 947–951.

[113] Hahnloser, R., Seung, H.S. (2001). Permitted and Forbidden Sets in Symmetric Threshold-Linear Networks. NIPS. Neural Comput. 2003 Mar; 15(3):621–38.

[114] Haloi, M. (2015). Improved microaneurysm detection using deep neural networks. arXiv:1505.04424.

[115] Haraty, R.A. Dimishkieh, M., Masud, M. (2015). An Enhanced k-Means Clustering Algorithm for Pattern Discovery in Healthcare Data. International Journal of Distributed Sensor Networks, June.

[116] Harnad, S. (1990). The Symbol Grounding Problem. Physica D. 42: 335–346.

[117] Havaei, M., Davy, A., et al. (2017). Brain tumor segmentation with deep neural networks. Med. Image Anal. 35:18–31, 2017.

[118] Heath, M., Bowyer, K. (2000). The digital database for screening mammography. In: Proceedings of the 5th international workshop on digital mammography, pp 212–218.

[119] Hebb, D.O. (1949). The Organization of Behavior. New York: Wiley & Sons.

[120] Heidenreich, F. et al. (2002). Modern approach of diagnosis and management of acute flank pain: review of all imaging modalities, European urology, vol. 41, no. 4, pp. 351–362, 2002.

[121] Hochreiter, Sepp; Schmidhuber, Jürgen (1997-11-01). "Long Short-Term Memory". Neural Computation. 9 (8): 1735–1780.

[122] Horvath, S. (2011). Weighted Network Analysis. Applications in Genomics and Systems Biology. Springer, New York, N.Y.

[123] Huang, Z., Dong, W. & Duan, H. (2015). A probabilistic topic model for clinical risk stratification from electronic health records. J. Biomed. Inform. 58, 28–36 (2015).

[124] Hutchins, J. (1999). Retrospect and prospect in computer-based translation. Proceedings of MT Summit VII, 1999, pp. 30–44.

[125] Hubel, D.H., and Torsten, N. W. (1962). Receptive fields, binocular interaction and functional architecture in the cat's visual cortex. The Journal of Physiology 160.1 (1962): 106–154.

[126] Hussain, S. M. A. (2017). Brain Tumor Segmentation Using Cascaded Deep Convolutional Neural Network, presented at the EMBC, Seogwipo, South Korea, 2017.

[127] Hiller, S. et al. (1973) Cybernetic methods of drug design. I. Statement of the problem—the perceptron approach. Comput. Biomed. Res. 6, 411–421.

[128] Hinton, G.E. et al. (2006) A fast learning algorithm for deep belief nets. Neural Comput. 18, 1527–1554.

[129] Ho, K.C., Speier, W., et al. (2014). Predicting discharge mortality after acute ischemic stroke using balanced data. AMIA Annu Symp Proc 2014:1787–96.

[130] Hope, T.M., Seghier, M.L., et al. (2013). Predicting outcome and recovery after stroke with lesions extracted from MRI images. Neuroimage Clin 2013; 2:424–33.

[131] Hoh, J.K., Cha, K.J., et al. (2012). Estimating time to full uterine cervical dilation using genetic algorithm. Kaohsiung J Med Sci 2012 Aug;28(8):423–428.

[132] Hou, Q. et al. 2018. RankProd Combined with Genetic Algorithm Optimized Artificial Neural Network Establishes a Diagnostic and Prognostic Prediction Model that Revealed C1QTNF3 as a Biomarker for Prostate Cancer. EBioMedicine (2018).

[133] Hsu, C.C., Chao, C.K., et al. (2006). Multiobjective optimization of tibial locking screw design using a genetic algorithm: Evaluation of mechanical performance. J Orthop Res. May;24(5):908–916.

[134] Hughes, T.B. et al. (2015) Modeling epoxidation of drug-like molecules with a deep machine learning network. ACS Central Sci. 1, 168–180.

[135] Hussain, S., Anwar, S. M., and Majid, M., (2018). Segmentation of glioma tumors in brain using deep convolutional neural network. Neurocomputing. 282:248–261, 2018.

[136] Hwang, S. (2019). Similarity-based machine learning and applications in rare disease trials. Vertex externship presentation. Vertex Inc. Boston, MA. August 15, 2019.

[137] Hwang, S. and Chang, M. (2019). Similarity-Based Artificial Intelligence for Adaptive Clinical Trial and Beyond, JSM presentation. Denver, Colorado.

[138] ICH (2017). General principles for planning and design of multi-regional clinical trials. ICH harmonised guideline, e17, final version adopted on 16 November.

[139] Ihmaid, S.K. et al. (2016). Self Organizing Map-Based Classification of Cathepsin k and S Inhibitors with Different Selectivity Profiles Using Different Structural Molecular Fingerprints: Design and Application for Discovery of Novel Hits. Molecules (Basel, Switzerland), 30 January, Vol. 21(2), pp. 175.

[140] Intrator, O. and Intrator, N. (1993) Using Neural Nets for Interpretation of Nonlinear Models. Proceedings of the Statistical Computing Section, 244–249. San Francisco: American Statistical Society (eds).

[141] Ishida, T., Nishimura, I. (2011). Use of a genetic algorithm for multiobjective design optimization of the femoral stem of a cemented total hip arthroplasty. Artif Organs. Apr;35(4):404–410.

[142] Itskowitz, P., Tropsha, A. (2005). K nearest neighbors QSAR modeling as a variational problem: theory and applications, J. Chem. Inf. Model. 45: 777–785.

[143] James, G., Witten, D., et al. (2013). An Introduction to Statistical Learning: with Applications in R (Springer Texts in Statistics). Springer, New York.

[144] Jaradat, N.J. et al. (2015) Combining docking-based comparative intermolecular contacts analysis and k-nearest neighbor correlation for the discovery of new check point kinase 1 inhibitors. J. Comput. Aid. Mol. Des. 29, 561–581.

[145] Jemal, A., Siegel, R., et al. (2008). Cancer statistics, 2008. CA Cancer J Clin 58(2):71–96.

[146] Jefferson, M.F., Pendleton, N., et al. (1997). Comparison of a genetic algorithm neural network with ogistic regression for predicting outcome after surgery for patients with nonsmall cell lung carcinoma. Cancer 1997 Apr;79(7):1338–1342.

[147] Jiang, F. et al. (2019). Artificial intelligence in healthcare: past, present and future. Stroke and Vascular Neurology 2017;2:e000101.

[148] Jingqin, L.V. and Fang, J. (2018). A Color Distance Model Based on Visual Recognition, Mathematical Problems in Engineering. Volume 2018, Article ID 4652526, 7 pages Jodoin, A. C., Larochelle, H., et al. (2017). Brain tumor segmentation with deep neural networks. Med Image Anal. 2017 Jan; 35:18–31.

[149] Kadurin, M., et al. (2017). druGAN: An Advanced Generative Adversarial Autoencoder Model for de Novo Generation of New Molecules with Desired Molecular Properties in Silico. Mol. Pharmaceutics, 2017, 14 (9): 3098–3104.

[150] Kadurin, A. et al. (2017). The cornucopia of meaningful leads: Applying deep adversarial autoencoders for new molecule development in oncology. Oncotarget. Feb 14; 8(7): 10883–10890.

[151] Kamnitsas, K., Ledig, C., et al. (2017). Efficient multi-scale 3D CNN with fully connected CRF for accurate brain lesion segmentation. Med Image Anal 2017;36:61–78.

[152] Karnan, M., Thangavel, K. (2007). Automatic detection of the breast border and nipple position on digital mammograms using genetic algorithm for asymmetry approach to detection of microcalcifications. Comput Methods Programs Biomed. Jul;87(1):12–20.

[153] Kennedy, J. and Eberhart, R. C. (1995). Particle swarm optimization. Proceedings of IEEE International Conference on Neural Networks, Piscataway, NJ. pp. 1942–1948.

[154] Khalil, A.S., Bouma, B.E., et a;. (2006). A combined FEM/genetic algorithm for vascular soft tissue elasticity estimation. Cardiovasc Eng 2006 Sep;6(3):93–102.

[155] Khotanlou, H., Afrasiabi, M. (2012). Feature selection in order to extract multiple sclerosis lesions automatically in 3D brain magnetic resonance images using combination of support vector machine and genetic algorithm. J Med Signals Sens 2012 Oct;2(4):211–218.

[156] Kim, J., Kang, U., Lee, Y. (2017). Statistics and deep belief network-based cardiovascular risk prediction, Healthc Inform Res. 2017 July;23(3):169–175.

[157] Kim, S., Liu, H., et al. (2015). Extracting drug-drug interactions from literature using a rich feature-based linear kernel approach. J. Biomed. Inform. 58, 23–30 (2015).

[158] Koer, S., Canal, M.R. (2008). Classifying epilepsy diseases using artificial neural networks and genetic algorithm. J Med Syst Aug;35(4):489–498.

[159] Koh, Y., Yap, C.W., Li, S.C. (2008). A quantitative approach of using genetic algorithm in designing a probability scoring system of an adverse drug reaction assessment system. Int J Med Inform 2008 Jun;77(6):421–430.

[160] Korb, O. et al. (2006). PLANTS: Application of ant colony optimization to structure-based drug design. ANTS 2006: Ant Colony Optimization and Swarm Intelligence pp 247–258.

[161] Kourou, K., Exarchos, T.P. et al. (2014). Machine learning applications in cancer prognosis and prediction. Comput Struct Biotechnol J. 2014 Nov 15;13:8–17.

[162] Koutsoukas, A. et al. (2013). In silico target predictions: defining a benchmarking data set and comparison of performance of the multiclass naïve bayes and parzen-rosenblatt window, J. Chem. Inf. Model. 53: 1957–1966.

[163] Kovačič, M. (2009). Genetic Programming and Jominy Test Modeling. Materials and Manufacturing Processes Volume 24, 2009 - Issue 7-8.

[164] Koza, J. et al. (1992). Genetic Programming: On the Programming of Computers by Means of Natural Selection (Complex Adaptive Systems). A Bradford Book.

[165] Koza, J. et al. (1994). Genetic Programming II: Automatic Discovery of Reusable Programs (Complex Adaptive Systems). A Bradford Book; First edition, New York, N.Y.

[166] Koza, J. et al. (1999). Genetic Programming III: Darwinian Invention and Problem Solving (Vol 3). Morgan Kaufmann; 1st edition.

[167] Koza, J. et al. (2003). Genetic Programming IV: Routine Human-Competitive Machine Intelligence. Springer.

[168] Koza, J. (2010a). Genetic Programming and Evolvable Machines, September 2010, Volume 11, Issue 3–4, pp. 251–284.

[169] Koza, J. (2010b). Human-competitive results produced by genetic programming. Genet Program Evolvable Mach (2010) 11:251–284.

[170] Kozegar, E., Soryani, M., et al. (2013) Assessment of a novel mass detection algorithm in mammograms. J Cancer Res Ther 9(4):592.

[171] Kumar, D.A. (2018). Artificial neural network model for effective cancer classification using microarray gene expression data. Neural Comput Appl 2018;29(12):1545–54.

[172] Kumardeep, C. et al. (2017). Deep learning based multi-omics integration robustly predicts survival in liver cancer. bioRxiv 2017(2017). 114892.

[173] Kuo, M.H., Kushniruk, A.W., et al. (2009). Application of the Apriori algorithm for adverse drug reaction detection. Stud Health Technol Inform. 2009;148:95–101.

[174] Kwon, S. and Yoon, S. (2017) DeepCCI: end-to-end deep learning for chemicalchemical interaction prediction. arXiv: 1704.08432.

[175] Laaksonen, J. et al. (2009). Cellular automata model for drug release from binary matrix and reservoir polymeric devices. Biomaterials 30(10):1978–87.

[176] Lampros, C. et al. (2014). Assessment of optimized markov models in protein-fold classification. J. Bioinform. Comput. Biol. 2014, 12, 1450016.

[177] Langdon, W. B. and Buxton, B. F. (2004). Genetic Programming and Evolvable Machines. Genetical program for mining DNA, Volume 5, Issue 3, pp 251–257.

[178] Langdon, W. B. and Buxton, B. F. (2004). Genetic Programming for Mining DNA Chip data from Cancer Patients. Genetic Programming and Evolvable Machines, 5 (3): 251–257.

[179] Langfelder, P., Horvath, S. (2008). WGCNA: an R package for weighted correlation network analysis. BMC Bioinformatics. 9: 559.

[180] Latkowski, T., Osowski, S. (2015). Computerized system for recognition of autism on the basis of gene expression microarray data. Comput Biol Med 2015 Jan;56:82–88.

[181] Lauby-Secretan, B. Scoccianti, C. et al. (2015) Breast-cancer screening–viewpoint of the IARC working group. New Engl J Med 372(24):2353–2358.

[182] Learning Dexterity. OpenAI Blog. July 30, 2018. Retrieved 2019-01-15. https://openai.com/blog/learning-dexterity/.

[183] LeCun, Y., Bengio, Y., Hinton, G. (2015). Deep learning. Nature. 521 (7553): 436–444. Bibcode: 2015Natur.521-436L.

[184] Li, J., Wu, J., Chen, K. (2013). PFP-RFSM: Protein-fold prediction by using random forests and sequence motifs. J. Biomed. Sci. Eng. 2013, 6, 1161–1170.

[185] Li, J., Wu, J., Chen, K. (2016). PFP-RFSM: Protein-fold prediction by using random forests and sequence motifs. J. Biomed. Sci. Eng. 2013, 6, 1161–1170.

[186] Li, X., Wu, X. (2014). Constructing Long Short-Term Memory based Deep Recurrent Neural Networks for Large Vocabulary Speech Recognition. 2014-10-15. arXiv:1410.4281.

[187] LIAng, M. et al. (2015) Integrative data analysis of multi-platform cancer data with a multimodal deep learning approach: IEEE/ACM Trans. Comp. Biol. Bioinformatics 12, 928–937.

[188] Lin, T.C., Liu, R.S., et al. (2013). Classifying subtypes of acute lymphoblastic leukemia using silhouette statistics and genetic algorithms. Gene 2013 Apr;518(1):159–163.

[189] Liu, J., Wang, X., et al. (2017). Tumor gene expression data classification via sample expansion-based deep learning. Oncotarget 2017;8(65):109646.

[190] Liu, X. and Chen, H. (2015). A research framework for pharmacovigilance in health social media: identification and evaluation of patient adverse drug event reports. J. Biomed. Inform. 58, 268–279 (2015).

[191] Lloyd-Jones, D., Adams, R., Carnethon, M. et al. (2009). Heart disease and stroke statistics – 2009 update. Circulation 119(3):21–181.

[192] Loprinzi, C.L., Laurie, J.A. et al. (1994). Prospective evaluation of prognostic variables from patient-completed questionnaires. North Central Cancer Treatment Group. Journal of Clinical Oncology. 12(3):601–7, 1994.

[193] Lohr, S. (2016). IBM is counting on its bet on Watson, and Paying Big Money for It. https://www.nytimes.com/2016/10/17/ technology/ibm-is-counting-on-its-bet-on-watson-and-paying-big-money-for-it.html.

[194] Long, E., Lin, H., Liu, Z. et al. An artificial intelligence platform for the multihospital collaborative management of congenital cataracts. Nat Biomed Eng 1, 0024 (2017). https://doi.org/10.1038/s41551-016-0024.

[195] Love, A., Arnold, C.W., et al. (2013). Unifying acute stroke treatment guidelines for a Bayesian belief network. Stud Health Technol Inform 2013; 192:1012.

[196] Lowe, R. et al. (2011) Classifying molecules using a sparse probabilistic kernel binary classifier. J. Chem. Inf. Model. 51, 1539–1544.

[197] Lu, L. Zhang, Y., et al. (2017). Deep Learning and Convolutional Neural Networks for Medical Image Computing. Springer, Switzland, 2017.

[198] Luczkowich, J.J., Borgatti, S.P. (2003). Defining and measuring trophic role similarity in food webs using regular equivalence. Journal of Theoretical Biology. 220 (3): 303–321.

[199] Lynn, S. (2019). https://www.shanelynn.ie/self-organising-maps-for-customer-segmentation-using-r/.

[200] Macías-García, L. et al. (2017). A Study of the Suitability of Autoencoders for Preprocessing Data in Breast Cancer Experimentation, J Biomed Inform. 2017 Aug;72:33-44.

[201] Mandal, S., Saha, G., Pal, R.K. (2013). Reconstruction of dominant gene regulatory network from microarray data using rough set and bayesian approach. J Comput Sci Syst Biol 2013; 6(5): 262–70.

[202] Mannini, A., Trojaniello, D., et al. (2016). A machine Learning Framework for Gait classification using inertial sensors: application to Elderly, Post-Stroke and Huntington's Disease Patients. Sensors 2016;16:134.

[203] Marinakis, Y., Dounias, G., Jantzen, J. (2009). Pap smear diagnosis using a hybrid intelligent scheme focusing on genetic algorithm based feature selection and nearest neighbor classification. Comput Biol Med 2009 Jan;39(1):69–78.

[204] Marsland, S. (2014). Machine Learning - an algorithm perspective, 2nd Ed. CRC. Boca Raton, FL.

[205] McCallum, E. and Weston, S. (2011) Parallel R. O'Reilly Media.

[206] Menche, J. et al. (2015) Uncovering disease–disease relationships through the incomplete interactome. Science 347, 1257601.

[207] Miller, T.P., Li. Y., et al. (2017). Using electronic medical record data to report laboratory adverse events. Br J Haematol 2017; 177:283–6.

[208] Mistry, P. et al. (2016) Using random forest and decision tree models for a new vehicle prediction approach in computational toxicology. Soft. Comput. 20, 2967–2979.

[209] Mnih, V. et al. (2012) Conditional restricted Boltzmann machines for structured output prediction. arXiv 2012 12023748.

[210] Moeskops, P. et al. (2016). Deep learning for multi-task medical image segmentation in multiple modalities, in International Conference on Medical Image Computing and Computer-Assisted Intervention, 2016, pp. 478–486.

[211] Moeskops, P. et al. (2016). Automatic segmentation of MR brain images with a convolutional neural network. IEEE transactions on medical imaging, vol. 35, no. 5, pp. 1252–1261, 2016.

[212] Mojsilovi, C. (2005). A computational model for color naming and describing color composition of images," IEEE Transactions on Image Processing, vol. 14, no. 5, pp. 690–699, 2005.

[213] Moravec, H. (1988), Mind Children, Harvard University Press.

[214] Nazareth, D.P., Brunner, S., et al. (2009). Optimization of beam angles for intensity modulated radiation therapy treatment planning using genetic algorithm on a distributed computing platform. J Med Phys 2009 Jul;34(3):129–132.

[215] Nelder, J., Wedderburn, R. (1972). Generalized Linear Models. Journal of the Royal Statistical Society. Series A (General). Blackwell Publishing. 135 (3): 370–384.

[216] Newman, D.J. et al. (1998). UCI Repository of machine learning databases [http://www.ics.uci.edu/~mlearn/MLRepository.html]. Irvine, CA: University of California, Department of Information and Computer Science.

[217] Newman, M.E.J. (2001). Scientific collaboration networks: II. Shortest paths, weighted networks, and centrality. Physical Review E. 64 (1): 016132.

[218] Nguyen, L.B., Nguyen, A.V., et al. (2013). Combining genetic algorithm and Levenberg-Marquardt algorithm in training neural network for hypoglycemia detection using EEG signals. Conf Proc IEEE Eng Med Biol Soc. 2013;2013:5386–5389.

[219] Ngo, T.A., Carneiro, G. (2014). Fully automated non-rigid segmentationwith distance regularized level set evolution initialization and constrained by deep-structured inference. In: Proceedings of IEEE conference computer vision and pattern recognition, pp 1–8.

[220] Ngo, T.A., Lu, Z., Carneiro, G. (2016). Combining deep learning and level set for the automated segmentation of the left ventricle of the heart from cardiac cine magnetic resonance. Med Image Anal 35:159–171.

[221] Nigsch, F., Bender, A., et al. (2006). Melting point prediction employing k-nearest neighbor algorithms and genetic parameter optimization, J. Chem. Inf. Model. 46: 2412–2422.

[222] Nikfarjam, A., Sarker, A., et al. (2015). Pharmacovigilance from social media: mining adverse drug reaction mentions using sequence labeling with word embedding cluster features. J. Am. Med. Inform. Assoc. 22, 671–681.

[223] Ocak, H. 2013). A medical decision support system based on support vector machines and the genetic algorithm for the evaluation of fetal well-being. J Med Syst 2013 Apr;37(2):9913.

[224] Ooi, C.H., Tan, P. (2003). Genetic algorithms applied to multiclass prediction for the analysis of gene expression data. Bioinformatics 2003 Jan;19(1):37–44.

[225] Oliver, A., Freixenet, J. et al. (2010). A review of automatic mass detection and segmentation in mammographic images. Med Image Anal 14(2):87–110.

[226] Opsahl, T., Colizza, V. et al. (2008). Prominence and control: The weighted rich-club effect. Physical Review Letters. 101 (16): 168702.

[227] Opsahl, T. and Panzarasa, P. (2009). Clustering in Weighted Networks. Social Networks. 31 (2): 155–163.

[228] Otake, T. (2016). IBM Big Data used for rapid diagnosis of rare leukemia case in Japan. 2016. https://disruptivetechasean.com/big_news/ibm-big-data-used-for-rapid-diagnosis-of-rare-leukemia-case-in-japan/.

[229] Padideh, D. (2016). A deep learning approach for cancer detection and relevant gene identification. In Pacific Symposium on Biocomputing. 2016. p. 219.

[230] Panesar, A. (2019). Machine Learning and AI for Healthcare Big Data for Improved Health Outcomes. Berkeley, CA : Apress.

[231] Pei, Y., Kim, Y., et al. (2011). Trajectory planning of a robot for lower limb rehabilitation. Conf Proc IEEE Eng Med Biol Soc. 1259–1263.

[232] Pineda, L., Ye, A. et al. (2015). Comparison of machine learning classifiers for influenza detection from emergency department free-text reports. J. Biomed. Inform. 58, 60–69 (2015).

[233] Pereira, J.C. et al. (2016). Boosting docking-based virtual screening with deep learning. J. Chem. Inf. Model. 56, 2495–2506.

[234] Pereira, S., Pinto, A., et al. (2016). Brain tumor segmentation using convolutional neural networks in mri images. IEEE Trans. Med. Imaging 35(5): 1240–1251, 2016.

[235] Prasoon, A., Petersen, K., et al. (2013) Deep feature learning for knee cartilage segmentation using a triplanar convolutional neural network. Med Image Comput Comput Assist Interv 16(Pt 2): 246–253.

[236] Putin, E. et al. (2018). Reinforced Adversarial Neural Computer for de Novo Molecular Design. J. Chem. Inf. Model., 2018, 58 (6), pp 1194–1204.

[237] Qayyuma, A. et al. (2018.) Medical Image Analysis using Convolutional Neural Networks: A Review, Journal of Medical Systems. November: 42:226.

[238] Rai, B. (2018). *https://www.youtube.com/watch?v=LxTDLEdetmI*.

[239] Rajendra, R.B., Vivek, V., Li, X. (2017). DeepCancer: Detecting Cancer through Gene Expressions via Deep Generative Learning. IEEE 2017 International Conference on Big Data Intelligence and Computing, Orlando, FL, Nov 6–10, 2017.

[240] Rajeshkumar, J., Kousalya K. (2017). Applicat5ions of swarm based intelligence algorithm in pharmaceutical industry: areview. Int. Res. J. Pharm. 2017, 8 (11).

[241] Ramachandran, P., Barret, Z., and Quoc, V.L. (October 16, 2017). Searching for Activation Functions. arXiv:1710.05941 [cs.NE].

[242] Rasool, F, et al. (2013). Using deep learning to enhance cancer diagnosis and classification. In Proceedings of the International Conference on Machine Learning 2013.

[243] Rehme, A.K., Volz, L.J., et al. (2015). Identifying neuroimaging markers of Motor Disability in acute stroke by machine Learning Techniques. Cereb Cortex 2015; 25:3046–56.

[244] Rekera, D. et al. (2014). Identifying the macromolecular targets of de novo-designed chemical entities through self-organizing map consensus. vol. 111: no. 11: 4067–4072.

[245] Riedmiller, M. (1994). Rprop - Description and Implementation Details. Technical Report. University of Karlsruhe.

[246] Riedmiller, M. and Braun, H. (1993) A direct adaptive method for faster backpropagation learning: the RPROP algorithm. Proceedings of the IEEE International Conference on Neural Networks (ICNN), pages 586-591. San Francisco.

[247] Rodriguez, J. (July 2, 2018). The Science Behind OpenAI Five that just Produced One of the Greatest Breakthrough in the History of AI. Towards Data Science. Retrieved 2019-01-15.

[248] Rohrer, B., 2019, https://brohrer.github.io/how_convolutional_neural_networks_work.html).

[249] Rondina, J.M., Filippone, M., et al. (2016). Decoding post-stroke motor function from structural brain imaging. Neuroimage Clin 2016;12:372–80.

[250] Rose, V.S. et al. (1991) An application of unsupervised neural network methodology Kohenen topology-preserving mapping to QSAR analysis. Quant. Struct. Act. Relat. 10, 6–15.

[251] Rosenbaum, P. R., Rubin, D. B. (1983). The Central Role of the Propensity Score in Observational Studies for Causal Effects. Biometrika. 70 (1): 41–55. doi:10.1093/biomet/70.1.41.

[252] Roth, H.R., Lu, L., et al. (2016) Improving computer-aided detection using convolutional neural networks and random view aggregation. IEEE TMI 35(5): 1170–1181.

[253] Roth, H.R., Lu, L. et al. (2015). DeepOrgan: multi-level deep convolutional networks for automated pancreas segmentation. In: Navab N, Hornegger J, Wells WM, Frangi AF (eds) Medical image computing and computer-assisted intervention – MICCAI 2015, Part I, vol 9349. LNCS. Springer, Heidelberg, pp 556–564.

[254] Rubner, Y., Tomasi, C, Guibas, L.J. (1998). A Metric for Distributions with Applications to Image Databases. Proceedings ICCV 1998: 59–66.

[255] Sahiner, B., Chan, H.P., et al. (1996). Image feature selection by a genetic algorithm: application to classification of mass and normal breast tissue. Med Phys 1996 Oct;23(10):1671–1684.

[256] Sak, H. et al. (2014). Long Short-Term Memory recurrent neural network architectures for large scale acoustic modeling.

[257] Schmidhuber, Jürgen (1993). Habilitation thesis: System modeling and optimization (PDF). Page 150 ff demonstrates credit assignment across the equivalent of 1,200 layers in an unfolded RNN.

[258] Schneider, P., Tanrikulu, Y., Schneider, G. (2009). Self-organizing maps in drug discovery: compound library design, scaffold-hopping, repurposing. Curr Med Chem. 2009;16(3):258–66.

[259] Schmider, J., Kumar, K., et al. (2019). Innovation in Pharmacovigilance: Use of Artificial Intelligence in Adverse Event Case Processing. Clinical pharmacology & therapeutics | volume 105 number 4 | April 2019.

[260] Scott Gottlieb (2019). Statement from FDA Commissioner Scott Gottlieb, M.D. on steps toward a new, tailored review framework for artificial intelligence-based medical devices.

[261] Scrucca, L. (2013) GA: A Package for Genetic Algorithms in R. Journal of Statistical Software, 53/4, 1-37. URL https://www.jstatsoft.org/v53/i04/.

[262] Scrucca, L. (2017) On some extensions to GA package: hybrid optimisation, parallelisation and islands evolution. The R Journal, 9/1, 187–206.

[263] Segler, M.H.S. et al. (2017) Generating focussed molecule libraries for drug discovery with recurrent neural networks. arXiv 1701, 01329.

[264] Segura-Bedmar, I. and Martinez, P. (2015). Pharmacovigilance through the development of text mining and natural language processing techniques. J. Biomed. Inform. 58, 288–291 (2015).

[265] Sennaar, K. (2018). https://www.techemergence.com/machine-learning-medical-diagnostics-4-current-applications/, 2018.

[266] Shahlaei, C.Y.M. (2017). The applications of PCA in QSAR studies: A case study on CCR5 antagonists. Chem Biol Drug Des. 2018;91:137–152.

[267] Shamim, M.T.A. et al. (2007). Support vector machine-based classification of protein-folds using the structural properties of amino acid residues and amino acid residue pairs. Bioinformatics 2007, 23, 3320–3327.

[268] Sharma, G., Wu, W. and Dalal, E. N. (2005). The CIEDE2000 color difference formula: implementation notes, supplementary test data, and mathematical observations, Color Research & Application, vol. 30, no. 1, pp. 21–30, 2005.

[269] Sharma, A. et al. (2013). A feature extraction technique using bi-gram probabilities of position specific scoring matrix for protein-fold recognition. J. Theor. Biol. 2013, 320, 41–46.

[270] Shen, H.B., Chou, K.C. (2009). Predicting protein-fold pattern with functional domain and sequential evolution information. J. Theor. Biol. 2009, 256, 441–446.

[271] Shi, L.M., et al. 1998. Mining the National Cancer Institute Anticancer Drug Discovery Database: cluster analysis of ellipticine analogs with p53-inverse and central nervous system-selective patterns of activity. Mol Pharmacol. 1998 Feb;53(2):241–51.

[272] Shi, Y. and Eberhart, R. C. (1998). A modified particle swarm optimizer. Proceedings of the IEEE Congress on Evolutionary Computation (CEC 1998), Piscataway, NJ. pp. 69–73, 1998.

[273] Siegel. J.S., Ramsey, L.E., et al. (2016). Disruptions of network connectivity predict impairment in multiple behavioral domains after stroke. Proc Natl Acad Sci USA 2016;113:E4367–E4376.

[274] Simon, S. (1965), p. 96 quoted in Crevier 1993, p. 109.

[275] Simon Smith (2019). https://blog.benchsci.com/startups-using-artificial-intelligence-in-drug-discovery.

[276] Singh, H. et al. (2015) QSAR based model for discriminating EGFR inhibitors and non-inhibitors using Random forest. Biol Direct 10, 1745–6150.

[277] Sirota, M. et al. (2011). Discovery and Preclinical Validation of Drug Indications Using Compendia of Public Gene Expression Data. Science translational medicine 3(96): 96ra77.

[278] Sloot, P., Chen, F., and Boucher, C. (2002). Cellular Automata Model of Drug Therapy for HIV Infection. in S. Bandini, B. Chopard, and M. Tomassini (Eds.): ACRI 2002, LNCS 2493, pp. 282–293, 2002. Springer-Verlag Berlin Heidelberg 2002.

[279] Smith, S.L., Cagnoni, S. (2011). Genetic and evolutionary computation medical applications; John Wiley & Sons, 2011.

[280] Smolensky, P (1986). Chapter 6: Information Processing in Dynamical Systems: Foundations of Harmony Theory, in Rumelhart, David E.; McLelland, James L. Parallel Distributed Processing: Explorations in the Microstructure of Cognition, Volume 1: Foundations. MIT Press. pp. 194–281. ISBN 0-262-68053-X.

[281] Soulami, K.S. et al. (2017). A CAD System for the Detection of Abnormalities in the Mammograms Using the Metaheuristic Algorithm Particle Swarm Optimization (PSO), in Advances in Ubiquitous Networking 2, ed: Springer, 2017, pp. 505–517.

[282] Sparkes, S. (2018). The role of artificial intelligence within pharmacovigilance and medical information. PIPELINE Issue 56; September 2018.

[283] Stanford, S. (January 25, 2019). DeepMind's AI, AlphaStar Showcases Significant Progress Towards AGI. Medium ML Memoirs. Retrieved 2019-01-15.

[284] Steve, L. et al. (1997). Face recognition: A convolutional neural-network approach. IEEE Trans Neural Netw 1997;8(1):98–113.

[285] Sudip, M. and Indrojit, B. (2015). Cancer classification using neural network. Int J 2015; 172.

[286] Tan, J. et al. (2015). Unsupervised feature construction and knowledge extraction from genome-wide assays of breast cancer with denoising autoencoders. In Pacific Symposium on Biocomputing. Pacific Symposium on Biocomputing. 2015. p. 132.

[287] Tan, C., Chen, H., et al. (2011). Modeling the relationship between cervical cancer mortality and trace elements based on genetic algorithm-partial least squares and support vector machines. Biol Trace Elem Res 2011 Apr;140(1):24–34.

[288] Tang, J., Rangayyan, R.M., et al. (2009) Computer-aided detection and diagnosis of breast cancer with mammography: recent advances. IEEE Trans Inf Technol Biomed 13(2):236–251.

[289] Thomas, N. and Mathew, D. (2016). KNN Based ECG Pattern Analysis and Classification. International Journal of Science, Engineering and Technology Research (IJSETR) Volume 5, Issue 5, May 2016.

[290] Thornhill, R.E., Lum, C., et al. (2014). Can shape analysis differentiate free-floating internal carotid artery Thrombus from atherosclerotic plaque in patients evaluated with CTA? for stroke or transient ischemic attack? Acad Radiol 2014;21:345–54.

[291] Trichopoulos et al. (1976) Br. J. of Obst. and Gynaec. 83, 645–650.

[292] Tsamardinos, I. et al. (2006). The max-min hill-climbing Bayesian network structure learning algorithm. Machine learning 65.1 (2006): 31–78.

[293] Tseng, K.L., Lin, Y.L., et al. (2017). Joint sequence learning and cross-modality convolution for 3d biomedical segmentation. arXiv:1704.07754.

[294] Tulder, G.V. and Bruijne, M.D. (2016). Combining Generative and Discriminative Representation Learning for Lung CT Analysis with Convolutional Restricted Boltzmann Machines. IEEE Transactions on Medical Imaging, May 2016, Vol.35(5), pp.1262–1272.

[295] Tzeng, E., Hoffman, J., et al. (2017). Adversarial discriminative domain adaptation. In: Computer Vision and Pattern Recognition (CVPR), Vol. 1, p. 4, 2017.

[296] Tzezana, R. (2017). https://curatingthefuture.com/2017/02/28/ai_future_darpa/ autoencoders.

[297] Villar, J.R., González, S., et al. (2015). Improving human activity recognition and its application in early stroke diagnosis. Int J Neural Syst 2015; 25:1450036.

[298] Vinterbo. S., Ohno-Machado, L. (1999). A genetic algorithm to select variables in logistic regression: example in the domain of myocardial infarction. Proc AMIA Symp 1999;984–988.

[299] Wallner, F.K. et al. (2010). Correlation and cluster analysis of immunomodulatory drugs based on cytokine profiles. Pharmacological Research, 2018. 244–251.

[300] Wang, Y. et al. (2015) A comparative study of family-specific protein-ligand complex affinity prediction based on random forest approach. J. Comput. Aid. Mol. Des. 29, 349–360.

[301] Wang, S., Yin, Y., et al. (2015) Hierarchical retinal blood vessel segmentation based on feature and ensemble learning. Neruocomputing 149:708–717.

[302] Warden, P. (2011). http://radar.oreilly.com/2011/05/data-science-terminology.html.

[303] Watts, D. J. and Strogatz, D. (June 1998). Collective dynamics of 'small-world' networks. Nature. 393 (6684): 440&ndash, 442.

[304] Way, G.P. and Greene, C.S. (2017). Extracting a biologically relevant latent space from cancer transcriptomes with variational autoencoders. bioRxiv (2017), 174474.

[305] Wei, L. and Zou, Q. (2016). Recent Progress in Machine Learning-Based Methods for Protein-Fold Recognition. Int. J. Mol. Sci. 2016, 17, 2118.

[306] Weidlich, I.E. et al. (2013) Inhibitors for the hepatitis C virus RNA polymerase explored by SAR with advanced machine learning methods. Bioorg. Med. Chem. 21, 3127–3137.

[307] Wen, M. et al. (2017) Deep-learning-based drug-target interaction prediction. J. Proteome. Res. 16, 1401–1409.

[308] Williams, R. J., Hinton, G. E., Rumelhart, D. E. (October 1986). Learning representations by back-propagating errors. Nature. 323 (6088): 533–536.

[309] Willems, J.L. et al. (1991). The diagnostic performance of computer programs for the interpretation of electrocardiograms. N Engl J Med 325(25): 1767–1773.

[310] Wu, A., Xu, Z., et al. (2016). Deep vessel tracking: a generalized probabilistic approach via deep learning. In: Proceedings of IEEE international symposium on biomedical, imaging, pp 1363–1367.

[311] Wu, X., Zhu, Y. (2000). A mixed-encoding genetic algorithm with beam constraint for conformal radiotherapy treatment planning. Med Phys 2000 Nov;27(11):2508–2516.

[312] Xavier, G., Bordes, A. and Bengio, Y. (2011). Deep sparse rectifier neural networks (PDF). AISTATS. Rectifier and softplus activation functions. The second one is a smooth version of the first.

[313] Xu, Y. et al. (2017). Demystifying Multitask Deep Neural Networks for Quantitative Structure–Activity Relationships. J. Chem. Inf. Model.201757102490-2504.

[314] Xu, Y. (2017). Demystifying Multitask Deep Neural Networks for Quantitative Structure–Activity Relationships. J. Chem. Inf. Model. 201757102490-2504.

[315] Shamim, M.T.A., Anwaruddin, M., Nagarajaram, H.A. (2007). Support vector.

[316] Xing, F., Yang, L. (2016) Robust nucleus/cell detection and segmentation in digital pathology and microscopy images: a comprehensive review. IEEE Rev Biomed Eng 9:234–263.

[317] Xiao, Y. et al. (2018). A deep learning-based multimodel ensemble method for cancer prediction. Comput Methods Programs Biomed 2018;153:1–9.

[318] Yahi, A., et al. (2017). Generative Adversarial Networks for Electronic Health Records: A Framework for Exploring and Evaluating Methods for Predicting Drug-Induced Laboratory Test Trajectories.

[319] Yang, M., Kiang, M., Shang, W. (2015). Filtering big data from social media: building an early warning system for adverse drug reactions. J. Biomed. Inform. 54, 230–240 (2015).

[320] Yang, J.Y., Chen, X. (2011). Improving taxonomy-based protein-fold recognition by using global and local features. Proteins Struct. Funct. Bioinform. 2011, 79, 2053–2064.

[321] Ye, H., Shen, H., et al. (2017). Using Evidence-Based medicine through Advanced Data Analytics to work toward a National Standard for Hospital-based acute ischemic Stroke treatment. Mainland China, 2017.

[322] Yildirim, P. et al. (2014). Knowledge Discovery and Interactive Data Mining in Bioinformatics. BMC Bioinformatics 2014 15 (Suppl 6) :S7.

[323] Yu, X., Yu, G., Wang, J. (2017). Clustering cancer gene expression data by projective clustering ensemble. PloS One 2017; 12(2):e0171429.

[324] Yu, Y., Schell, M.C., Zhang, J.B. (1997). Decision theoretic steering and genetic algorithm optimization: application to stereotactic radiosurgery treatment planning. Med Phys 1997 Nov;24(11):1742–1750.

[325] Yuan Y., Shi, Y., et al. (2016). DeepGene: an advanced cancer type classifier based on deep learning and somatic point mutations. BMC Bioinformatics 2016;17 (17):476.

[326] Zandkarimi, M., Shafiei, M., et al. (2014). Prediction of pharmacokinetic parameters using a genetic algorithm combined with an artificial neural network for a series of alkaloid drugs. Sci Pharm 2014 Mar;82(1):53–70.

[327] Zhang, B., Horvath, S. (2005). A general framework for weighted gene co-expression network analysis. Statistical Applications in Genetics and Molecular Biology. 4: Article17.

[328] Zhang, L.M., Chang, H.Y., Xu, R.T. (2013). The Patient Admission Scheduling of an Ophthalmic Hospital Using Genetic Algorithm. Adv Mat Res 2013;756:1423–1432.

[329] Zhang, W. et al. (2015). Deep convolutional neural networks for multi-modality isointense infant brain image segmentation, NeuroImage, vol. 108, pp. 214–224, 2015.

[330] Zhang, Z. et al. (2017). Hierarchical cluster analysis in clinical research with heterogeneous study population: highlighting its visualization with R. Ann Transl Med. 2017 Feb; 5(4): 75.

[331] Zhang, L., et al. (2017). From machine learning to deep learning: progress in machine intelligence for rational drug discovery. Drug Discovery Today Volume 22, Number 11 November 2017.

[332] Zhang, D, Lu, Z., et al. (2018). Integrating Feature Selection and Feature Extraction Methods with Deep Learning to Predict Clinical Outcome of Breast Cancer. IEEE.

[333] Zhang, Q., Xie, Y., et al. (2013). Acute ischaemic stroke prediction from physiological time series patterns. Australas Med J 2013;6:280–6.

[334] Zhang, Y., Padnia, R. & Patel, N. Paving the COWpath: learning and visualizing clinical pathways from electronic health records. J. Biomed. Inform. 58, 186–197 (2015).

[335] Zhen, X., Wang, Z., et al. (2016). Multi-scale deep networks and regression forests for direct bi-ventricular volume estimation. Med Image Anal 30:120–129.

[336] Zheng, Y., Loziczonek, M. (2011). Machine learning based vesselness measurement for coronary artery segmentation in cardiac CT volumes. In: Proceedings of SPIE medical imaging, vol 7962, pp 1–12.

[337] Zheng, Y., Tek, H., Funka-Lea, G. (2013) Robust and accurate coronary artery centerline extraction in CTA by combining model-driven and data-driven approaches. In: Proceedings of international conference medical image computing and computer assisted intervention, pp 74–81.

[338] Zimek, A., Schubert, E. (2017), Outlier Detection, Encyclopedia of Database Systems, Springer New York, pp. 1–5.

[339] Zhou, T. et al. (2010). Solving the apparent diversity-accuracy dilemma of recommender systems. Proc. Natl. Acad. Sci. U. S. A. 107, 4511–4515.

Index

Bayes, Thomas, 189
Bayes' factor (BF), 201, 202, 209
Bayes' formula, 197–198
Bayes' law, 238
Bayes' model, hierarchical, 206–207
Bayes' rule, 186, 190
Bayes' theorem, 190, 197–198
Bayesian Q-learning. *See* Q-learning
Bayesian credible interval (BCI),
    200–201
Bayesian decision-making
    decision theory, applying, 208
    defining, 208
    overview, 208
Bayesian inference, 193–195, 208
Bayesian networks
    beta posterior distribution, 198
    binomial endpoint case example,
        199
    clinical trial, 200
    coronary heart disease case
        example, 194–197
    hypothesis testing, 201–202
    implementing, 195–197
    inferences, 197
    layers, 192
    model selection, 205–206
    normal posterior distribution,
        198–199
    overview, 192
    probabilities, 193, 194
    similarity searches, 192
    variables, 192
Bayesian paradigms, 190
    nuisance parameters,
        eliminating, 207
Bayesian point estimate, 200
Bayesian probability, 191
Bayesian stochastic decision process
    (BSDP), 253, 254, 255
Bayesianism, 288, 289
Bellman equations, 241, 242, 246,
    247, 250, 254, 255
Bellman's optimality principle, 240,
    257, 287

Best matching units (BMUs), 223
Bias. *See also* confirmation bias;
    selection bias
    defining, 21
    drug effectiveness, in predicting,
        53–54
Binary outcome variables, 87, 88
Bioinformatics
    data mining, 213
    description, 3
    drug discovery, use in, 216
    industrial uptake, 7
    kernel methods, 167
    Needleman-Wunsch algorithm
        use of, 64
    word embedding, 145–146
Biological networks (pathways), 228
Blyth, Colin R., 52
Boole, George, 1
Boosting, 14, 89, 174, 183, 184,
    185–186, 278
Brooks, Rodney, 3

C
Capek, Karel, 1
Carpenter, Rollo, 2
CART. *See* Gini index
CC-Cruiser, 290
Cellular automata (CA), 266–267,
    267–268
Chebyshev distance, 63
Clark, Sally, case of, 50–51, 191
Classical conditioning, 99–100
Classical statistics, 17, 28, 44, 53, 289
Classification trees
    case example, 175–177
    constructing, 175–176
    cross-entropy, 176, 177
    deviance, 177
    Gini index (*see* Gini index)
    goodness of fit, 179
    ID3, 177, 178
    overview, 175
    threshold determination, 178
    tree pruning, 178–179